S0-CBH-008

Host Plant Resistance to Pests

Host Plant Resistance to Pests

Paul A. Hedin, EDITOR
USDA, Boll Weevil Research Laboratory

A symposium sponsored by the Division of Pesticide Chemistry at the 174th Meeting of the American Chemical Society, Chicago, IL, Aug. 31–Sept. 1, 1977.

ACS SYMPOSIUM SERIES **62**

AMERICAN CHEMICAL SOCIETY
WASHINGTON, D. C. 1977

Library of Congress CIP Data

Host plant resistance to pests.
(ACS symposium series; 62 ISSN 0097-6156)

Includes bibliographical references and index.

1. Plants—Disease and pest resistance—Congresses.
2. Plant immunochemistry—Congresses.
I. Hedin, Paul Arthur. II. American Chemical Society. Division of Pesticide Chemistry. III. Series: American Chemical Society. ACS symposium series; 62.

SB750.H67 632'.94 77-13823
ISBN 0-8412-0389-X ACSMC8 62 1-286 1977

PRINTED IN THE UNITED STATES OF AMERICA

FOREWORD

The ACS SYMPOSIUM SERIES was founded in 1974 to provide a medium for publishing symposia quickly in book form. The format of the SERIES parallels that of the continuing ADVANCES IN CHEMISTRY SERIES except that in order to save time the papers are not typeset but are reproduced as they are submitted by the authors in camera-ready form. As a further means of saving time, the papers are not edited or reviewed except by the symposium chairman, who becomes editor of the book. Papers published in the ACS SYMPOSIUM SERIES are original contributions not published elsewhere in whole or major part and include reports of research as well as reviews since symposia may embrace both types of presentation.

CONTENTS

Preface ix

1. **Phytoalexins and Chemicals That Elicit Their Production in Plants** 1
   N. T. Keen and B. Bruegger

2. **Role of Elicitors of Phytoalexin Accumulation in Plant Disease Resistance** 27
   Barbara S. Valent and Peter Albersheim

3. **Biochemical Aspects of Plant Disease Resistance and Susceptibility** 35
   Doug S. Kenfield and Gary A. Strobel

4. **Interactions between *Phytophthora Infestans* and Potato Host** 47
   Donald D. Bills

5. **Biosynthetic Relationships of Sesquiterpenoidal Stress Compounds from the Solanaceae** 61
   Albert Stoessl, J. B. Stothers, and E. W. B. Ward

6. **Activated Coordinated Chemical Defense against Disease in Plants** 78
   Joseph Kuć and Frank L. Caruso

7. **Biochemical and Ultrastructural Aspects of Southern Corn Leaf Blight Disease** 90
   Peter Gregory, Elizabeth D. Earle, and Vernon E. Gracen

8. **Host Plant Resistance to Insects** 115
   A. C. Waiss, Jr., B. G. Chan, and C. A. Elliger

9. **The Effects of Plant Biochemicals on Insect Growth and Nutritional Physiology** 129
   John C. Reese

10. **Isolation and Identification of Toxic Agents from Plants** 153
    Martin Jacobson

11. **Insect Antifeedants and Repellents from African Plants** 165
    I. Kubo and K. Nakanishi

12. **Antifeedant Sesquiterpene Lactones in the Compositae** 179
    Tom J. Mabry, William C. Burnett, Jr., Samuel B. Jones, Jr., and James E. Gill

13. **Insect Antifeedants of *Spodoptera Litura* in Plants** 185
    Katsura Munakata

14. **Natural Insecticides from Cotton *(Gossypium)*** 197
    Robert D. Stipanovic, Alois A. Bell, and Maurice J. Lukefahr

15. Role of Repellents and Deterrents in the Feeding of *Scolytus Multistriatus* **215**
Dale M. Norris

16. Behavioral and Developmental Factors Affecting Host Plant Resistance to Insects **231**
P. A. Hedin, J. N. Jenkins, and F. G. Maxwell

Index **277**

PREFACE

For the past several decades, chemical pesticides including insecticides and herbicides have been the foundation of agricultural pest control. However, the use of these pesticides has come under increasing regulatory control because of their side effects on nontarget species and on the environment. Moreover, there is evidence of increasing resistance through genetic selection of these pests against chemicals that still are authorized for use. Yet the urgency to control pests becomes greater because man, largely because of his own rampant increase in population, seems to be running a collision course with insects and plant diseases in competition for foods and fibers.

During the past 20 years, several new concepts of pest control have evolved. For insects, these concepts include the agents that modify behavior and development, such as pheromones, kairomones, and hormones. Other approaches encompass the biological agents such as viruses, bacteria, and protozoa. Genetic manipulation has been investigated also. There have been efforts to combine various cultural techniques with these aforementioned approaches to develop integrated pest management systems.

The control of plant diseases has concentrated primarily on prophylaxis, with a lesser emphasis on immunization. The three major prophylactic approaches include: (1) exclusion of the pathogen by quarantine, inspection, and certification of seed; (2) eradication of the pathogen by crop rotation, sanitation of equipment, elimination of overwintering hosts, eradication of host plant parts, and eradication of alternate hosts; and (3) direct protection by regulating the environment; i.e., moisture, temperature, insect vectors, spraying and dusting, and treating seed and the soil. The control of plant diseases by immunization, as with viruses and host disease resistance, has focused on acquired resistance. This host resistance may be attributed to the plant's morphology, undefined physiological factors, and/or chemicals that are either present initially or are biosynthesized in response to attack by the disease organism.

These alternative approaches to pest control generally have not achieved the successes predicted by their original expectations, however. In some instances, the degree of control is not adequate. Often, the procedures are too complicated for the grower or pest control agent to use correctly. Cost is usually higher and, in some cases, excessive. Also, because these control programs are directed toward individual crops, there may be disincentives for industrial commercial development.

The development of host plant varieties has several advantages compared with the other alternative approaches to pest control. There is no complicated technology for the grower to understand, and it is inexpensive, since any increased seed costs will be more than offset by decreased costs for pesticides and their application. Decreased use of pesticides is also desirable for environmental purposes since the spread of pollutants is decreased, and energy requirements to manufacture, distribute, and apply pesticides is lessened. The major disadvantage associated with plant protection by resistant varieties is that because new races arise, germ plasm must be available from which to select new varieties. However, this is not insurmountable because plant breeders have provided successfully a series of rust resistant wheats in response to the new races of the pathogen for at least four decades.

Recently, the interest in elucidating some of the chemical aspects of resistance has increased because of: (1) greater interest in the field of host plant resistance as an alternative to our current dependence on pesticides and their role in developing effective integrated pest management programs; (2) some success in recent programs; (3) information derived from basic plant–insect and plant–plant interactions; (4) greater interest by private and public agencies in supplying the support necessary to conduct more basic chemical studies; (5) a greater awareness and interest by natural product chemists of the opportunities for research contributions; (6) better methodology and technology in microchemical techniques; (7) probably most important for the future, the potential Food and Drug Administration regulation that will require chemical research to document that changes (toxins, nutritional, etc.) in the plant from development of resistance to a pest will not be health or nutrition hazards to humans or to other animals that might consume the resistant crop.

All but two of the chapters in this volume were presented originally at the symposium on "The Chemical Basis for Host Plant Resistance to Pests" at the 174th National Meeting of the American Chemical Society. Seven chapters on plant diseases are followed by nine chapters on insects. It is hoped that this volume will initiate and stimulate work by cooperating groups of chemists, biochemists, entomologists, plant pathologists, plant breeders, and other related scientists. Defining the chemical basis for plant resistance to pests requires a multidisciplinary approach.

U.S. Department of Agriculture
Mississippi State, MS
August 22, 1977

PAUL A. HEDIN

1

Phytoalexins and Chemicals That Elicit Their Production in Plants

N. T. KEEN and B. BRUEGGER

Department of Plant Pathology, University of California, Riverside, CA 92521

Since the discovery of phytoalexins by Müller and Börger (1), their number has steadily grown until we now know of at least 64 different chemicals for which structures have been offered and which meet at least some of the criteria for phytoalexins. Although phytoalexins have been isolated from at least 75 species representing 20 families, it is too early to say whether their production is a general feature of higher plants. With the possible exception of *Ginkgo* (2), phytoalexins have not been reported from plants other than angiosperms and conifers with only a few from the latter. At present, they are most studied in vegetative plants and most often in the Leguminosae and Solanaceae. There are plants in which phytoalexins have been looked for and not found (see 3), but such negative data of course does not establish that they are not made; generally, when intensively searched, plants have been observed to make phytoalexins in response to the proper challenge and with the use of proper extraction techniques. This review cannot be comprehensive and will not recite the known list of phytoalexins since this has been done many time (4-9); instead it will hazard to deduce some generalizations about phytoalexins, their properties, biosynthesis, and, of paramount importance, to assess their presumed association with the resistance of plants to infectious plant pathogens. We will also discuss what is known about phytoalexin elicitors--chemicals or stimuli that initiate the biochemical steps ultimately resulting in phytoalexin accumulation.

All the known phytoalexins are low molecular weight products of plant biosynthesis that have antibiotic properties to one or several groups of microorganisms. The preformed levels of phytoalexins are generally very low or non-detectable in healthy plant tissue, but they accumulate to high levels at the site of attack of the plant by an invading microorganism. Phytoalexins have at times been called "abnormal metabolites" or "stress metabolites". It is our opinion that these terms are, generally speaking, synonyms, and as such need not be used. Phytoalexins are prime examples of secondary metabolites in plants; they would appear

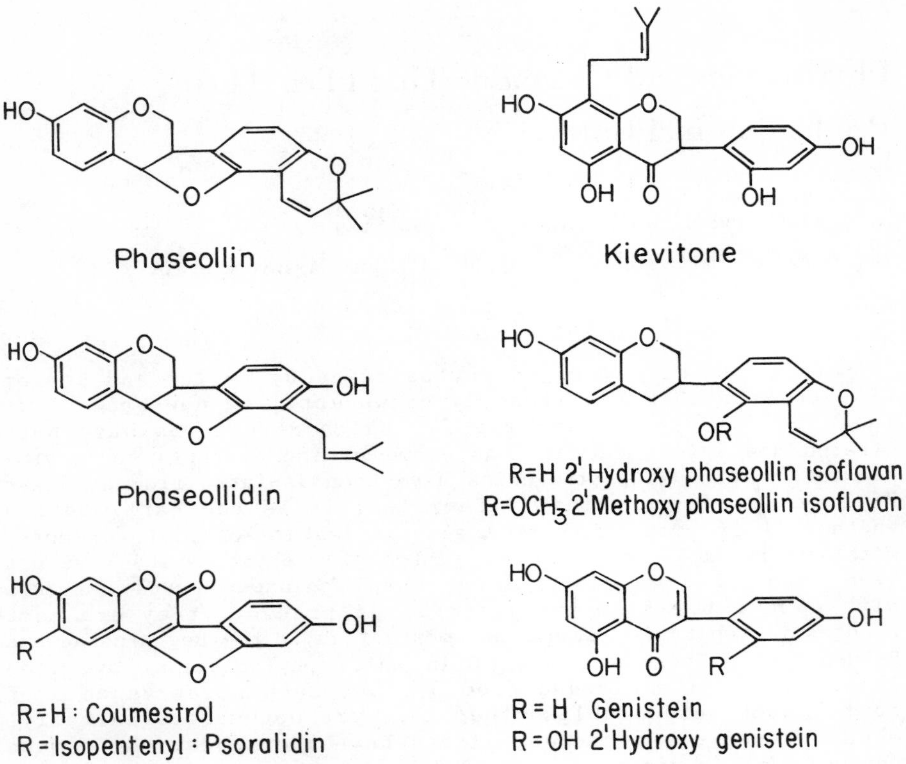

Phaseollin

Kievitone

Phaseollidin

R=H 2' Hydroxy phaseollin isoflavan
R=OCH_3 2' Methoxy phaseollin isoflavan

R=H : Coumestrol
R=Isopentenyl : Psoralidin

R=H Genistein
R=OH 2' Hydroxy genistein

Figure 1. Phytoalexins from Phaseolus vulgaris

to provide one likely answer to the long-standing search for physiologic functions of these products--defense against parasite attack (see 10). To summarize this into a simple definition offering criteria for phytoalexins, we consider them to be inducibly formed higher plant metabolites that are antibiotic to certain potential plant pathogens.

We define here some terms appearing in this paper. Disease resistance in plants is the restriction of development of a pathogenic agent or parasite and it may vary in degree from immunity (no development) to only slight retardation relative to a so-called susceptible reaction. We employ the term incompatible to denote host-parasite combinations in which the plant exhibits resistance and compatible to refer to susceptible reactions of the host. Similarly, an incompatible pathogen is one leading to a resistant plant reaction, while a compatible one leads to a susceptible host reaction.

The burst in phytoalexin research in the last 10 years has established that many plants simultaneously make several distinct but chemically related phytoalexins. For instance, the green bean plant, Phaseolus vulgaris, produces at least 9 isoflavonoid compounds (Figure 1) in response to various microorganisms or elicitors. Phaseollin is the best known and has pronounced activity against a broad spectrum of fungi, except for those that metabolize it to less toxic products. Most of the other bean phytoalexins are antifungal; however, genistein and hydroxygenistein accumulate along with the other compounds (11), but have only weak antifungal or antibacterial properties. Do they accumulate as a mere coincidence of the general activation of isoflavonoid biosynthesis or do they have antibiotic activity to organisms (nematodes? protozoans?) that we don't know about? Wyman and Van Etten (12) reported that kievitone and phaseollin isoflavan possessed pronounced antibacterial activity against several pathogenic bacteria, but coumestrol and phaseollin were generally without effect. This, however, has been disputed by Lyon and Wood (13) and Patil and Gnanamanickam (14) who observed antibacterial activity for coumestrol and phaseollin, respectively.

Another plant producing multiple, related phytoalexins is Solanum tuberosum, where at least eight sesquiterpenoid compounds are known (see Dr. Stoessl's paper in this series). It should be noted that some of these compounds may originate due to direct chemical modification of others by the challenge microorganism, as Ward and Stoessl (15) have recently demonstrated for the production of 15-dihydrolubimin from lubimin by fungi. However, rishitin, the best known potato phytoalexin, as well as phytuberin, lubimin and phytuberol can be elicited in the absence of living microorganisms (16), thus negating the possibility that they represent structure modifications. Rishitin is antibiotic to several fungi (17,18), having a ED_{50} value of about 50 μg/ml in vitro against Phytophthora infestans, with somewhat higher concentrations required against other fungi. Phytuberin, on the other hand, exhibited activity only against P. infestans. Solavetivone and anhydro-β-rotunol were active against all fungi tested (17), but lubimin was inactive against P. infestans.

In potato, green bean, soybean, and a few other plants producing multiple phytoalexins, the relative proportions of the various phytoalexins are not always the same (19,20 and see 21). Price et al. (22) noticed this for the potato sesquiterpenes and Rich et al. (23) observed that lima bean roots produced coumestrol and psoralidin in response to the nematode pathogen Pratylenchus scribneri, but did not accumulate detectable levels of kievitone, earlier found to be made by the same plant in response to bacteria (24).

The antibiotic properties of phytoalexins are routinely determined in in vitro assays, and several technical and interpretational problems can occur in these (25,26). Since highly purified phytoalexins have relatively low solubility in pure water, solubilizers such as ethanol or other organic solvents are generally added such that their final concentration is sufficiently low as not to affect the assay organism. Some have argued that this requirement detracts from the possible activity of phytoalexins in the plant, but we believe that the opposite is more likely to occur--co-dissolution of phytoalexins in the plant tissues by related compounds, fatty acids, phospholipids, saponins, etc. likely increases their _in vivo_ solubility and perhaps their activity as well. Indeed, evidence is available (_27_) indicating that the pea phytoalexin pisatin may be more active against infecting hyphae of _Erysiphe graminis in vivo_ than against the same organism _in vitro_.

As already mentioned, phytoalexins are often active against fungi and at least some of them affect bacteria, but what about other plant pests? We know of no case of phytoalexin involvement with resistance to insects, and this is perhaps to be expected with non-sedentary pests. However, what about aphids, mites, and other more stationary feeders? In recent work, Rich et al. (_23_) demonstrated that the lima bean (_Phaseolus lunatus_ L.) phytoalexin coumestrol exhibited a paralytic effect on the nematode pathogen _Pratylenchus scribneri_ and as such could account for the resistance of lima beans to this pathogen. Perhaps this portends other cases of phytoalexin mediated resistance to nematodes, but we do not have information. The resistant or so-called "local lesion" reaction of several plants to viruses results in substantial phytoalexin production (_28,29_), and although it is appealing to speculate causality, we have no direct evidence that they deleteriously affect virus replication, protein synthesis and/or virion assembly. Pterocarpinoid compounds have been reported to possess antitumor activity in animals (_30_), a finding of possible pharmacological significance.

Despite considerable information on the antimicrobial of activity phytoalexins, we know relatively little concerning the mechanism of their toxicity. Unlike some microbial antibiotics and synthetic fungicides, phytoalexins do not appear to have discreet "biochemical targets," but instead, as Van Etten and Bateman first showed (_31,32_), they likely function as nonspecific pleiotropic membrane antagonists. For instance, phaseollin (_32_) caused ion leakage from treated fungus cells and glyceollin, phaseollin, and medicarpin also caused bursting of red blood cells (_31_). Similarly, phytuberin and other sesquiterpene phytoalexins from _Solanum tuberosum_ caused rapid lysis of _Phytophthora infestans_ zoospores (_17_), an effect that would seem again to imply membrane antagonism. There is also evidence that phaseollin and other phytoalexins incite the same effects on permeability of the plant cells where they accumulate (_33,34_). It is therefore appealing to think that the disorganization noted to occur in the hypersensitive reaction to be discussed could be caused by the extremely high levels of phytoalexins acting on the plant cells. It was suspected that the aplanar configuration of the pterocarpan phytoalexins might be essential for their fungitoxicity (35), but Van Etten (36) subsequently showed that several 6a-hydroxy pterocarp-6a-enes, which do not show the same configuration as the corresponding parent 6a-hydroxy pterocarpans, had essentially unchanged activity. As noted, several isoflavones (viz. genistein and hydroxygenistein), do not possess pronounced antifungal activity, but isopentenyl substitution makes them considerably more active (37,38). This leads to the suspicion that lipophilicity is an important contributor to activity.

Bell and Presley (39), obtained data with the cotton-*Verticillium dahliae* host-parasite system that satisfactorily summarizes our view of phytoalexins and their contribution to compatibility and incompatibility of plants to certain potential pathogens. We call this relationship the "equilibrium hypothesis," the salient point being that any modulator of the host-parasite complex positively influencing pathogen colonization rates but independent of effects on host phytoalexin production will promote disease susceptibility. Similarly, any factor that positively influences the local rate of phytoalexin production independently of effects on pathogen growth will promote disease resistance. The converse of both tenets produces the corresponding effects. To illustrate, Bell and Presley (39), found that increasing temperature from 15 to 28°C increased the relative ability of cotton plants to produce phytoalexins, and phytoalexin production was constant from 28-35°C. However, increasing temperature above 25° deleteriously affected the growth rate of the fungus pathogen *Verticillium dahliae*. This was consistent with the observation that disease becomes less severe above *ca*. 25°C in nature. Conversely, lowering temperatures caused less phytoalexin production but did not adversely affect the fungus and disease became more severe. Similar to these results, many investigations have shown that treatment of infected plants with metabolic inhibitors that affect only the plant block phytoalexin production and cause more susceptible reactions. On the other hand, treatment of plants with agents that specifically inhibit development of the pathogen would be expected to result in a more resistant reaction and this too has been observed experimentally.

Phytoalexin production is often associated with a widespread but poorly understood plant disease defense reaction called the "hypersensitive reaction" (HR). Classically, the hypersensitive reaction is observed after a few hours or a few days following infection by an incompatible race or species of plant pathogen as the death or disorganization of the plant cells immediately adjacent to the infection site concomitant with or preceeding the restriction of pathogen development. Despite the fact that the HR operates in response to non-pathogenic members of all the known types of plant pathogens--fungi, bacteria, nematodes, and viruses--no plausible explanation as to why it worked was offered until phytoalexins appeared on the scene. As will be discussed in more detail later, many hypersensitive reactions result in the accumulation of large quantities of phytoalexins in the "hypersensitive area," up to the phenomenal concentration of 10% of the dry weight of tissue (40)! It would seem implausible that the cells in the immediate infection site could themselves account for the relatively massive biosynthesis required for such high concentrations. There are many non-infected host cells surrounding hypersensitive areas, however, and recent evidence suggests that these metabolically active neighboring cells both export biosynthetic intermediates to cells immediately adjacent to the infection site (41,43) and biosynthesize phytoalexins and export them into the host cells and intercellular spaces that are actually colonized by the pathogen (41,42,43). This "dumping" of phytoalexins into the hypersensitive area may account for the very high local concentrations present and would presumably constitute a very hostile local environment for the pathogen. There is good evidence that phytoalexins are in fact made in living, metabolically active plant cells (44,45,46) and the hypersensitive death of host cells may result from the toxicity of the accumulated phytoalexins (44). The observed extracellular export of phytoalexins is also consistent with the common observation that phytoalexins can be efficiently eluted from intact plant tissues by merely placing them into contact with water or aqueous ethanol (4,47,48). These so-called diffusion techniques are useful for obtaining phytoalexins free from most of the intracellular plant metabolites, and in one case the use of this technique was shown to elute the phytoalexin from host tissue at a sufficient rate that the normally resistant tissue became susceptible to the pathogen (47).

Kiraly, Barna, and Ersek (49,50,134) applied various metabolic inhibitors to three normally compatible host-parasite combinations and reported that the inhibitors, presumed to affect only the parasite, caused occurrence of the same hypersensitive defense reaction as in genetically incompatible plants; in one case--the potato-*Phytophthora infestans* system--they reported that the inhibitor treated plants accumulated the phytoalexin rishitin just as did normally incompatible plants. The authors interpreted their data as indicating that the hypersensitive reaction and phytoalexin production were therefore only consequences but not causes of resistance, and that some unknown mechanism actually accounted for resistance expression. Despite the acceptance of this reasoning by some (51), the logic of Kiraly and co-workers is flawed. Disregarding arguments (52,53) of the necessary assumption that the inhibitors were only affecting the pathogens, it is plausible that if the pathogens were in fact specifically inhibited by the treatments, the release of phytoalexin eliciting substances from them could account for the observations. By applying the anti-pathogen antibiotics, the authors were then quite simply swinging the equilibrium relationship of the last paragraph grossly in favor of the host. Were the authors' interpretation correct, we would similarly have to dismiss the immune system in higher animals as a defense mechanism, since attenuated pathogens and antigens from them trigger it, just as Kiraly _et al_. reported in plants (_49,50,134_).

Are Phytoalexins Really a Mechanism for Disease Resistance in Plants?

Since Cruickshank's classic review (_4_) brought phytoalexins to the attention of the scientific community and forcefully proposed their role in disease defense, they have inexplicably been under attack by some workers generally not involved in phytoalexin research on the principle presumption that they are merely spurious products or formed coincidentally to some other as yet undescribed defense mechanism. Enough evidence is now at hand to attempt a reasonable assessment of the proposed role of phytoalexins as a defense mechanism in plants but not to deduce the generality of this role. There are two types of resistance to be considered, _general resistance_ of a plant species to various organisms that are non-pathogenic on it and _specific resistance_, occurring in certain cultivars of a plant species to specialized biotypes, or races, of a single pathogen species (gene-for-gene systems [_54,55,56_]).

General Resistance

As with all parasities, those of plants must possess certain characteristics that enable them to satisfactorily reach the host, penetrate and infect, colonize host tissue, escaped defense mechanisms, and finally to reproduce and disseminate. Failure to do any of these properly results in inability to parasitize the plant and we say that the plant is resistant or is incompatible with the parasite. Many mechanisms clearly can account for this that have nothing to do with phytoalexins--structural features of the plant such as cell wall composition, cuticular features, leaf hairs, stomatal behavior, lectins, induced lignification and suberization, preformed chemical inhibitors and others are undoubtedly important in excluding a great many potentially pathogenic microorganisms (_see 3,10_). But there is substantial reason to think that phytoalexins are one of the mechanisms that confer general resistance, either singly or in concert with other mechanisms (_57_).

Cruickshank (_4_) first noticed that successful pathogens on pea and bean plants were generally less sensitive to the phytoalexins produced by those plants than were non-pathogens and this has been confirmed by Van Etten (_58_). This indicated that phytoalexins were perhaps responsible for resistance to the non-pathogens, and that the failure of the host to produce them in sufficient quantity or the ability of the parasite to degrade them could account for the successful parasites. We now know of several examples where successful parasites are able to degrade the plant phytoalexins, while taxonomically closely related microorganisms cannot. For instance van den Heuvel (59) showed that three strains of the pathogen *Botrytis cinerea* that were pathogenic on green beans had the ability to degrade phaseollin, while two strains of the same fungus that were not pathogenic lacked ability to significantly degrade it. Similar observations were made by Mansfield and Widdowson (60), who found that *Botrytis fabae*, a pathogen of *Vicia fabae* plants, degraded the major plant phytoalexin wyerone acid much more rapidly than the related non-pathogen *B. cinerea*. The principle degradation product was subsequently identified as reduced wyerone acid and it had little or no antifungal activity. Hargreaves *et al.* (61) also observed the degradation of another broad bean phytoalexin, wyerone epoxide, to wyerol epoxide by *B. cinerea* and to dihydro-dihydroxy wyerol epoxide by *B. fabae*. Again, the degradation products had less antifungal activity than the parent compound, with the *B. fabae* product being least active.

Jones *et al.* (62) showed that the pepper pathogen *Phytophthora capsici* was much less sensitive *in vitro* to inhibition by the pepper phytoalexin capsidiol than was the non-pathogen of peppers, *P. infestans*. No evidence was obtained for degradation of capsidiol by *P. capsici*; instead the fungus was apparently more tolerant of the compound. In addition, *P. infestans* also elicited much greater amounts of capsidiol in peppers than did *P. capsici*. Thus, it would seem that two factors (differential sensitivity and differential elicitor activity) distinguish the two fungi. The work by the Canadian group offers considerable support for the role of capsidiol in determining the general resistance of peppers to *P. infestans* and their susceptibility to *P. capsici*.

A major line of evidence linking phytoalexins with general resistance is that certain treatments break the normal resistance of plant tissues to non-pathogens and concomitantly negate phytoalexin biosynthesis after inoculation (53,63). For instance, Chamberlain (64,65) found that treatment of soybean hypocotyls at 44 C for 1 hr produced no apparent damage to the plants, but it destroyed resistance to several non-pathogenic fungi along with phytoalexin production, providing that the plants were inoculated within about 24 hr after treatment. After 1-2 days, however, resistance and phytoalexin production were concomitantly restored. A key point (64) was the fact that heat treatment broke resistance to some non-pathogens of soybean but not all of them, thus implying that phytoalexins are associated with general resistance to some but not all non-pathogens. Yoshikawa (69) followed this with the finding that the protein synthesis inhibitor blasticidin S also rendered soybean plants susceptible to several normally non-pathogenic *Phytophthora* sp. and that the treated plants did not produce the soybean phytoalexin glyceollin after inoculation.

Specific Resistance

Plant pathologists and plant breeders have known for almost 80 years that some cultivars within a plant species have resistance to certain plant pathogens and that this resistance to disease is frequently inherited as a dominant, single gene character. Predictably, the incorporation of these single genes into agronomically desirable cultivars of crop plants has become a mainstay of practical disease control. The curse of the methodology is that the pathogens involved frequently mutate to biotypes or races that are pathogenic on the resistant plant genotype. This reacquisition of pathogenicity or virulence has also been found to be inherited as a single gene character in several parasites, and thus led Flor (see 56) to first formally recognize the complementarity of host resistance genes and parasite virulence genes; this relationship is called gene-for-gene complementarity and exists in many plant-pest relationships (54,55,56). Aside from being a detriment to the permanent use of disease resistance genes in practical agriculture, the gene-for-gene relationship offers a splendid system for the critical testing of mechanisms involved in the biochemical function of resistance genes in plants and virulence genes in parasites. Rowell *et al*. (55) first presented this as the "quadratic check," shown in Figure 2 for comparing one resistance gene in the plant and one complementary virulence gene in the parasite. Since it is possible to develop host and parasite lines that are "near-isogenic" except for the genes in question, use of these systems clearly minimizes the problem of interference by genes other than those under study (55).

Two terms originally introduced by Talboys (66) are useful in conceptualizing events occurring in plant-parasite interactions, especially gene-for-gene systems. The *determinative phase* denotes those initial recognitional events occurring between host and pathogen that determine or dictate whether the plant will subsequently react in a compatible or incompatible way. The *expressive phase* constitutes those events that are, as we will see, irreversibly predicated by the determinative phase, and may involve rapid or slow phytoalexin production which in turn may constitute resistance or susceptibility, respectively.

Gene-for-gene host-parasite systems have generally offered the most critical and convincing proof that rapid phytoalexin production constitutes a mechanism for restriction of pathogen growth in resistant plant tissue--conversely, phytoalexins are the only mechanism that is currently known to explain resistance in gene-for-gene systems, other proposed mechanisms having negligible experimental support. The experimental points that support phytoalexin involvement with resistance expression in gene-for-gene systems may be summarized as follows: (i) incompatible reactions result in more rapid and higher rates of phytoalexin production than do compatible ones (13,34,41,67,68,70) and all races of the respective pathogens seem equally sensitive to the

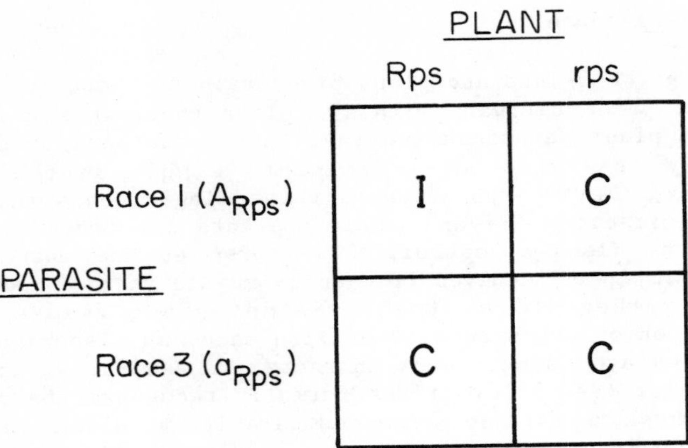

Figure 2. The quadratic check (55); gene-for-gene complementarity as illustrated by the reactions of two alleles of the Rps resistance locus in soybeans to Phytophthora megasperma *var.* sojae *and two races of the fungus shown with their presumed but unproved virulence genotypes. Reactions: I = incompatible; C = compatible. Resistance in the plant and avirulence in the fungus are shown as dominant characters, a general observation (54).*

phytoalexins in vitro. (ii) in the relatively few cases in which localized levels of phytoalexins have been assessed either by extraction from specific host cells layers or histochemistry, they have been shown to reach highly inhibitory levels to the pathogen at the infection sites in incompatible reaction (34,40,41,68,70); (iii) application of metabolic inhibitors or heat treatments simultaneously block both resistance and phytoalexin production (53,71); (iv) although technically difficult, limited data suggest that application of purified phytoalexins to normally compatible host-parasite infection sites results in incompatible reactions (27,72).

The most extensively researched and best understood gene-for-gene plant-pathogen system is the soybean-Phytophthora megasperma var. sojae interaction, and the evidence strongly supports a role for phytoalexins in the expression of resistance. Plant breeders have found several different single genes for resistance to the root and hypocotyl infecting fungus, but only one, the dominant Rps gene, has been incorporated into near-isogenic lines. As with all gene-for-gene systems, several races of the fungus have been found (9 at last count), and two of these, races 1 and 3, together with the rps and Rps soybean genotypes, fill out the quadratic check for the host-parasite system (Figure 2). Klarman and

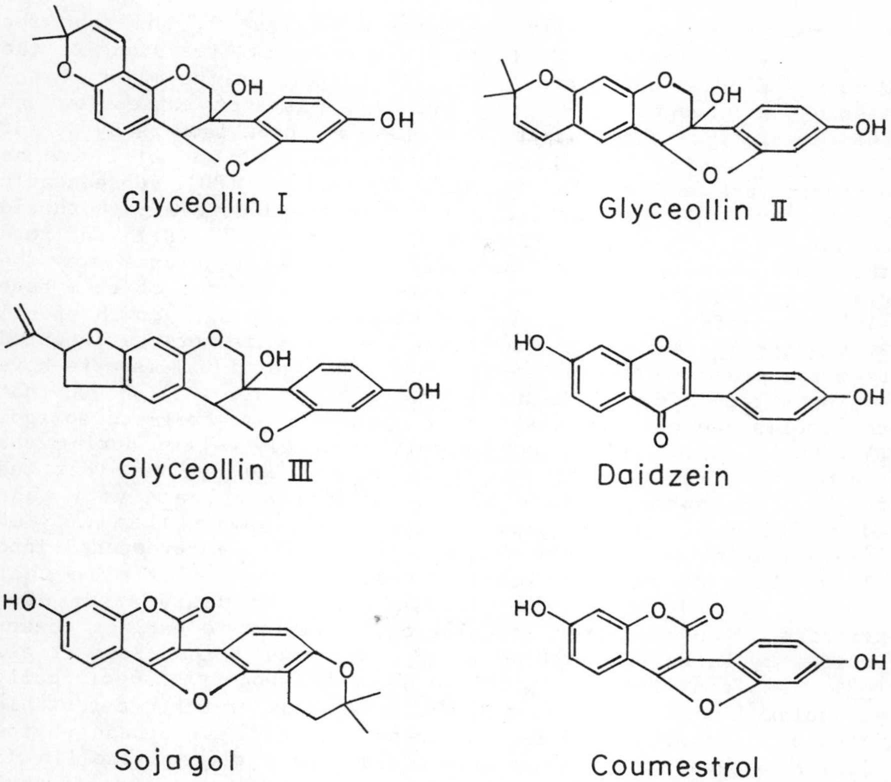

Glyceollin I

Glyceollin II

Glyceollin III

Daidzein

Sojagol

Coumestrol

Figure 3. Structures of soybean (Glycine max) *phytoalexins*

Gerdemann (47,74) first showed the possible involvement of phytoalexins in the expression of resistance due to the Rps gene. They demonstrated that elution of the infection site of incompatible inoculated soybean hypocotyls with water removed the phytoalexins and made the plants susceptible. They also chromatographically showed the presence of two phytoalexins in extracts from incompatible reacting plants that were not recovered from compatible plants. Paxton and associates showed that phytoalexins were associated with resistance conferred by one additional resistance gene (75) and found that inoculation with an incompatible fungus race led to "cross-protection" of the plants against subsequent re-inoculation with a normally compatible race. They also noted that administration of two of the soybean phytoalexins (now known to be glyceollin and PA_k) to compatible host-parasite combinations resulted in a resistant reaction (72). Bridge and Klarman (76) followed this same logic by showing that ultraviolet light elicited glyceollin production in the rps genotype, normally susceptible to race 1, and that the cortical tissue of such UV treated plants became resistant to the fungus. Klarman and Sanford (77) first isolated the major phytoalexin now thought to be involved in resistance expression and Keen et al. (78) showed that it existed as three isomers by GC-MS. Sims et al. (79) formulated the structure of one of these as 6a-hydroxyphaseollin, but Burden and Bailey (80) subsequently showed that this was incorrect and that the isoprene unit should be placed on the A ring (Figure 3). Lyne et al. (81) confirmed this and formulated the currently accepted structures for the other two glyceollin isomers (Figure 3). All three of them have similar antifungal properties and similarly inhibit growth of all P. megasperma races (78). Although the mixed isomers as isolated from the plant are antibacterial (82), the individual isomers have not been tested for their activity. Keen et al. (83) showed that the isoflavone daidzein and the coumestans coumestrol and sojagol (Figure 3) accumulated coordinately with glyceollin during the incompatible reaction to P. megasperma var. sojae, but that the former compounds posessed much less antifungal activity than glyceollin. In the same paper, ^{14}C-phenylalanine and ^{14}C-isoliquiritigenin were found to be readily incorporated into glyceollin, thereby indicating that the phytoalexin in fact results from de novo synthesis and not from degradation of a preformed compound after inoculation. Similar to earlier observations by Rathmell and Bendall (84) in green beans, Keen et al. (83) also found that incompatible soybean hypocotyls specifically accumulated isoflavonoids, but not flavonoids or other detectable phenols. Genistein and PA_k are also regarded as soybean phytoalexins because they accumulate coordinately with glyceollin in resistant reactions (85) and possess at least weak antifungal activity. Keen (40) showed that glyceollin was produced from 10-100 times faster in incompatible than in compatible reactions, demonstrated that production was strictly localized at the infection site in incompatible hypocotyls, and found that concentrations reached exceedingly high levels (up to 10% of the tissue dry weight by 48 hr). In common with all other known phytoalexins, there was a ca. 10 hr lag period after inoculation before glyceollin was detected in the plants, but then it accumulated rapidly in the incompatible reactions until 48 hr, at which time the fungus seems dead (86).

Yoshikawa et al. (71) have recently made a critical and elegant study of the levels of glyceollin occurring at very early times after inoculation with P. megasperma var. sojae by making freeze microtome sections of soybean hypocotyls and in this way quantitating levels of the phytoalexin occurring at various cell depths from the inoculation site on the hypocotyls. Their data conclusively establishes that incompatible reactions are distinguishable from compatible reactions at the earliest time of glyceollin detection at about 8 hrs after inoculation when the fungus has only invaded a few host cell layers. Glyceollin

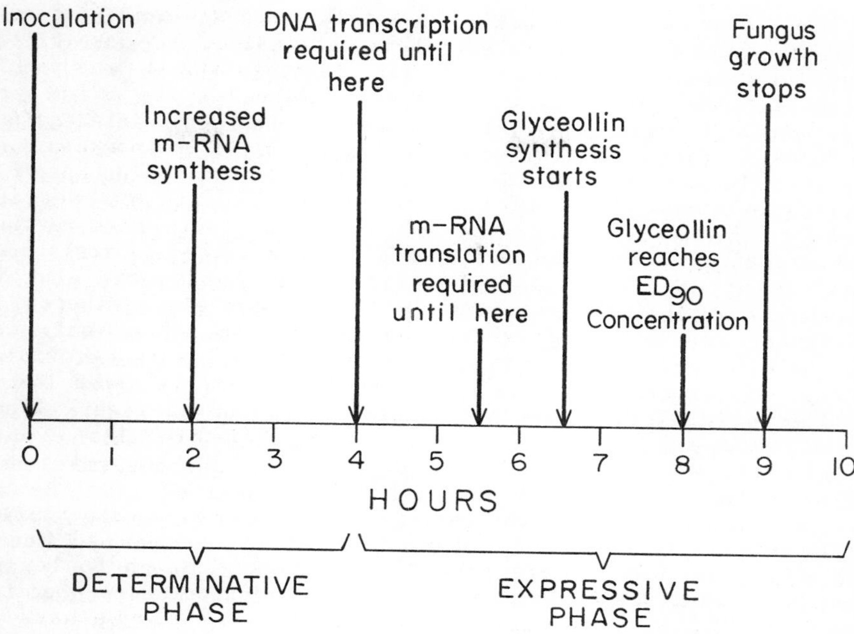

Figure 4. Timing of events occurring in incompatible-reacting soybean hypocotyls inoculated wtih Phytophthora megasperma *var.* sojae. *The determinative phase is defined as that time required for recognitional events between parasite and plant and for DNA transcription; the expressive phase encompasses translational events and phytoalexin biosynthesis.*

production was strictly localized in the host cells immediately adjacent to the fungus in the incompatible reactions and it accumulated to strongly inhibitory levels (ED_{90} concentrations for in vitro fungus growth) immediately before the time when fungus growth stopped at ca. 9 hrs after inoculation (see Figure 4). In compatible reacting plants, however, levels of glyceollin never exceeded the ED_{50} value for inhibition of fungus growth in host cells that were near the growing margin of the invading fungus. This data strongly supports the idea that glyceollin constitutes an inhibitory mechanism in incompatible host-parasite combinations and that the failure of the compatible host to make the chemical fast enough accounts for its susceptibility to restrict the fungus.

In other important work Yoshikawa et al. (87) found that incompatible reacting soybean hypocotyls synthesized poly(A)-containing mRNA at 6 times the rate of non-inoculated hypocotyls, but that inoculated compatible hypocotyls synthesized mRNA at only 2 times that of the controls. These data raised the possibility that the expression of incompatibility is biochemically the result of derepressed transcription of genes coding for the enzymes of phytoalexin biosynthesis. This has been further substantiated (69) with the finding that the transcription inhibitor actinomycin D at 10 µg/ml blocked phytoalexin production in normally incompatible plants and concomitantly resulted in a compatible reaction. However, application of actinomycin D to the plants at more than 4 hours after inoculation was ineffective in reversing resistance expression or glyceollin production. Similarly, the protein synthesis inhibitor blasticidin S blocked protein synthesis in resistant inoculated hypocotyls and this resulted in only low phytoalexin levels and susceptibility of the plants, providing that the inhibitor was applied in the first 6 hours after inoculation (69). These observations are summarized in Figure 4 and constitute an elegant elucidation of the molecular events associated with resistance expression to *P. megasperma* var. *sojae* in the soybean plant.

If, as appears to be the case in the soybean-*P. megasperma* var. *sojae* host-parasite system, incompatibility of the host to fungus infection is due to derepressed biosynthesis of phytoalexins resulting from specific *de novo* DNA transcription, does specific induction of biosynthetic enzymes in fact occur? We have no evidence on this except for the work of Sitton and West (73) with the antifungal and antibacterial phytoalexin casbene. Sitton and West obtained a cell-free system from fungus-challenged castor bean tissue that synthesized casbene 20-40 times faster from labelled mevalonic acid than comparable extracts from healthy plant tissue. Of considerable interest, rates of synthesis of other diterpene hydrocarbons was essentially the same with both cell-free systems. This suggests that casbene biosynthesis is a consequence of specific induction of the appropriate biosynthetic enzymes and does not result from a general acceleration of terpenoid biosynthesis. With the legume isoflavonoid phytoalexins, early enzymes in the presumed biosynthetic pathways including phenylalanine-ammonia lyase, chalcone-flavonone isomerase, peroxidase and others appear to have no specific role (88,93), and later enzymes in the presumed pathways have not been obtained in cell-free extracts.

A prime remaining question in plant pathology is elucidation of the specificity mechanism(s) in plant-parasite systems that operates between parasite and host plant and dictates whether the plant will respond in a compatible or incompatible way. It is most plausible that the host in such cases specifically recognizes some chemical feature(s) of the incompatible parasite (but not the compatible) and that this in a yet unknown manner interacts with the host to ultimately result in derepressed phytoalexin production and resistance. This suspicion has generated a great deal of interest and will involve the next section.

Phytoalexin elicitors

Phytoalexin production may be elicited in plant tissues in the absence of living microorganisms by such agents as heavy metal ions (102), UV light (76), 3′,5′ cyclic AMP (97), various peptides, proteins, antibiotics (95) and chemicals of pathogen origin. These chemical agents have been called "inducers" of phytoalexin production, but we prefer to call them "elicitors," because all we currently know is that they initiate or elicit the production of phytoalexins. The non-specificity of phytoalexin response to such an array of agents has been used as a detracting argument against the presumed role of phytoalexins in disease resistance, despite the obvious fact that pathogens do not routinely liberate heavy metal ions, antibiotics, etc. Further, many of the elicitors from pathogens are obtained by such harsh treatments as autoclaving or homogenization, leading one to doubt whether they play any physiologic role in host-parasite interactions. Few attempts have been made to critically test the association of elicitors with the differential phytoalexin production that occurs in gene-for-gene host-parasite systems, but logic dictates that such specificity imparting factors must function during the determinative phase. We will discuss non-specific elicitors, those with no differential effects of various genotypes of a plant species, and specific elicitors, those which have differential elicitor efficiency, depending on the disease resistance genotype. Among non-specific elicitors, we recognize biotic elicitors as those of pathogen origin, and abiotic elicitors as those from other sources.

In attempts to understand the biochemical processes by which plants initiate phytoalexin production, many abiotic chemicals have been used to elicit phytoalexin production in several plant systems such as pea pods, bean pods and hypocotyls, sweet potato root discs, soybean cotyledons, hypocotyls, suspension cells, and leaves, jack bean callus, potato tuber slices and cowpea hypocotyls (92,93,94,95,96). Some of the non-specific abiotic chemicals that elicit phytoalexin production in various of these plants systems are inorganic ions as those of mercury, copper and silver, metabolic inhibitors such as sodium azide and cyanide, DNA intercalating agents such as actinomycin D, mitomycin C, and acridine orange, basic proteins and peptides such as protamine, poly-L-lysine, and spermine, and protein synthesis inhibitors such as cycloheximide and puromycin (89,94,101,102). Some of the compounds such as cycloheximide and actinomycin D elicit phytoalexin production at low concentrations but at higher concentrations are inhibitory (94,95). One of the problems in formalizing a general hypothesis for phytoalexin induction from data obtained with abiotic elicitors is that not all plant species respond to the same chemicals (96); even different tissues within the same plant respond differently--for example glyceollin production can be elicited in soybean cotyledons with $K_2Cr_2O_7$, iodoacetate, SDS,

$CuCl_2$ or $HgCl_2$, but only $K_2Cr_2O_7$, iodoacetate, or $CuCl_2$ elicit phytoalexin production when injected into soybean leaves--SDS or $HgCl_2$, while causing necrosis, do not elicit detectable quantities of phytoalexins. Since many abiotic elicitors affect DNA structure and metabolism, specific gene derepression has been hypothesized by Hadwiger as the mechanism by which phytoalexin production is elicited (89). However, since most elicitors have not been shown to act exclusively on DNA, "side affects" could account for their activity. Another idea is that the abiotic elicitors may put the cell under "stress"--that is, change the normal physiological state of the plant cell environment and thereby cause phytoalexin production. Although specific mechanisms are as yet indefinite and perhaps diverse, the previously discussed work of Yoshikawa and associates (69,71,87) points to involvement of specific DNA transcription with glyceollin elicitation in the *Phytophthora*--soybean system. Reilly (100) also supported this in experiments with soybean suspension culture cells with UV light as the eliciting agent. Exposure of cells to increasing doses of UV light resulted in concomitant increases in thymine dimer formation and glyceollin synthesis. Photoreversal experiments on UV-treated cells using visible light showed that thymine dimers and glyceollin synthesis also declined in parallel. The observed simultaneous effects of UV-light and photoreversal on phytoalexin elicitation and physical DNA permutation point to DNA as the primary elicitor target of UV light.

A number of apparently non-specific biotic elicitors have now been reported. Polysaccharides from the culture fluid of 3 species of *Colletotrichum* (two of which were non-pathogens) elicited phytoalexin production on cotyledons of bean plants (98). Also, the culture fluid of *Penicillium expansum* (non-pathogen) elicited phytoalexin production on bean hypocotyls and pea pods (84) and *Nectria galligena* Bres. released proteases that elicited benzoic acid production in apples (99).

Culture fluids as well as soluble extracts from the mycelia and conidia of compatible and incompatible strains of *Ceratocystis fimbriata* elicited ipomeamarone, ipomeamoronol, and dehydroipomeamarone in sweet potato root tissue. The elicitor was soluble in water or 0.02M KCl, heat stable, dialyzable, had no anionic or cationic properties and caused cellular injury to root tissue (103).

Incompatible and compatible races of *Colletotrichum lindemuthianum* released an elicitor into culture fluids that was largely glucan in nature (104) and efficiently elicited phaseollin production in green beans. A similar elicitor was also isolated from purified mycelial cell wall preparations of the fungus.

Incompatible and compatible races of *Phytophthora megasperma* var. *sojae* produce a heat-stable 1,3-glucan that is an extremely potent non-specific elicitor of the soybean phytoalexin glyceollin

(105,106,107). Valent and Albersheim discuss the work elsewhere in this volume.

Stekoll and West (108) recently obtained an elicitor of casbene synthesis in castor bean from the non-pathogenic fungus *Rhizopus stolonifer* that, unlike the glucan elicitors, was heat labile. The partially purified elicitor appeared to be a glycoprotein of about 23,000 molecular weight and was active at 0.15 μg/ml.

Varns *et al*. (96) showed that sonicates and homogenates of *Phytophthora infestans* elicited production of phytoalexins in potato tuber tissues, but other chemicals that caused necrosis did not necessarily function as elicitors. Lisker and Kuc (16) also observed that potatoes were relatively insensitive to some of the abiotic and biotic elicitors that were effective in leguminous plants such as poly-L-lysine and polysaccharides such as the *P. megasperma* var. *sojae* glucan (105). Only autoclaved sonicates and heat-killed fungi having glucan-cellulose walls were active as elicitors of the potato phytoalexins. This perhaps indicates that potato responds to a narrower range of elicitors than leguminous plants.

Among non-specific phytoalexin elicitors, the fact is emerging that biotic elicitors are generally much more active than abiotic elicitors--by several orders of magnitude in some cases (105,108). Although this may argue for a physiologic role for the former, they are produced both by compatible and incompatible pathogen races and strains, are often only recovered from fungus mycelium by harsh extraction treatments, and show no host specificity (98,105). All these factors would appear to exclude a role for non-specific elicitors in the determination of gene-for-gene specificity (see also 109) but they should nonetheless be useful in studies of elicitation mechanisms in plants and could be involved in the conferral of general resistance.

There is a paucity of data on specific phytoalexin elicitors. The best known is a genus specific elicitor isolated by Cruickshank and colleagues (110). A polypeptide, monilicolin A, was purified from the culture fluid of *Monilinia fructicola* (a non-pathogen of *Phaseolus vulgaris*) and was shown to elicit phaseollin accumulation in green bean, but neither elicited pisatin in peas (*Pisum sativum*) nor phytoalexins in broad bean (*Vicia faba*) (110).

An acetone precipitated supernatant of the homogenate of *Rhizoctonia versicolar* (a non-pathogenic fungus asociated with roots of the orchid *Loroglossum hircinum*) elicited hircinol synthesis in sterile tissue of the orchid bulb. A similar homogenate from *Rhizoctonia solani*, a fungus which destroys the bulbs of *L. hircinum*, did not elicit the phytoalexin--a case of fungal species specificity (111).

Strain specificity was observed in the *Fusarium solani*-pea system (112). The culture fluid of avirulent *Fusarium solani* strains elicited larger quantities of pistin than did the culture fluid from virulent strains; however, the difference was apparently due to quantities of the same rather than different types of elicitors.

Keen (113,114) detected a race specific phytoalexin elicitor in culture fluids of race 1 of Phytophthora megasperma var. sojae but not from race 3 of the fungus (see Figure 2). The partially purified race 1 factor elicited greater production of the soybean phytoalexin glyceollin in Rps genotype soybeans than in rps--that is, the same phytoalexin specificity on these genotypes as by the living fungus. Despite this initial encouragement, the race 1 specific elicitor has proved insufficiently stable to allow isolation and identification and a genetic cross of the two fungus races designed to critically test the association of the race 1 specific elicitor with the race 1 phenotype has failed on technical grounds (115,116). Proof of involvement of the metabolite in pathogenesis is therefore lacking. Nevertheless, the initial success prompted the hypothesis that, in gene-for-gene systems, the determinative phase involves production of specific elicitors by the parasite that are unique for each dominant avirulence gene and specific receptors by the plant that are unique for each dominant resistance gene (Figure 5). This is, or course, a molecular conceptualization of the gene-for-gene concept (Figure 2) and there is nothing biologically unusual about it. Similar versions have been proposed previously by several workers (104, 109,117,118), and it should be noted that its involvement in the determinative phase could occur irrespective of whether the expressive phase involved differential phytoalexin production or some other, as yet unrecognized mechanism(s) for restriction of pathogen growth. The specific elicitor-specific receptor mechanism also is virtually identical as a recognition system to that currently proposed for certain pollen-style incompatibility systems in higher plants governed by single genes (119,120), for Rhizobium-legume specificity (see 121), and for receptor systems for certain host-specific plant toxins (122). Also in common with these systems, the elicitors and receptors most likely are constitutive metabolites that are either present on the cell wall or plasmalemma surfaces or, in the case of elicitors, might be extracellular metabolites. This is due to the relatively short times available for recognition; for example in the soybean-Phytophthora system, events of the determinative phase are concluded by about 4 hours after initial contact of host and parasite (Figure 4).

Important recent research by Goodman and co-workers (123), Sing and Schroth (124) and Sequeira's group (125) strongly suggests that specific host receptor substances, possibly lectins (104,120), may confer specificity in tobacco, green bean and potatoes to incompatible strains of Pseudomonas spp. Although these are not regarded as gene-for-gene systems and the possible role of phytoalexins in the expressive phase has not been well investigated, the results demonstrate the specific recognition and "attachment"

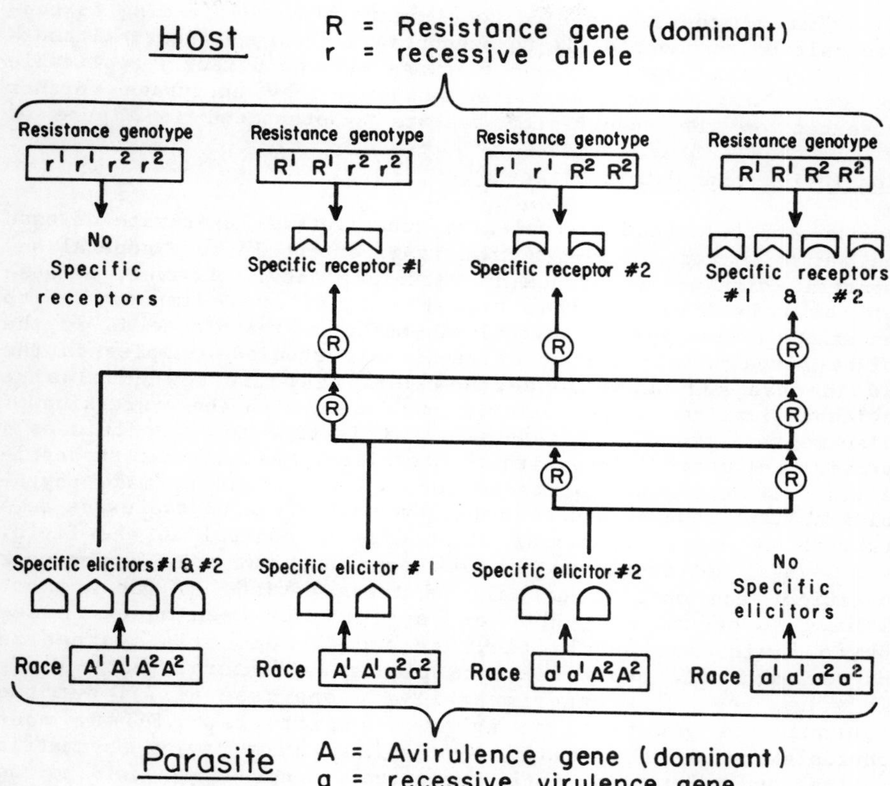

Figure 5. Scheme of the specific elicitor-specific receptor hypothesis for determination of the hypersensitive reaction in gene-for-gene plant–parasite interactions, considering two resistance loci in the plant and two complementary virulence loci in the parasite.

Host resistance genes R^1 and R^2 code either directly or indirectly for specific and unique receptor substances 1 and 2, possibly proteins, that specifically recognize and bind specific phytoalexin elicitors 1 or 2 from the pathogen. The hypothesis assumes that only the dominant plant resistance alleles make functional receptors. Correspondingly, for the parasite, the virulence genes code for unique specific elicitors that are specifically recognized by the complementary plant receptors. Only the dominant parasite avirulence alleles make functional gene products. Plant disease resistance (R) (rapid phytoalexin production) will be expressed in any combination that involves production of either specific elicitor and its complementary receptor. On the other hand, absence of a specific elicitor or its specific receptor will lead to repressed phytoalexin production and a susceptible plant response. Thus, as the gene-for-gene relationship dictates, in the absence of functional resistance genes in the host or functional avirulence genes in the parasite, susceptible reactions will be observed, regardless of the genotype of the complementary partner.

of incompatible but not compatible bacteria. With the potato system, Sequeira and Graham (126) have shown that both compatible and incompatible strains of *Pseudomonas solanacearum* possess structural features in their walls that bind to a potato lectin, but only the compatible strains possess an extracellular polysaccharide that "masks" the lectin binding elements, thereby leading to non-recognition and eventually to susceptibility of the host. Although it is not clear whether these findings will be directly applicable to gene-for-gene host-parasite systems, they encourage further research on the specificity factors hypothesized in Figure 5.

Conclusions and Future Directions

We conclude that phytoalexins constitute a legitimate defense mechanism in certain plants that is important in both general and specific disease resistance in certain plants. However, because of the relatively limited research done, it is impossible to assess how wide-spread phytoalexin-mediated resistance is in the plant kingdom. In the more extensively studied examples in the Solanaceae and Leguminosae, however, there is now conclusive evidence implicating phytoalexin production with the expression of disease resistance. We think that it is time to take this as a provisional dogma, along with the corollary stating that susceptibility is often due to the failure of the plant to make phytoalexins fast enough, and consider how we can use phytoalexins as a vehicle to obtain practical plant disease control in the field.

Several approaches to this have surfaced but none is yet near practical and most if not all of those devised so far are not likely to become so. However, a start has been made. Since phytoalexins are antibiotic, one could chemically synthesize phytoalexins and apply these to plants as "natural pesticides". This has been done experimentally in one case (127), but the approach is generally fraught with pitfalls. First, most phytoalexins are not readily synthesized, many having asymmetric carbons and relatively labile functional groups--they would not be cheap to make. Most disturbing, however, is the fact that phytoalexins are not innocuous chemicals, despite being "natural pesticides". As discussed earlier, at least some phytoalexins are membrane antagonists and several of them have been shown to be potent toxicants in mammals (for example, 32,128,129); clearly they are not the sort of thing to be spread around the environment, especially onto food crop plants.

Instead of spraying phytoalexins on plants, Albersheim (see 130) suggests applying phytoalexin elicitors, anticipating that the elicited phytoalexins would ward off possible pathogens. The approach may have merit but would be plagued by methods for applying the elicitors and by dosage requirements. The danger is that as elicitor doses are increased to levels possibly providing disease control, one will likely observe phytotoxicity as a result of the phytoalexins produced. As discussed, phytoalexins are normally associated with the necrotic, hypersensitive reaction in plants; in the normal HR, a plant can easily afford to lose a few of its cells to contain pathogens, but what happens if an applied elicitor causes wholesale hypersensitive necrosis? or if foodstuffs contain significant quantities of the elicited phytoalexins?

A. A. Bell (unpublished) has suggested a potentially useful approach to harnessing phytoalexins based on the observed fact that several plants make multiple, related phytoalexins. Bell suggests that disease resistance plant breeding programs could attempt to select plants that preferentially made favorable phytoalexins toward the pathogen in question--for instance, against a bacterial pathogen, breed for genotypes that preferentially produce potent antibacterial phytoalexins; against fungus diseases, select for preferential production of the most antifungal phytoalexins; perhaps it would even be possible to select plants making phytoalexins that a pathogen could not degrade, thereby conferring general disease resistance.

A final approach to harnessing phytoalexins in disease control involves applying chemicals to susceptible plants that are neither antibiotic nor phytoalexin elicitors, but interact with the plant-parasite interaction to cause higher production of phytoalexins by the plant cells and thus impart resistance. Because of this behavior we propose to call such chemicals "sensitizers". There are suggestions that such chemicals already exist. For instance, Pring and Richmond (131) found that bean plants inoculated with a normally compatible race of *Uromyces phaseoli* responded to treatment with the weakly antifungal fungicide oxycarboxin in the same way as genetically resistant plants, thus implying that oxycarboxin somehow caused invocation of a normal defense reaction. Langcake and Wickins (132,133) have shown that 2,2-dichloro-3,3-dimethyl cyclopropane carboxylic acid controls the blast disease caused by *Piricularia oryzae* on compatible rice varieties, but acts neither as a fungicide nor as a phytoalexin elicitor. However, inoculated plants that were treated with the cyclopropane were fully resistant to the disease, and, unlike untreated inoculated plants or treated but noninoculated plants, accumulated momilactones A and B, which appear to be phytoalexins in rice. There are several other less well-defined but possible examples of sensitizers in the literature, but we do not yet know if the concept is sound or if the required chemicals can be found. It is an idea worthy of consideration, however, since chemicals such as Langcake's cyclopropane appear relatively innocuous ecologically, and do not appear to have deleterious effects on plants that are not challenged by pathogens.

The authors' research is supported by National Science Foundation grant BMS 75-03319.

Literature Cited

1. Müller, K. O. and Börger, H. (1940). Arb. Biol. Reichsanst. Land-u. Forstwiss. 23, 189-231.
2. Christensen, T. G. and Sproston, T. (1971). Phytopathology 62, 493-494.
3. Deverall, B. J. "Defence mechanisms of plants." Cambridge Univ. Press. London. (1977),
4. Cruickshank, I. A. M. (1963). Annual Rev. Phytopathol. 1, 351-374.
5. Deverall, B. J. (1972). Proc. Roy Soc. Lond. B181, 233-246.
6. Ingham, J. L. (1972). Botan. Rev. 38, 343-424.
7. Kuc, J. "Phytoalexins", p. 632-652, In Physiological Plant Pathology, Edit. by Heitefuss, R. and Williams, P. H. (1976).
8. Kuc, J. and Currier, W. (1976). Adv. Chem. Ser. 149, 356-368.
9. Wood, R. K. S. (1973). Mitteil. Biologischen Bund. fur Land. u. Fortschaft 154, 95-105.
10. Swain, T. (1977). Annual Rev. Plt. Physiol. 28, 479-501.
11. Biggs, R. T. (1975). Austr. J. Chem. 28, 1389-1391.
12. Wyman, J. and Van Etten, H. D. (1975). Proc. Amer. Phytopath. Soc. 2, 110.
13. Lyon, F. M. and Wood, R. K. S. (1975). Physiol. Plant Pathol. 6, 117-124.
14. Patil, S. S. and Gnanamanickam, S. S. (1976). Nature 259, 486-487.
15. Ward, E. W. B. and Stoessl, A. (1977). Phytopathology 67, 468-471.
16. Lisker, N. and Kuc, J. (1977). Phytopathology (in press).
17. Harris, J. E. and Dennis, C. (1976). Physiol. Plant Pathol. 9, 155-165.
18. Ishizaka, N., Tomiyama, K., Katsui, N., Murai, A. and Masamune, T. (1969). Plt. Cell Physiol. 10, 183-192.
19. Elliston, J., Kuc, J., Williams, E. B., and Rahe, J. E. (1977). Phytopathol. Z. 88, 114-130.
20. Keen, N. T. p. 268. In "Specificity in Plant Diseases." Ed. by Wood, R. K. S. and Graniti, A. Plenum Press, New York (1976).
21. Stoessl, A., Stothers, J. B., and Ward, E. W. B. (1976). Phytochemistry 15, 855-872.
22. Price, K. R., Howard, B. and Coxon, D. T. (1976). Physiol. Plant Pathol. 9, 189-197.
23. Rich, J. R., Keen, N. T. and Thomason, I. J. (1977). Physiol. Plant Pathol. 10, 105-116.
24. Keen, N. T. (1975). Phytopathology 65, 91-92.
25. Skipp, R. A. and Bailey, J. A. (1976). Physiol. Plant Pathol. 9, 253-263.
26. Bailey, J. A., Carter, G. A., and Skipp, R. A. (1976).
27. Oku, H., Shiraishi, T., and Ouchi, S. (1976). Ann. Phytopath. Soc. Japan 42, 597-600.

28. Klarman, W. L. and F. Hammerschlag. (1972). Phytopathology 62, 719-721.
29. Bailey, J. A. and Ingham, J. L. (1971). Physiol. Plant Pathol. 1, 451-456.
30. Kojima, R., Fukushima, S., Neno, A. and Saiki, Y. (1970). Chem. Pharm. Bull. Japan 18, 2555.
31. Van Etten, H. D. (1972). Phytopathology 62, 795.
32. Van Etten, H. D. and Bateman, D. F. (1971). Phytopathology 51, 1363-1372.
33. Shiraishi, T., Oku, H., Isono, M. and Ouchi, S. (1975) Plt. Cell Physiol. 16, 939-942.
34. El Naghy, M. A. and Heitefuss, R. (1976). Physiol. Plant Pathol. 8, 269-277.
35. Perrin, D. R. and Cruickshank, I. A. M. (1969). Phytochemistry 8, 971-978.
36. Van Etten, H. D. (1976). Phytochemistry 15, 655-659.
37. Fukui, H., Egawa, H., Koshimizu, K. and Mitsui, T. (1973). Agr. Biol. Chem. 37, 417-421.
38. Ingham, J. L., Keen, N. T., and Hymowitz, T. (1977). Phytochemistry (in press).
39. Bell, A. A. and Presley, J. T. (1969). Phytopathology 59, 1141-1146.
40. Keen, N. T. (1971). Physiol. Plant Pathol. 1, 265-276.
41. Nakajima, T., Tomiyama, K. and Kinukawa, M. (1975). Ann. Phytopath. Soc. Japan 41, 49-55.
42. Tomiyama, K., and Fukaya, M. (1975). Ann. Phytopathol. Soc. Japan 41, 418-420.
43. Imaseki, H. and Uritani, I. (1964). Plt. Cell Physiol. 5, 133-143.
44. Shiraishi, T., Oku, H., Ouchi, S. and Isono, M. (1976). Ann. Phytopath. Soc. Japan 42, 609-612.
45. Mansfield, J. W., Hargreaves, J. A., and Boyle, F. C. (1974). Nature 252, 316-317.
46. Paxton, J., Goodchild, D. J., and Cruickshank, I. A. M. (1974). Physiol. Plant Pathol. 4, 167-171.
47. Klarman, W. L. and Gerdemann, J. W. (1963). Phytopathology 53, 863-864.
48. Higgins, V. J. and Millar, R. L. (1968). Phytopathology 58, 1377-1383.
49. Ersek, T., Barna, B., and Kiraly, Z. (1973). Acta Phytopath. Acad. Sci. Hung. 8, 3-12.
50. Barna, B., Ersek, T., and Mashaal, S. F. (1974). Acta Phytopathol. Acad. Sci. Hung. 9, 293-300.
51. van der Plank, J. E. "Principles of Plant Infection". Academic Press, New York. (1975).
52. Kim, W. K., Rohringer, R., Samborski, D. J. and Howes, N. K. (1977). Can. J. Bot 55: 568-573.
53. Doke, N., Nakae, Y., and Tomiyama, K. (1976). Phytopathol. Z. 87, 337-344.

54. Day, P. R. "Genetics of host-parasite interaction". Freeman, San Francisco (1974).
55. Rowell, J. B., Loegering, W. Q., and Powers, H. R. Jr. (1963). Phytopathology 53, 932-937.
56. Flor, H. H. (1971). Annual Rev. Phytopathol. 9, 275-296.
57. Heale, J. B. and Sharman, S. (1977). Physiol. Plant Pathol. 10, 51-61.
58. Van Etten, H. D. (1973). Phytopathology 63, 1477-1482.
59. Heuvel, J. van den (1976). Neth. J. Plant Pathol. 82, 153-160.
60. Mansfield, J. W. and Widdowson, D. A. (1973). Physiol. Plant Pathol. 3, 393-404.
61. Hargreaves, J. A., Mansfield, J. W., Coxon, D. T., and Price K. R. (1976). Phytochemistry 15, 1119-1121.
62. Jones, D. R., Unwin, C. H. and Ward, E. W. B. (1975). Phytopathology 65, 1286-1288.
63. Kojima, M. and Uritani, I. (1976). Physiol. Plant Pathol. 8, 97-111.
64. Chamberlain, D. W. (1972). Phytopathology 62, 645-646.
65. Chamberlain, D. W. and Gerdemann, J. W. (1966). Phytopathology 56, 70-73.
66. Talboys, P. W. (1957). Trans. Brit. Mycol. Soc. 40, 415-427.
67. Bailey, J. A. and Deverall, B. J. (1971) Physiol. Plant Pathol. 1, 435-449.
68. Rahe, J. E. (1973). Can. J. Bot. 51, 2423-2430.
69. Yoshikawa, M. (1977). Plt. Physiol. (submitted).
70. Shiraishi, T., Oku, H., Ouchi, S., and Tsuji, Y. (1977). Phytopathol. Z. 88, 131-135.
71. Yoshikawa, M., Yamauchi, K., and Masago, H. (1977). Physiol. Plant Pathol. (Submitted).
72. Chamberlain, D. W. and Paxton, J. D. (1968). Phytopathology 58, 1349-1350.
73. Sitton, D. and West, C. A. (1975). Phytochemistry 14, 1921-1925.
74. Klarman, W. L. and Gerdemann, J. W. (1963). Phytopathology 53, 1317-1320.
75. Paxton, J. D. and Chamberlain, D. W. (1967). Phytopathology 57, 352-353.
76. Bridge, M. A. and Klarman, W. L. (1973). Phytopathology 63, 606-609.
77. Klarman, W. L. and Sanford, J. B. (1968). Life Sci. 7, 1095-1103.
78. Keen, N. T., Sims, J. J., Erwin, D. C., Rice, E. and Partridge, J. E. (1971). Phytopathology 61, 1084-1089.
79. Sims, J. J., Keen, N. T., and Honwad, V. K. (1972). Phytochemistry 11, 827-828.
80. Burden, R. S. and Bailey, J. A. (1975). Phytochemistry 14, 1389-1390.

81. Lyne, R. L., Mulheirn, L. J. and LeWorthy, D. P. (1976). JCS Chem. Comm. p. 497-498.
82. Keen, N. T. and Kennedy, B. W. (1974). Physiol. Plant Pathol. 4, 173-185.
83. Keen, N. T., Zaki, A. I., and Sims, J. J. (1972). Phytochemistry, 11, 1031-1039.
84. Rathmell, W. G. and Bendall, D. S. (1971) Physiol. Plant Pathol. 1, 351-362.
85. Keen, N. T. and Paxton, J. D. (1975). Phytopathology 65, 635-637.
86. Frank, J. A. and Paxton, J. D. (1970) Phytopathology 60, 315-318.
87. Yoshikawa, M., Masago, H., and Keen, N. T. (1977). Physiol. Plant Pathol. 10, 125-138.
88. Partridge, J. E. and Keen, N. T. (1977). Phytopathology 67, 50-55.
89. Schwochau, M. E., and Hadwiger, L. A. (1969). Arch. Biochem. Biophys. 134, 34-41.
90. Biggs, D. R. (1972). Plt. Physiol. 50, 660-666.
91. Uritani, I., Uritani, M. and Yamada, H. (1960). Phytopathology 50, 30-34.
92. Gustine, D. L. (1976). Fed. Proc. 35, 1552.
93. Munn, C. B., and Drysdale, R. B. (1975). Phytochemistry 14, 1303-1307.
94. Bailey, J. A. (1969). Phytochemistry. 8, 1393-1395.
95. Hadwiger, L. A., and Schwochau, M. E. (1971). Plt. Physiol. 47, 346-351.
96. Varns, J. L., Currier, W. W. and Kuc, J. (1971). Phytopathology 61, 968-971.
97. Oguni, I., Suzuki, K., and Uritani, I. (1976). Agr. Biol. Chem. 40, 1251-1252.
98. Anderson, A. (1976). Proc. Amer. Phytopath. Soc. 3, 314.
99. Swinburne, T. R. (1975). Phytopathol. Z. 82, 152-162.
100. Reilly, J. J. (1975). Diss. Abstr. B36, 2552.
101. Hadwiger, L. A. & Schwochau, M. E. (1970). Biochem. Biophys. Res. Comm. 38, 683-691.
102. Perrin, D. R., and Cruickshank, I. A. M. (1965). Austr. J. Biol. Sci. 18, 803-816.
103. Kim, W. K., and Uritani, I. (1974). Plt. Cell Physiol. 15, 1093-1098.
104. Albersheim, P., and Anderson-Prouty, A. J. (1975). Ann. Rev. Plt. Physiol. 26, 31-52.
105. Ayers, A. R., Ebel, J., Finelli, F., Berger, N. and Albersheim, P. (1976). Plt. Physiol. 57, 751-759.
106. Ayers, A. R., Ebel, J., Valent, B. and Albersheim, P. (1976). Plt. Physiol. 57, 760-765.
107. Ayers, A. R., Valent, B., Ebel, J., and Albersheim, P. (1976). Plt Physiol. 57, 766-774.
108. Stekoll, M. (1976). Ph.D. Dissertation, Univ. of California, Los Angeles.

109. Wood, R. K. S. p. 327, _In_ "Specificity in Plant Diseases". Edit. by Wood, R. K. S. and Graniti, A. Plenum, New York (1976).
110. Cruickshank, I. A. M., and Perrin, D. R. (1968). Life Sci. _7_, part II 449-458.
111. Nuesch, J. (1963). Soc. Gen. Microbiol. Symp. _13_, 335-343.
112. Daniels, D. L., and Hadwiger, L. A. (1976). Physiol. Plt. Pathol. _8_, 9-19.
113. Keen, N. T. (1975). Science _187_, 74-75.
114. Keen, N. T., Partridge, J. E., and Zaki, A. I. (1972). Phytopathology _62_, 768.
115. Long, M. and Keen, N. T. (1977). Phytopathology _67_, 670-674.
116. Long, M. and Keen, N. T. (1977). Phytopathology _67_, 675-677.
117. Ercolani, G. L. (1970). Phytopathol. Med. _9_, 151-159.
118. Hadwiger, L. A. and Schwochau, M. E. (1969). Phytopathology _59_, 223-227.
119. Heslop-Harrison, J., Heslop-Harrison, Y., and Barker, J. (1975). Proc. Roy. Soc. _B188_, 287-297.
120. Linskens, H. F. p. 311, _In_ "Specificity in Plant Diseases." Edit. by Wood, R. K. S. and Graniti, A. Plenum, New York. (1976).
121. Dazzo, F. B. and Brill, W. J. (1977). Appl. Environ. Microbiol. _33_, 132-136.
122. Strobel, G. A. (1975). Sci. Amer. _232_, 80-89.
123. Goodman, R. N., Huang, P. and White, J. A. (1976). Phytopathology _66_, 754-764.
124. Sing, V. and Schroth, M. N. (1977). Science (In press).
125. Sequeira, L., Gaard, G. and De Zoeten, G. A. (1977). Physiol. Plant Pathol. _10_, 43-50.
126. Sequeira, L. and Graham, T. C. (1977). Physiol. Plant Pathol. (in press).
127. Ward, E. W. B., Unwin, C. H. and Stoessl, A. (1975). Phytopathology _65_, 168-169.
128. Uritani, I. (1967). J. Assoc. Agric. Chemists _50_, 105-113.
129. Beradi, L. C. and Goldblatt, L. A. p. 212. _In_ "Org. Const. Plt. Foodstuffs." Liener, I. E. (ed.). Academic Press, New York. (1969).
130. Maugh, T. H. (1976). Science _192_, 874-876.
131. Pring, R. J. and Richmond, D. V. (1976). Physiol. Plant. Pathol. _8_, 155-162.
132. Langcake, P. and Wickins, S. G. A. (1975). Physiol. Plant Pathol. _7_, 113-126.
133. Langcake, P. (1977). Nature _267_, 511.
134. Kiraly, Z., Barna, B., and T. Ersek. (1972) Nature 239, 456-458.

2

Role of Elicitors of Phytoalexin Accumulation in Plant Disease Resistance

BARBARA S. VALENT and PETER ALBERSHEIM

Department of Chemistry, University of Colorado, Boulder, CO 80309

Plants are exposed to attack by an immense array of microorganisms, and yet plants are resistant to almost all of these potential pests. Many plant tissues have been observed to respond to an invasion by a pathogenic or nonpathogenic microorganism, whether a fungus, a bacterium or a virus, by accumulating phytoalexins, low molecular weight compounds which inhibit the growth of microorganisms. The production of phytoalexins appears to be a widespread mechanism by which plants attempt to defend themselves against pests (1, 2, 3). The molecules of microbial origin which trigger phytoalexin accumulation in plants have been called elicitors (4). Plants recognize and respond to elicitors as foreign molecules. It is highly improbable that plants have evolved separate recognition systems for every bacterial species and strain and every fungal race and every virus that plants are exposed to. Thus, elicitors are likely to be molecules common to many microbes and, in fact, the one to be described in this paper is a fungal polysaccharide, a polysaccharide which is a structural component of the mycelial walls of many fungi.

Most plants produce several structurally related phytoalexins. The most studied phytoalexin of soybeans is glyceollin (5). Lyne *et al.* (6) have characterized two additional soybean phytoalexins which are structural isomers of glyceollin and which appear to have similar antibiotic characteristics. The synthesis of glyceollin, a phenylpropanoid derivative, is probably initiated from phenylalanine via the reaction catalyzed by phenylalanine ammonia-lyase, but, as yet, no biosynthetic pathway for the production of a phytoalexin has been completely described.

Steven Thomas, in our laboratory, has been studying the effect of glyceollin on a variety of microorganisms, for the mechanism by which phytoalexins work is unknown. Glyceollin is a static agent rather than a toxic agent, a trait which seems to be common to many phytoalexins. Glyceollin inhibits the growth, *in vitro*, of the soybean pathogen, *Phytophthora megasperma* var. *sojae* (Pms), the causal agent of root and stem rot. In addition,

Steven has found that glyceollin will stop the growth of three Gram-negative bacteria, *Pseudomonas glycinea*, *Rhizobium trifolii*, and *Rhizobium japonicum*, of the Gram-positive bacterium, *Bacillus subtilis*, and of baker's yeast, *Saccharomyces cerevisiae*. Interestingly, it requires about 25 μg/ml of glyceollin to inhibit by 50% and 100 μg/ml to inhibit by 100% the growth of all of these different organisms. Thus, it appears that a plant's phytoalexins can potentially protect the plant from a broad spectrum of microorganisms.

Glyceollin is accumulated by soybean tissues in response to infection by Pms, the soybean pathogen. Glyceollin accumulates in soybean hypocotyls, within 9 hours of infection with Pms mycelia, to levels which are inhibitory to the growth of Pms *in vitro*. A component of Pms mycelial walls has been demonstrated to stimulate glyceollin accumulation at the same rate as live Pms mycelia. This observation and other data have convinced us that this mycelial wall component is responsible for triggering glyceollin accumulation during infection by the live fungus; and, therefore, we believe that the mycelial wall component is the natural elicitor of this system.

Three different soybean tissues respond to the Pms elicitor by accumulating glyceollin, and these have been used for biological assays of elicitor activity. An assay using 8-day old cotyledons (seed leaves) was used for the purification of the elicitor since this was the least laborious assay developed (*7*). A second bioassay uses the hypocotyls (upper stems) of 5-day old soybean seedlings (*7*) and a third assay uses suspension-cultured soybean cells (*8*). In all three assays, the production of glyceollin is proportional to the amount of elicitor applied. The time course of elicitor-stimulated glyceollin accumulation and the amount of elicitor required is very similar in all three soybean tissues.

Soybean tissues are sensitive to extremely small amounts of Pms elicitor. About 10^{-12} moles of elicitor applied to a single hypocotyl stimulates quantities of glyceollin sufficient to prevent the growth of Pms and other microorganisms *in vitro*. It is impressive, too, to observe the effects on the growing suspension-cultured soybean cells caused by the addition of submicromolar quantities of the polysaccharide elicitor. These cells respond to the small amount of elicitor even though the cells are growing in the presence of 50 mM sucrose. Within a few hours, the cells turn light brown. At the same time, the activity in the cells of at least one of the enzyme believed to be involved in the synthesis of glyceollin, phenylalanine ammonia-lyase, is greatly increased. The increase in activity of the phenylalanine ammonia-lyase precedes the accumulation of glyceollin both in the cells and in the culture medium. The growth of the suspension-cultured cells, as measured by fresh weight, stops upon addition of the elicitor. The cells also stop taking up ions from the media, which is another indication of the lack of growth of these cells (8).

The Chemical Nature of the *Phytophthora megasperma* var. *sojae* Elicitor.

The component of the Pms mycelial walls which stimulates glyceollin accumulation by soybean tissues is a structural polysaccharide. The elicitor was first found in the fluid of old cultures of Pms, probably being released into the culture fluid by autolysis. It was later demonstrated that elicitor-active molecules with the same properties as the culture fluid elicitor could be isolated from the mycelial walls of Pms by a heat treatment similar to that used to solubilize the surface antigens from the cell walls of *S. cerevisiae* (9). The best method for obtaining large amounts of Pms elicitor is partial acid hydrolysis of the mycelial walls. The series of oligosaccharides so obtained are extremely active as elicitors and possess characteristics identical to the culture fluid elicitor. All elicitor-active molecules examined have been found to be glucans. Methylation analysis of the purified elicitor has demonstrated that this glucan is largely a 3-linked polymer with glucosyl branches to carbon 6 of about one out of every three of the backbone glucosyl residues. The elicitor-active glucan is susceptible to hydrolysis by an exo-β-1,3-glucanase isolated from *Euglena gracilis* (10), indicating that the Pms mycelial wall glucan is a β-linked polymer. Optical rotation and NMR studies have confirmed that the glucan is β-linked. This is not surprising as other *Phytophthora* cell walls have a quantitatively dominant component which is a β-1,3-linked glucan with some branches to carbon 6 (11). Indeed, it appears that as much as 60% of the mycelial wall of the Pms is composed of this polymer.

The Pms mycelial wall elicitor has been well characterized (12) and unpublished results of this laboratory). That portion of the elicitor, which is released from the walls by aqueous extraction at 121 C, is heterogeneous in size with an average molecular weight of approximately 100,000. The *E. gracilis* enzyme hydrolyzes glucans from the non-reducing end and is capable of hydrolyzing the the glycosidic bond of 3-linked glucosyl residues that have other glucosyl residues attached to carbon 6. The product of the exoglucanase-degraded mycelial wall-released elicitor is still size heterogeneous, but has an average molecular weight of approximately 10,000. This highly branched glucan fragment retains as much activity as the undegraded elicitor. The predominant glycosidic linkages remaining after extensive exoglucanase treatment are 3-linked, 3,6-linked and terminal glucosyl linkages in a ratio of 1:1:1. Small amounts of 4-linked and 6-linked glucosyl residues are also present.

The evidence that demonstrated that the Pms elicitor is a polysaccharide included the findings that the elicitor is stable to autoclaving at 121 C for several hours, lacks affinity for both anion and cation exchange resins, is completely stable to treatment by pronase, and is size heterogeneous. Periodate treatment of the wall-released elicitor confirms the polysaccharide nature of the active component and demonstrates the essential role of a branched oligosaccharide having terminal glycosyl residues. Exposing the elicitor to periodate eliminates almost all of the elicitor activity. On the other hand, if the periodate-degraded polymers are reduced with sodium borohydride and then are subjected to mild acid hydrolysis, a considerable portion of the elicitor activity is regained. Since the 3- and 3,6-linked glucosyl residues lack vicinyl hydroxyls and are, therefore, resistant to periodate degradation, it seems likely that the periodate has destroyed the elicitor activity by modifying the terminal glucosyl residues of the elicitor. The elicitor activity may be recovered after partial acid hydrolysis of the periodate-treated elicitor because new terminal glucosyl residues have been exposed and are able to provide the proper structure of an active elicitor.

The requirement for elicitor activity of a branched oligosaccharide is supported by our observation that 3-linked glucans which lack branches to carbon 6 or have only a single carbon 6-branched glucosyl residue, such as laminarin, have little or no elicitor activity (less than one thousandth of the Pms elicitor). Indeed, a series of commercially available polysaccharides, oligosaccharides, methylglycosides, and simple sugars have been tested for elicitor activity, and, besides laminarin, the only commercially available product found with elicitor activity was nigeran, a mycelial wall component from the fungus, _Aspergillus niger_.

A major goal of our research is the determination of the detailed molecular structure of the active-site of the Pms elicitor. It is expected that this goal will be achieved by the isolation and structural characterization of the smallest possible elicitor-active oligosaccharide which can be derived from the glucan elicitor. This smallest elicitor-active oligosaccharide has been produced by partial acid hydrolysis of Pms mycelial walls. The series of oligosaccharides obtained by this partial hydrolysis have been partially resolved by high resolution Bio-Gel P-2 gel permeation chromatography. It was found that oligosaccharides containing as few as 7 or 8 glucosyl residues still retain elicitor activity. Glucose is the only detected component of these oligosaccharides.

The smallest elicitor-active oligosaccharide-containing fractions from the P-2 column have been fractionated by high pressure liquid chromatography into at least 5 oligosaccharides. Two of the 5 oligosaccharides obtained by high pressure liquid chromatography can be eliminated by treatment of the mixture of oligosaccharides with the _E. gracilis_ exoglucanase. There appears to be little loss of elicitor activity by treatment with the exoglucanase. Of the three oligosaccharides remaining (there may still be one or more additional oligosaccharides which have not been detected by the fractionation procedures used), two actively elicit soybean tissues to accumulate glyceollin. Sufficient quantities of these oligomers are now being produced to test for chemical purity and to permit structural characterization.

Elicitors Lack Species Specificity

Our experiments have shown that the elicitors of phytoalexin accumulation are not the specificity determining factors in the Pms-soybean system. Three Pms races (races 1, 2, and 3) are distinguished by their differing abilities to infect various soybean cultivars. The elicitor obtained from each of the three Pms races purifies in exactly the same manner, and at least the major structural features of the elicitors from the three races are identical (12). The activities of the elicitors purified from the three Pms races were carefully examined using the three separate bioassays: the cotyledon assay (7), the hypocotyl assay (7), and the cell suspension culture assay (8). All three assays gave the same results, that is, the activities of the elicitors from different Pms races are identical. These findings indicate that the three races of Pms are equally effective at stimulating phytoalexin accumulation in their host soybean tissues.

The results of another type of experiment support our conclusion that elicitors are not responsible for race-specific resistance in the Pms-soybean system. Soybean hypocotyls accumulate glyceollin when inoculated with living mycelia of Pms. The response which is characteristic of natural infections with either an infective or a non-infective race of Pms is retained with this inoculation technique. We have compared the relative effectiveness of live mycelia and purified elicitor in stimulating glyceollin accumulation. The result is the following: the onset and the rate of glyceollin accumulation in seedlings inoculated with infective mycelia was indistinguishable from the onset and rate of glyceollin accumulation in seedlings inoculated with either non-infective mycelia or purified elicitor. These results demonstrate that differences in rates of glyceollin accumulation in response to different races of Pms do not account for the resistance or susceptibility of various soybean cultivars to the Pms races.

The available evidence does indicate that elicitors have a role in resistance even though they are not determinants of race specificity. The elicitor isolated from Pms is capable of protecting soybean hypocotyls from infection by an infective race of Pms if the elicitor is applied to the hypotocyls 6 hours prior to inoculation with Pms. The elicitor cannot protect soybean tissue when applied simultaneously with an infective race of Pms.

Phytoalexins are not Capable of Protecting Plants from Their Pathogens

A microorganism which has evolved the ability to grow successfully on a plant and thus become pathogenic to that plant must also have evolved a mechanism of avoiding the toxic effects of phytoalexins. There are several plausible mechanisms for such avoidance by successful pathogens. One such mechanism might be simply the ability of an infective strain of a pathogen to grow away from the area in which the plant accumulates toxic levels of phytoalexin. This possibility seems likely as an explanation for the avoidance of the effects of glyceollin in soybean plants by infective races of Pms.

There are other mechanisms by which a pathogen might prevent a plant from stopping the growth of the pathogen by accumulation of phytoalexins. For example, a pathogen might kill the plant cells in the region of the pathogen before those cells are capable of synthesizing the enzymes necessary for synthesis of the phytoalexin. Still another mechanism by which a successful pathogen might prevent a plant from accumulating sufficient phytoalexins might be to repress synthesis of enzymes involved in phytoalexin synthesis or else to inhibit the enzymes once they are synthesized. A known mechanism by which pathogens overcome phytoalexin inhibition is metabolizing the phytoalexins to less toxic or unstable compounds (13, 14, 15, 16). A final possible mechanism might be the production by the pathogen of proteins or other molecules which specifically inhibit the enzymes of the host which solubilize elicitors from the mycelial walls of the pathogen. Evidence suggestive of this type of mechanism has also been obtained (17).

Elicitors are Widespread in Nature

Soybean plants have evolved the ability to recognize and respond to the structural β-glucan of *Phytophthora* mycelia walls. Similar β-glucans are found in the walls of a wide range of fungi. One fungus containing such β-glucans is brewer's yeast, *S. cerevisiae*, a non-pathogen of plants. An elicitor has now been purified from a commercially available extract of brewer's yeast (Difco) (M. Hahn, unpublished results). The 80% ethanol insoluble fraction of the yeast extract contains a very active elicitor of glyceollin accumulation in soybeans. Most of the polysaccharide in this 80% ethanol insoluble fraction is a mannan. However, yeast extract does contain small amounts of the β-glucan. The glucan can be almost completely separated from the mannan by binding the mannan to an affinity column consisting of Concanavalin A covalently attached to Sepharose. The glucan can be separated from glycoproteins by binding the proteins to sulfopropyl-Sephadex. Both the purified mannan and purified glucan remain contaminated by small amounts (≃2%) of arabinogalactan. Ribose, which contaminates the 80% ethanol insoluble fraction, is removed on a DEAE-cellulose column. The elicitor activity of the crude 80% ethanol yeast extract precipitate resides in the glucan component. The small amount of residual activity remaining in the mannan fraction can be attributed to the minor contamination of this fraction by glucan. The glucan is composed of the same glucosyl linkages found in the Pms elicitor. The same quantities of the yeast and Pms elicitor are required to stimulate glyceollin accumulation in soybeans.

Our laboratory has obtained other evidence that the elicitor-phytoalexin story is a general one. For example, the Pms elicitor stimulates suspension-cultured cells of sycamore and parsley to produce large amounts of phenylalanine ammonia lyase activity (8). In addition, we have obtained evidence that the Pms elicitor stimulates *Phaseolus vulgaris*, the true bean, to accumulate its phytoalexins (K. Cline, unpublished results). We have already reported (18) that a wall glucan from *Colletotrichum lindemuthianum*, a pathogen of *P. vulgaris*, stimulates *P. vulgaris* to produce its phytoalexin. The *C. lindemuthianum* glucan also stimulates soybeans to produce glyceollin (K. Cline, unpublished results). And, finally, we have recently demonstrated that the Pms elicitor stimulates potato tubers to accumulate their phytoalexins (M. Wade, unpublished results).

In summary, elicitors appear to be general in nature, and diverse plants are able to respond to a single elicitor. Elicitors may therefore provide a new way of protecting plants against their pests, for elicitors may activate the plant's own defense mechanism and thereby eliminate some of the need for spraying agricultural crops with poisonous pesticides. Several industrial firms have recognized this potential and have initiated their own research or are supporting out-of-house research on elicitors.

Literature Cited

1. Ingham, J.L., Botanical Rev. (1972) 38, 343.
2. Kuč, J., Ann. Rev. Phytopathol. (1972) 10, 207.
3. Deverall, B.J., Proc. R. Soc. Lond. B. (1972) 181, 233.
4. Keen, N.T., Partridge, J.E. & Zaki, A.I., Phytopathology (1972) 62, 768.
5. Burden, R.S. & Bailey, J.A., Phytochemistry (1975) 14, 1389.
6. Lyne, R.L., Mulheirn, L.J. & Leworthy, D.P., J.C.S. Chem. Commun. (1976) 497.
7. Ayers, A.R., Ebel, J., Finelli, F., Berger, N. & Albersheim, P. Plant Physiol. (1976) 57, 751.
8. Ebel, J., Ayers, A.R. & Albersheim, P. Plant Physiol. (1976) 57, 775.
9. Ayers, A.R., Ebel, J., Valent, B. & Albersheim, P., Plant Physiol. (1976) 57, 760.
10. Barras, D.R. & Stone, B.A., Biochim. Biophys. Acta (1969) 191, 342.

11. Bartnicki-Garcia, S., J. Gen. Microbiol. (1966) 42, 57.
12. Ayers, A.R., Valent, B., Ebel, J. & Albersheim, P. Plant Physiol. (1976) 57, 766.
13. Higgins, V.J., Physiol. Plant Pathol. (1975) 6, 5.
14. Higgins, V.J., Stoessl, A. & Heath, M.C., Phytopathology (1974) 64, 105.
15. Van Den Huevel, J. & VanEtten, H.D., Physiol. Plant Pathol. (1973) 3, 327.
16. Van Den Huevel, J., VanEtten, H.D., Coffen, D.L. & Williams, T.H., Phytochemistry (1974) 13, 1129.
17. Albersheim, P. & Valent, B.S. Plant Physiol. (1974) 53, 684.
18. Anderson-Prouty, A.J. & Albersheim, P. Plant Physiol. (1975) 56, 286.

3

Biochemical Aspects of Plant Disease Resistance and Susceptibility

DOUG S. KENFIELD and GARY A. STROBEL

Department of Plant Pathology, Montana State University, Bozeman, MT 59715

The events that either permit or preclude a pathogenic association between a host and its pathogen can manifest themselves at many different levels of the interaction (1). A complete knowledge of the interaction necessitates elucidation of the molecular events in connection with and pertinent to the genetics of pathogenicity and resistance. The purpose of this report is both to propose and illustrate a few of the possible molecular interactions between host and parasite and to discuss how these interactions relate to resistance or susceptibility in plant diseases.

Molecular Interactions

The molecular interplay that can occur between a plant and its pathogen can be divided into five categories. The first two categories arise when small molecules from either the pathogen or host interact with one or more structures in the other. The second two occur when the active constituents are macromolecules from either source. Such molecules may exert either positive or negative effects on their targets, and more than one type of interaction can and often does occur in a given disease. A fifth type of interaction may involve the decrease or elevation of the concentration of a normal constituent in either the host or the pathogen as influenced by the other. This is best exemplified in the case of pathogens that may utilize some constituent which ultimately results in an imbalance in metabolism in the host (2).

Parasite-Produced Small Molecules

It is conceivable that a small molecule produced by a pathogen could act selectively on a plant and actuate either a susceptible or resistant reaction. In the latter case, Keen (3) reported on the presence of a dialyzable compound in culture filtrates of an incompatible race of *Phytophthora megasperma* var. *sojae* that specifically elicited the increased production of the antifungal phytoalexin hydroxy-phaseollin in a resistant cultivar of soybean but not in a near-isogenic, susceptible cultivar. Compatible races of this fungus did not produce this elicitor. While attempts by some to repeat these interesting observations have not been successful (4), it seems likely that such a phenomenon does exist somewhere in nature.

Pathogens are also capable of producing small molecules which act as toxins in the plant, and numerous instances are known where these toxins exhibit host selectivity (5). Characterization of such toxins is a vital step that can ultimately lead to the discovery of their mode of action, biosynthesis, preparation by organic techniques, and, most importantly, to serve as tools in discovering the molecular basis of disease specificity. A toxin which exhibits host specificity may also be of value in screening for resistant plants in a genetics or breeding program.

Highly virulent strains of *Alternaria mali*, the causal agent of apple blotch, produce as many as seven host-specific toxins in culture while less virulent strains lack the ability to produce some of the toxins (6,7). The two major toxins have been characterized as alternariolide (cyclo-[α-hydroxyisovaleryl-α-amino-p-methoxyphenylvaleryl-α-amino-acryl-alanyl-lactone]) and its demethoxy derivative (8). Resistance in apple trees has been shown to be controlled by multiple dominant genes, thus this system offers excellent opportunities to elucidate the attack-counterattack hypotheses of species evolving in close association with one another.

Alternaria tenuis produces a toxin called tentoxin. This molecule is also a cyclic tetrapeptide with the structure cyclo (L-leucyl-N-methyl-(Z)-dehydrophenylalanyl-glycyl-N-methyl-L-alanyl) (9). The toxin exhibits some species specificity as it causes chlorosis in seedlings of certain species but not others, notably the Cruciferae (10). Steele, et al. (11) reported that the active site for this toxin in sensitive plants is the chloroplast coupling factor 1(CF1). Evidence supporting this hypothesis is that, in a sensitive species, both CF1-ATPase and the coupled electron transport system (ETS) in chloroplasts are inhibited by tentoxin. Also, radio-labelled tentoxin appears to bind in a molar ratio to partially purified CF1 as measured by a time-release of radio-labelled tentoxin utilizing ultra-filtration techniques. An affinity constant of $10^8 M^{-1}$ was reported for the toxin to CF1 from a sensitive species. In an insensitive species, a 23-fold increase in tentoxin was required to inhibit CF1-ATPase, coupled-ETS was not inhibited, and tentoxin bound to the partially purified CF1 with an apparent affinity of less than $10^4 M^{-1}$. More convincing evidence of a direct involvement of CF1 is required. Coincident migration of radioactivity from bound, labelled tentoxin and CF1 during gel electrophoresis would be substantiative evidence, as would resolution of the differences between *in vivo* and *in vitro* results (12,13).

Tentoxin has been synthesized and a description of the active part of the molecule is being attempted (14). Results of this approach may aid in a more rigorous characterization of the active site in the plant and allow a fuller understanding of the biological activity of such cyclic tetrapeptides.

The Helminthosporium species of fungi constitute an important group in terms of specificity as many of them produce host-specific toxins. *H. victoriae* and *H. carbonum* produce peptidyl toxins which act selectively on the plasma membrane. The complete structures of neither of these toxins nor specific sites of activity have been characterized, however.

Southern corn leaf blight, the most devastating plant disease epidemic in recent history, is caused by *Helminthosporium maydis*. Differential susceptibility in corn is related to the presence of Texas male sterile (Tcms) or normal (N) cytoplasm. Pathogenicity of the fungus is determined by the ability to produce as many as five different host specific toxins (toxins I-V) which appear to be terpenoids and their derivatives (15,16). Toxin preparations have been shown to stimulate the oxidation of malate and pyruvate, inhibit phosphorylation, and induce swelling of mitochondria from susceptible but not resistant plants (17). Such preparations have also been shown to cause ion leakage, reduce ADP content, inhibit root growth, and directly affect stomatal functioning (18). Toxin preparations inhibit oxidative phosphorylation and stimulate ATPase activity in Tcms but not N mitochondria (19). While these diverse effects are interesting, crude toxin preparations were utilized in these studies which makes it difficult to evaluate the findings.

Watrud, et al. (20) demonstrated that malate dehydrogenase was inhibited in mitochondria from Tcms corn but not in those from N corn. Upon rupture or removal of the outer mitochondrial membrane, however, mitochondria from N cytoplasm became sensitive. Toxin-binding studies showed that mitochondria from both N and Tcms cytoplasm bound toxins II and I, but that the greatest binding activity was associated with the inner mitochondrial membrane of Tcms mitochondria. Toxin-binding in plant extracts was also demonstrated by Ireland and Strobel (21). Utilizing dextran-coated charcoal to adsorb unreacted toxin from assay mixtures, they found that extracts from a susceptible corn line (W64A Tcms) contained a protein localized primarily in the cytosol which binds toxins I and II with a K_A in the order of 10^{-4}M. The binding activity is relatively heat insensitive but is destroyed by treatment with papain or ficin. Binding activity was found in extracts of both sensitive and insensitive corn lines, as well as other plant species, thus allowing no direct correlation between binding and resistance or susceptibility.

Mertz, et al. (22) argue that the plasmalemma contains the site of specificity for toxin action. They found that a toxin preparation containing primarily toxin II (10^{-6} to 10^{-8}M) inhibited K^+ uptake in leaf discs and apical root segments of both Tcms and N corn; however, a 30-fold increase in toxin concentration was required to obtain a similar inhibition in N corn. In Tcms corn, increasing the concentration of toxin or time of treatment, decreasing pH, decreasing the extracellular K^+ concentration, and treating with light all increased the inhibitory activity. Further, six genotypes of corn which show different degrees of susceptibility to the fungus also showed a corresponding differential inhibition of K^+ uptake upon treatment with toxin. A separate study has also revealed a differential sensitivity of potassium uptake in roots to toxin preparations (23).

An accurate picture of the activity of these host-specific toxins is not yet possible. Definitive experiments await the identification of the individual toxins and their use as homogeneous preparations. It does appear that multiple reactions occur both at the plasmalemma and in the cytosol, but the site(s) of specificity remain obscure.

Helminthosporium sacchari causes eyespot on sugarcane. This fungus produces a glycoside called helminthosporoside with the proposed structure 2-hydroxycyclopropyl-α-D-galactoside (24). The pathogenicity of this fungus depends upon its ability to produce helminthosporoside. Multiple transfers of the pathogenic, toxin-producing fungus on synthetic media resulted in a loss of toxin production and a temporary inability of the fungus to cause typical symptoms on susceptible leaves (25). Susceptibility in the plant is governed by the presence of a protein in the plasmalemma having the ability to bind helminthosporoside (26,27,28), this binding being the first and specific step of the molecular events of pathogenesis. One mode of resistance in sugarcane is conferred by the inability of the plant to bind the toxin. An originally susceptible cane clone was mutagenized via irradiation and three resistant clones were obtained, each of which lacked the ability to bind the toxin (29). In another experiment, susceptibility to helminthosporoside was transferred *in vitro* to tobacco and resistant sugarcane protoplasts by incorporating the toxin-binding protein from susceptible sugarcane into the normally resistant protoplasts (30). Elucidation of this pathogenic interaction has had an impact on sugarcane breeding since geneticists currently spray clones of sugarcane seedlings with helminthosporoside in order to sort out varieties susceptible to this world-wide disease (31).

Plant-Produced Small Molecules

Phytoalexins, a heterogeneous array of substances having fungistatic or fungicidal properties, have received major emphasis of compounds in this category. The production of phytoalexins by plants upon treatment with various stimuli undoubtedly plays a role in the host-parasite interaction, and this response is dealt with elsewhere in this volume.

Specificity as a time related phenomenon appears to be

exhibited in the interaction of Fusarium graminearum and wheat in the disease known as headblight. Wheat is susceptible to headblight after the anthers have emerged and is resistant prior to anthesis or if the anthers have been removed. Two related compounds, betaine and choline, have been isolated from the anthers both of which stimulate the in vitro growth of the fungus and increase the virulence of the macroconidia (32). It is not known, however, if resistance to headblight is a specific response related to the absence of these compounds.

In the case of H. sacchari on sugarcane it appears as if a second mode of disease resistance could arise by the inability of the plant to "activate" toxin production in the fungus. Pinkerton and Strobel observed that the fungus lost its ability to produce the toxin if grown for an extended time on synthetic media (33). The fungus regained its pathogenicity if grown on agar derived from susceptible sugarcane or if crude material from a water wash of susceptible sugarcane leaves was added to the synthetic media. They isolated, identified, and synthesized one of the activators and it was shown to be serinol, 2-amino-1,3-propanediol. Evidence indicated that serinol was of plant origin and, subsequently, Babczinski et al. (34) demonstrated that the biosynthesis of serinol proceeds via serinol phosphate from dihydroxyacetone phosphate in a transamination reaction in extracts of susceptible sugarcane. Resistant cane did not possess serinol, nor the metabolic pathway for its production. A second inducer (molecular weight ≅ 700) is currently being characterized by U. Matern of this laboratory. It is not known how these compounds act to stimulate the synthesis of helminthosporoside. It would also appear likely that numerous other host-parasite combinations are operating in a similar mode, as loss of pathogenicity during extended culture is frequently observed.

The antithesis of a host-formed activator would be a host-formed inhibitor of toxin production. To our knowledge, no one has yet demonstrated the existence of such a compound.

Macromolecules from Pathogens

The production of a macromolecule by the pathogen that confers specificity in the host-parasite interaction is a viable hypothesis, but is not supported by strong evidence. Pathogens do produce numerous degradative enzymes such as cellulases, pectinases, and proteases, but there is no evidence for host specificity being prescribed by these enzymes.

A model system in which the pathogen apparently counters a possible defense reaction by the host occurs between Colletotrichum lindemuthianum and Phaseolis vulgaris (35). A major component of the fungal cell wall is a β-1,3 glucan, and P. vulgaris posseses a β-1,3 glucanase that could pose a threat to the pathogen since its wall material could be destroyed. Albersheim, et al., isolated a proteinaceous compound from the fungus that inhibits the activity of the host's β-1,3 glucanase. These observations are worthy of being tested in other systems and should be further studied to ascertain how important this phenomenon is to pathogen specificity and the defense reaction of the host.

Both bacteria and fungi produce large phytotoxic glycopeptides which have been shown to possess some host specificity (36,37). For instance, Ries and Strobel isolated and characterized a glycopeptide from *Corynebacterium insidiosum* that causes wilting in alfalfa (38). The toxin is a glycofucan with a molecular weight of 5×10^6. At a critical concentration of 0.05%, the toxin was effective as a screening tool in selecting alfalfa clones resistant to *C. insidiosum*.

Phoma tracheiphila, the causal agent of mal secco disease of citrus, produced a glycopeptide capable of inducing veinal necrosis and vein clearing followed by wilt and leaf shedding in susceptible citrus. The toxin has a molecular weight of 93,000. Amino acids contribute 36% of its weight, which is considerably greater than exhibited by other phytotoxic glycopeptides (39). The toxin causes death of susceptible lemon callus, but has no detrimental effect on resistant orange callus. This glycopeptide has enormous potential for screening citrus protoplasts and/or callus tissue for toxin resistance. This technological advance seems feasible since citrus trees have recently been regenerated from protoplasts.

A host-parasite interaction that may involve a transfer of macromolecules between species is the stem rust-wheat association. This disease is characterized by a high degree of specificity at the genetic level in that cultivars of wheat carry genes for resistance (R) and susceptibility (r) to races of the fungus containing corresponding genes for virulence (a) or avirulence (A). Of the four possible combinations R/A, R/a, r/A, and r/a, only the R/A interaction leads to a resistant reaction characterized by localized necrosis (40).

Rohringer, et al. (41,42) published evidence that a gene-specific RNA, possibly the transcription product of the gene for avirulence in the fungus, is directly involved in the resistant reaction of wheat to stem rust. More rigorous support of this hypothesis is needed, however, as these researchers recently reported their failure to consistently reproduce their original results (43).

One of the best illustrations of the involvement of a macromolecule in plant disease specificity is the phenomenon of crown gall, a tumorous disease of plants caused by *Agrobacterium tumefaciens*. A macromolecule (bacterial plasmid) has been implicated as the key factor in infectious *A. tumefaciens* that brings about transformation of plant cells (44). Nevertheless, the ultimate gene product(s) that bring about the abnormal behavior of the plant cell remains to be identified. It is apparent from these and other studies that the field of macromolecular interactions between pathogens and their plant hosts is in a primordial state.

Macromolecules from Plants

Possibilities for specificity at the macromolecular level in plants include unique or novel cutin or cell wall structures, isozymes that dictate specificity, or even the production of enzymes that destroy or alter specific toxins produced by the pathogen. Evidence for the direct, specific involvement of any of these macromolecules in the host-parasite interaction is to be forthcoming.

An example of specificity in plant-disease resistance established at the macromolecular level is the *H. sacchari*-sugarcane interaction wherein resistance is expressed as an altered protein receptor that cannot bind the host-specific fungal toxin (27). From additional work on this system, it has become apparent to us that it is *incorrect to believe that the molecular event that confers specificity also must constitute the crucial detrimental effect in the disease process*. A certain dogma in phytopathology is that a host-specific toxin must reproduce all of the symptoms of the disease incited by the pathogen. This concept formulates a workable basis for isolating host specific toxins, but it must not be misconstrued to mean that the specific interaction of a toxin with its active site comprises the entire molecular events of pathogenesis. The specific event may simply be a potentiator for the crucial event.

These points are illustrated in the *H. sacchari*-sugarcane system. The binding protein for the toxin is localized on the plasma membrane, and all susceptible clones of sugarcane tested thus far possess an active binding protein. Clones that are resistant possess a protein which lacks detectable toxin binding activity. Certain lines of evidence indicate that toxin binding is only a potentiating reaction that is crucial to specificity but may be just a key to other events in the cell leading to its demise.

Upon treatment with the toxin, protoplasts from susceptible cane swell and eventually burst, tissue from susceptible cane undergoes an immediate drop in membrane potential and shows leakiness of ions, tissue ATP levels drop, and photosynthesis declines (45). Pretreatment with cycloheximide, N_2, or elevated temperature cause susceptible leaves to be resistant (46). Such treatments, however, do not affect the toxin-binding activity of the protein. The toxin activates (K^+, Mg^{++})-ATPase activity in plasma membrane preparations from susceptible tissues but not from heat-treated tissues (47). Arrhenius plots of heat effects on ATPase activity and symptom expression reveal a sharp break at 32°C for both (unpublished). Also, common inhibitors of ATPase activity confer resistance to susceptible tissue.

It appears that a specific, heat-insensitive step of toxin

binding precedes a heat-sensitive event crucial for symptom expression. The second event may well be an activation of plasma-membrane (K^+, Mg^{++})-ATPase activity.

That specificity resides in the toxin-binding step is further supported by the fact that sensitivity to helminthosporoside can be transferred _in vitro_ to both tobacco and resistant sugarcane protoplasts via transfer of the binding protein. The binding activity alone is not sufficient to promote disease symptoms, however, because toxin-binding activity has been found in TCA-solubilized extracts of plasmalemma-enriched fractions from diverse plants (See Table I). Mint shows a nearly two-fold increase in specific binding activity over that from susceptible sugarcane, yet toxin applied to mint leaves fails to educe any observable symptoms. The toxin also fails to cause an activation of the plasma-membrane (K^+, Mg^{++})-ATPase from mint (unpublished). In fact, the toxin does not cause symptoms on any of the plants tested except susceptible sugarcane. Note also that while tobacco possesses toxin-binding ability, protoplasts were not killed by helminthosporoside until they had been treated with the binding protein from sugarcane. Clearly, binding activity is essential but insufficient to cause disease.

Yet another interesting observation is that the TCA extracts of all the plants tested can bind raffinose and galactinol; yet resistant sugarcane, potato, and corn lack the ability to bind helminthosporoside. If, in truth, what we are studying is a facet of α-galactoside transport in plants, it appears that alterations can occur which drastically affect ligand specificity yet do not interfere with apparently normal physiological functions. Currently in our laboratory, isolation and characterization of the binding proteins from sugarcane, mint, and tobacco are in progress in an attempt to discern the uniqueness of the protein from susceptible sugarcane and its involvement in disease specificity.

Implications

What are the implications of molecular research into the specificity of plant diseases? Certainly a primary goal of plant pathologists is the development of more effective control of plant diseases. A very important application leading to enhancement of control measures is the identification of biochemical markers for plant breeders thereby increasing the efficiency and celerity of their efforts to produce disease-resistant plants. Currently in this laboratory, J. P. Beltran is characterizing in barley cultivars differential binding activity for a glycosidic toxin produced by _Rhyncosporium secalis_, the causal agent of barley scald. If successful, this work will provide the first biochemical marker of toxin resistance in a plant, the genetics of which have been extensively characterized.

Alternatively, the identification of molecules active in

Table I. α-Galactoside binding activity of trichloroacetate-solubilized, plasma-membrane-enriched fractions.

Source	Binding activity (m mole substrate/g protein)		
	Raffinose	Galactinol	Helminthosporoside
susceptible sugarcane 51NG97	0.05	0.19	2.7
resistant sugarcane H50-7209	0.04	0.16	<0.005
Mint	1.07	0.47	4.7
Tobacco	0.08	0.06	0.24
Potato	0.06	0.13	<0.005
Beet	0.11	0.05	1.9
Wheat	0.07	0.10	0.012
Barley	0.10	0.04	0.020
Corn	0.08	0.03	<0.005

Note: A pellet enriched in plasma membranes was obtained as in (25). This pellet was homogenized in Tris-HCl, 0.05 M, pH 7.0 containing 0.5 M trichloroacetate (TCA) and incubated at 4C for 4 hours. After centrifugation at 48,000 g for 20 minutes, 0.5 ml of the supernatant was incubated at 25C for 30 minutes with ^{14}C-ligand (10^{-4}M). The assay mixture was then chromatographed over Bio-gel P-2 (1.5 x 45 cm), eluting with buffer. The fractions eluting at the void volume were collected, placed in 5 ml of Aquasol, and counted in a Packard Tri-Carb Liquid Scintillation Spectrometer. Total protein was measured according to Lowry, et al. (46). Specific radioactivities of the ligands (DPM/μmole) were: raffinose - 351,500; galactinol - 33,700; and helminthosporoside - 4,000. Bovine serum albumin and boiled TCA extracts were used as controls.

specificity may promote a rational approach to development and application of fungicides and herbicides, allowing design of more effective and more selective poisons. This approach is especially important in crop systems where weeds are closely related to the commercial plants (i.e. wild oats and wheat) and high degrees of selectivity are required for efficient control.

Research on specificity will also increase our knowledge on the basic processes of life. As examples: tentoxin studies may well reveal important events in the phenomenon of uncoupling of oxidation-phosphorylation; aspects of protein-protein interaction in membranes and their effect on active transport are possible extensions of the work on the activity of helminthosporoside; the study of southern corn leaf blight should increase our knowledge of interactions of nuclear and cytoplasmic genes.

Acknowledgements

Certain aspects of the research work presented in this report were supported in part by the National Science Foundation Grant PCM 76-19565, a grant from the Herman Frasch Foundation and the Montana Agricultural Experiment Station.

Literature Cited

1. Brian, P. W., "Specificity in Plant Diseases", pp.15-26, Plenum Press, New York, 1976.
2. MacDonald, P. W. and Strobel, G. A., Plant Physiol., (1970), 46, 126-135.
3. Keen, N. T., Science, (1975), 187, 74-75.
4. Ayers, A. R., Ebel, J., Valent, B., and Albersheim, P., Plant Physiol., (1976), 57, 760-766.
5. Scheffer, R. P., "Physiological Plant Pathology", pp.247-269, Springer-Verlag, New York, 1976.
6. Kohmoto, K., Khan, I. D., Renbutso, Y., Taniguchi, T., and Nishimura, S., Physiol. Plant Pathol., (1976), 8, 141-153.
7. Nishimura, S., Kohmoto, K., Otani, H., Fukami, H., and Ueno, T., "Biochemistry and Cytology of Plant Parasite Interaction", pp.94-101, Elsevier Scientific Publishing Co., New York, 1976.
8. Okuno, T., Ishita, Y., Sawai, K., and Matsumoto, T., Chem Lett, (1974), 635-638.
9. Meyer, W. L., Templeton, G. E., Grable, C. I., Jones, R., Kuyper, L. F., Lewis, R. B., Sigel, C. W., and Woodhead, S. H., J. Am. Chem. Soc., (1975), 97, 3802-3809.
10. Durbin, R. D. and Uchytil, T. F., Phytopathology, (1977), 67, 602-603.
11. Steele, J. A., Uchytil, T. F., Durbin, R. D., Bhatnagar, P., and Rich, D. H., Proc. Nat. Acad. Sci., USA, (1976), 73, 2245-2248.
12. Arntzen, C. H., Biochem. Biophys. Acta, (1972), 283,

539-542.
13. Fulton, N. D., Bollenbacher, K., and Templeton, G. E., Phytopathology, (1965), 55, 49-51.
14. Rich, D. H., "Specificity in Plant Diseases," pp.169-184, Plenum Press, New York, 1976.
15. Karr, A. L., Karr, D. B., and Strobel, G. A., Plant Physiol., (1974), 53, 250-257.
16. Karr, D. B., Karr, A. L., and Strobel, G. A., Plant Physiol., (1975), 55, 727-730.
17. Miller, R. J. and Koeppe, D. E., Science, (1971), 173, 67-69.
18. Arntzen, C. J., Koeppe, D. E., Miller, R. J., and Peverly, J. H., Physiol. Plant Pathol., (1973), 3, 79-90.
19. Bednarski, M. A., Izawa, S., and Scheffer, R. P., Plant Physiol., (1977), 59, 540-545.
20. Watrud, L. S., Baldwin, J. K., Miller, R. J., and Koeppe, D. E., Plant Physiol., (1975), 56, 216-221.
21. Ireland, C. A. and Strobel, G. A., Plant Physiol., in press.
22. Mertz, S. M. and Arntzen, C. J., Plant Physiol., in press.
23. Frick, H., Bauman, L. F., Nicholson, R. L., and Hodges, T. K., Plant Physiol., (1977), 59, 103-106.
24. Steiner, G. W. and Strobel, G. A., J. Biol. Chem., (1971), 246, 4350-4357.
25. Pinkerton, F., Ph.D. Thesis, Montana State University, (1976).
26. Strobel, G. A., J. Biol. Chem., (1973), 248, 1321-1328.
27. Strobel, G. A., Proc. Nat. Acad. Sci., USA, (1973), 70, 1693-1696.
28. Strobel, G. A. and Hess, W. M., Proc. Nat. Acad. Sci. USA, (1974), 71, 1413-1417.
29. Strobel, G. A., Steiner, G. W., and Byther, R. S., Biochem. Genet., (1975), 13, 557-565.
30. Strobel, G. A. and Hapner, K. D., Biochem. Biophys. Res. Comm., (1975), 63, 1151-1156.
31. Byther, R. S. and Steiner, G. W., Phytopathology, (1972), 62, 466-470.
32. Strange, R. N., Majer, J. R., and Smith, H., Physiol. Plant Pathol., (1974), 4, 277-290.
33. Pinkerton, F. and Strobel, G. A., Proc. Nat. Acad. Sci. USA, (1976), 73, 4007-4011.
34. Babczinski, P., Matern, U., and Strobel, G. A., Plant Physiol., in press.
35. Albersheim, P. and Valent, B. S., Plant Physiol., (1974), 53, 684-687.
36. Strobel, G. A., Ann. Rev. Plant Physiol., (1974), 25, 541-566.
37. Strobel, G. A., Ann. Rev. Microbiol., (1977), 31, 205-224.
38. Ries, S. M. and Strobel, G. A., Physiol. Plant Pathol., (1972), 2, 133-142.
39. Nachmias, A., Barash, I., Solel, Z., and Strobel, G. A.,

Physiol. Plant Pathol., (1977), 10, 147-157.
40. Flor, H. H., J. Agric. Res., (1946), 73, 335-357.
41. Rohringer, R., Howes, N. K., Kim, W. K., and Samborski, D. J., Nature, (1974), 249, 585-588.
42. Rohringer, R., "Specificity in Plant Diseases", pp.185-198, Plenum Press, New York, 1976.
43. Rohringer, R., Howes, N. K., Kim, W. K., and Samborski, D. J., Can. J. Bot., (1977), 55, 851-852.
44. Kado, C. I., Ann. Rev. Phytopathol., (1976), 14, 265-308.
45. Strobel, G. A., Trends in Biochem. Sci., (1976), 1, 247-250.
46. Byther, R. S. and Steiner, G. W., Plant Physiol., (1975), 56, 415-419.
47. Strobel, G. A., Proc. Nat. Acad. Sci., USA, (1974), 71, 4232-4236.
48. Lowry, O. H., Rosebrough, N. J., Farr, A. L., and Randall, R. J., J. Biol. Chem., (1951), 193, 265-275.

4

Interactions between *Phytophthora Infestans* and Potato Host

DONALD D. BILLS

Eastern Regional Research Center, Agricultural Research Service, U.S. Department of Agriculture, Philadelphia, PA 19118

The white potato (Solanum tuberosum) was unknown outside of South America until the sixteenth century, but by the nineteenth century, it was widely cultivated in the British Isles, Europe, Russia, and the United States. Ireland, in particular, shifted its agricultural emphasis from cereal grains to the more productive potato with such success that the population of Ireland increased considerably to about 7 million by 1844. In 1845, climatic conditions and widespread growth of the potato in Ireland permitted catastrophic field infection by Phytophthora infestans and resulted in the loss of nearly the entire potato crop in that year and the subsequent year. The results for Ireland were devastating; over one million people died of starvation or diseases associated with malnutrition, and over one million more emigrated. P. infestans, the pathogen responsible for potato late blight, remains the most significant fungal parasite of the potato.

The Nature of Late Blight

The symptoms of late blight on the potato plant are visible first as dark brown patches on the leaflets, usually near the margin. On the leaflets of varieties that are incompatible with the infecting race of P. infestans, the necrosis is restricted to small spots or flecks that do not enlarge. In a susceptible interaction, the necrotic areas expand, and white mold growth is usually visible along their edges on the underside of the leaflet. If weather conditions remain sufficiently damp, the infection can spread and destroy the entire haulm in a period of several weeks. During the course of progressive infection, spores are spread to other plants by wind, rain, or insects and transferred to the soil and the tubers of infected plants by rain.

The means by which late blight is perpetuated are shown in Fig. 1. In the vicinity of potato fields, elimination of cull piles containing blighted tubers is a well-accepted practice for controlling the spread of infection from this source, and the selection of seed tubers as free from infection as possible is recognized as a necessity. One blighted seed tuber in 100,000 provides an adequate initial inoculum for a major epidemic (1). Control of the spread of infection between potato plants is dependent upon the use of fungicides such as the ethylenebisdithiocarbamates and the planting of potato varieties that have some degree of resistance. In the United States, the use of fungicides to prevent the spread of late blight generally is practiced in all areas except the western states where the summer climate is usually dry and unconducive to the development of late blight. Under current management practices in the United States, late blight decreases the yield of potatoes at harvest by about 4%. Further losses are encountered in storage when tubers with minor blighted areas become mixed inadvertently with sound tubers. *P. infestans* does not generally spread to healthy tubers during storage, but the loci of infection provide footholds for secondary bacterial infections that spread to other tubers and cause significant losses (2).

Potato Resistance to Fungal Infection

The potato is immune to parasitization by the vast majority of fungi; there are, including *P. infestans*, only six or seven fungal parasites of the plant and tuber that are economically significant (2). The term immunity is used to denote complete and apparently permanent protection against a given parasite, whereas resistance denotes either incomplete or impermanent protection. Parallels have been drawn between immunity and resistance (1, 3), but Robinson (1) suggests that immunity is outside of the conceptual bounds of parasitism. Perhaps the major difficulty in conceptualizing immunity is that no one yet has formulated a research approach that might lead to a description of the circumstances responsible for the immune relationship of plants to fungi. Resistance, on the other hand, provides chemically and physically observable and measurable phenomena as a basis for research.

Resistance to fungal parasites commonly is divided into two general categories, horizontal resistance and vertical resistance. Horizontal resistance is said to provide a constant level (high, moderate, low, or nearly zero) of infectability with all existing and potential races of a given fungal species. From the constant level of infectability, the term and concept of horizontal resistance was formulated by van der Plank (4). It is believed that horizontal resistance usually has a polygenic origin in the host and that mechanisms responsible for horizontal resistance are likely to be numerous and complex. If this were not the usual case, minor genetic changes in a fungal parasite would be more likely to result in new races capable of parasitizing the host at higher levels.

Horizontal resistance can be viewed speculatively as a barrier of physical or chemical defenses which slow fungal invasion. Unavailability in the host of adequate essential nutrients required for fungal growth would constitute a passive form of horizontal resistance. The concept of horizontal resistance can also be extended to chemicals applied externally. The effectiveness of new fungicides is often temporary because the fungi produce new races unaffected by them. Other fungicides, such as Bordeaux Mixture, appear to have "horizontal properties" and maintain their effectiveness permanently, indicating that the fungi are unable to produce new races able to tolerate them. The inability of fungi to circumvent the toxicity of the "horizontal" dithiocarbamate fungicides, which inhibit twenty or more enzymes (5), is easily understood. The presence of a "natural horizontal fungicide" in host tissue would confer some degree of resistance, and there is evidence, cited later, that the potato tuber contains indigenous compounds with fungitoxic properties.

While potato varieties can exhibit varying degrees of horizontal resistance to *P. infestans*, most commercial varieties have (and most breeding programs are aimed at producing) vertical resistance (1). The term vertical resistance is derived from the observation that in some host-parasite interactions infection of a given host variety by different fungal races can occur at either a high level (susceptible or compatible interaction) or a very low level (resistant, incompatible, or hypersensitive interaction) (4). The terms compatible and incompatible are used in the remainder of this paper. Vertical resistance is often referred to as a gene-for-gene relationship between host and fungal parasite (1). Obviously, it is not the genes but the chemical and physical characteristics whose inheritance is genetically controlled that either match or do not match each other. Albersheim and Anderson-Prouty (6) extensively reviewed the evidence for the existence of genetically controlled recognition mechanisms that result in an incompatible response between fungal parasite and plant host. A formal classification scheme has been established to designate the genetic relationship between potato varieties and *P. infestans* races in terms of compatibility and incompatibility (7). In this widely accepted scheme, potato varieties are described as possessing vertical resistance genes designated as R_1, R_2, R_3, etc. A single potato variety may possess zero, one, or multiple resistance genes. *P. infestans* races are described as possessing pathogenicity genes designated as v_1, v_2, v_3, etc. (or as race 1, race 2, race 3, etc.), and a single race may possess zero, one, or multiple pathogenicity genes. A potato variety having the single resistance gene R_2 interacts in a compatible manner with *P. infestans* races with the v_2 gene and in an incompatible manner with all other races. Such a scheme has great utility as a classification system, but its empirical nature must be kept in mind. The designations used have no real meaning apart from the observed interactions between host and parasite.

Not all investigators believe that resistance can be neatly categorized as either horizontal or vertical. Also, both types of resistance may exist simultaneously in a single cultivar, although strong vertical resistance usually tends to mask horizontal resistance.

Native Properties Related to Resistance

The potato plant and tuber have native physical and chemical properties related to resistance. The epidermis of the foliar portion of the plant and the skin of the tuber present physical barriers to fungal and bacterial infection. A moist environment is required to permit spore germination and fungal invasion across these primary defenses. The stomata of the leaflets and the lenticels of the tuber represent weak points in the primary defenses, but wounding provides a more immediate foothold for fungal invasion.

As described by Kuć and Currier (8), a number of indigenous compounds which have some degree of fungitoxicity are found at higher levels in the outer few millimeters of the tuber. These compounds are chlorogenic acid, caffeic acid, glycoalkaloids, and scopolin. The levels of chlorogenic acid (9), caffeic acid (10), and glycoalkaloids (11, 12) increase in mechanically wounded tuber tissue, thus adding, perhaps, some degree of protection in an otherwise vulnerable situation. Rapid suberization followed by formation of a wound periderm in as little as 24 hr (2) is probably more important as a defense.

Scopolin and its aglycone scopoletin are normally present in potato tubers. The scopoletin content of tubers varies with the physiological state, being highest in newly-harvested, dormant tubers (13). Mechanical disruption of tuber tissue does not increase the concentration of scopolin, but infection by *P. infestans*, other fungi, bacteria, and viruses causes significant increases (14, 15).

Compounds Produced in Infected Potato Tissue

The incompatible interaction of *P. infestans* with potato tissue leads to the production of a number of compounds that are not native to the host. Studies have been carried out predominantly with tuber tissue disrupted by slicing prior to inoculation and care must be exercised in attempting to extrapolate such findings to events that may transpire in intact foliar tissue during field infection. Zacharius *et al*. (16) have shown that tuber tissue may not even provide a corresponding compatible or incompatible response when compared to the leaflets of the same potato variety.

The terpenoid structures that have been isolated and characterized as products of an incompatible interaction between *P. infestans* and potato tuber tissue are shown in Fig. 2. Tomiyama

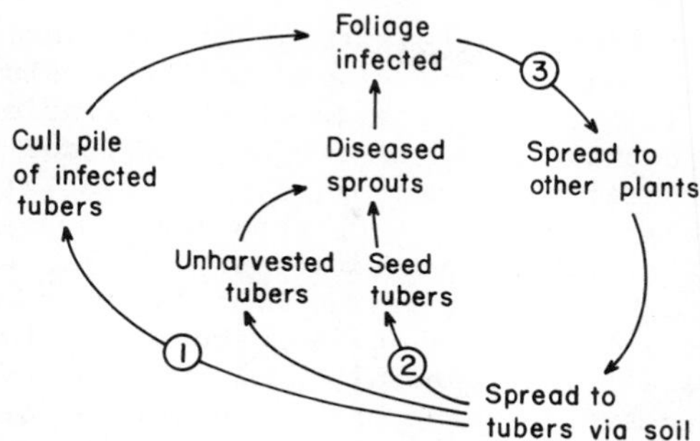

Figure 1. Cycle of Phytophthora infestans *infection of* Solanum tuberosum. *Points at which the spread of infection may be disrupted effectively are indicated at 1, 2, and 3.*

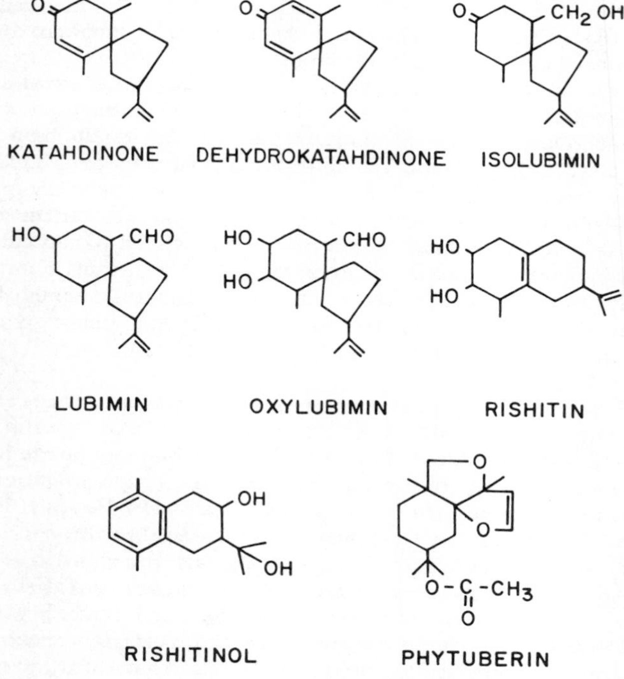

Figure 2. Compounds produced by the potato tuber host in response to infection. Most of these compounds have demonstrated fungitoxic properties and are often referred to as phytoalexins.

et al. isolated rishitin from infected Rishiri variety potatoes (17), and Katsui et al. characterized this compound (18). Katsui et al. (19, 20) isolated and characterized rishitinol. Varns et al. (21) isolated phytuberin, and Hughes and Coxon (22) established its structure. Metlitskii et al. (23) first isolated lubimin from a Soviet potato variety, Lyubimets, and Stoessl et al. (24) and Katsui et al. (25) determined its structure. Katsui et al. (25) isolated and identified oxylubimin; Kalan and Osman (26), isolubimin. Investigators in England and in the U.S. Department of Agriculture (27) simultaneously isolated and identified katahdinone and dehydrokatahdinone. The trivial name, katahdinone, was derived from the Katahdin potato variety from which the compound was isolated in the United States.

The use of potato tuber slices in the above studies is advantageous in that many potato cells are infected, inoculated slices can be incubated under controlled conditions, and the production of terpenoids is high per unit mass of potato tissue. The production of compounds in the incompatible interaction in intact foliar tissue is more difficult to study because the infection is restricted to very small areas. Metlitskii et al. reported lubimin production in potato leaflets inoculated with *P. infestans* (28). This suggests, at least, that terpenoid production may not be peculiar to tuber tissue.

While the above terpenoids have been isolated from incompatible interactions between *P. infestans* and potato tuber tissue, this is not the exclusive mechanism for their production. Zacharius et al. (29) reported that rishitin was produced in both a compatible interaction and an incompatible interaction at a similar rate during the first and second days following inoculation. Other reports document the production of rishitin, phytuberin, and lubimin in potato tuber tissue invaded by *Erwinia carotovora* var. *atroseptica* (30, 31, 32). The formation of terpenoids in potato tuber tissue chemically treated with NaF was reported by Metlitskii et al. (33), but this experiment could not be duplicated by Zacharius et al. (34).

In addition to the production of terpenoids, detectable changes in the soluble protein patterns of potato tissue interacting incompatibly with *P. infestans* have been reported. In infected tuber tissue, Tomiyama and Stahman found an increased number of electrophoretically separable protein bands (35). Yamamoto and Konno observed a new protein in leaflets of the potato plant 6 hr after inoculation with an incompatible race of *P. infestans* (36). It is possible that the incompatible interaction results in the induction of enzymes not normally present in the tuber and that such enzymes are the new, observable proteins. Further work is needed to determine the role, if any, of such proteins in resistance.

Elicitors

Cell-free fungal preparations that initiate an incompatible response in host tissue have been referred to as elicitors or inducers. Sonication of *P. infestans* mycelia yields cell-free, autoclavable preparations that initiate a typical incompatible response, including terpenoid production and necrosis, when applied to potato tuber slices. Varns *et al.* (37) found that sonicates of both compatible and incompatible races of *P. infestans* produced an incompatible response when applied to tuber tissue. Glucans isolated from *Phytophthora megasperma* var. *sojae*, a fungal parasite of the soy bean plant, were reported by Ayers *et al.* to initiate an incompatible response in host tissues (38).

Albersheim and Anderson-Prouty suggested that elicitors are carbohydrate-containing molecules for which receptor sites exist on the plasma membrane of host cells (6). They further conjectured that the specificity of the receptor site for its "matching" carbohydrate elicitor may be the basis for incompatible interactions as the phenotypic expression of vertical resistance. In other words, receptor sites and elicitors may represent the genetically determined primary recognition factor for a potato variety and an incompatible race of *P. infestans*.

In a recent study, Kota and Stelzig demonstrated that preparations which act as elicitors caused membrane depolarization of potato petiole cells in less than 1 min following exposure (39). These preparations were a sonicate, a β-1,3-glucan, and a lipopolysaccharide (all obtained from *P. infestans*) and a preparation from *P. megasperma*. Among eight other carbohydrates which are not elicitors, only pectin produced depolarization. This report is supportive of the elicitor theory of Albersheim and Anderson-Prouty (6) and the mechanism of action of fungal phytotoxins advanced by Strobel (40).

In contrast to the theory of Albersheim and Anderson-Prouty (6), Ward and Stoessl proposed that a recognition mechanism (or incompatibility suppression mechanism) is involved in an interaction between molecules produced by the fungus and the compatible host (41). Metlitskii *et al.* added another dimension to speculation about the incompatible interaction by proposing that the potato host produces compounds which induce the formation of compounds by *P. infestans* which in turn induce the incompatible interaction in the host cell (42). Keen referred to the elicitor phenomenon as a derepression of the production of an antifungal compound in the soy bean host cell (43). More work obviously is needed to determine the nature of the primary events that determine compatibility or incompatibility between host and fungal parasite.

The Phytoalexin Theory

Potato tuber tissue inoculated with an incompatible race of *P. infestans* was reported to display an incompatible interaction when later inoculated with a normally compatible race (44). This observation led to the "phytoalexin theory" (45), which holds that fungitoxic compounds, phytoalexins, are produced by host cells as a result of invasion by an incompatible fungus. Kuć and Currier (8) recently reviewed phytoalexin studies, most of which were conducted with members of the families Solanaceae (mainly the potato) and Leguminosae.

For the potato, the compounds shown in Fig. 2 are often referred to as phytoalexins and most of them have demonstrated fungitoxic properties. The fungitoxicity of rishitin, phytuberin, lubimin, katahdinone, and dehydrokatahdinone was tested against mycelial growth of *P. infestans* by Beczner and Érsek (46). At levels of 25 to 100 ppm in solid media in Petri dishes, all of the compounds exhibited moderate to strong inhibitory effects toward the further spread of 7-day-old cultures of *P. infestans* transferred to the media. Ward *et al*. (47) reported similar inhibition of mycelial growth with somewhat lower levels of rishitin. Germination of *P. infestans* zoospores was also inhibited by similarly low amounts of phytuberin (48), lubimin (49), and rishitin (50, 51). Since the levels of phytoalexins able to inhibit pathogen growth *in vitro* are attained in many incompatible interactions, investigators have suggested that phytoalexins play an important role in inhibiting fungal invasion (6, 52-57). However, ultrastructural studies (cited later) indicate observable differences between the compatible and incompatible interactions well in advance of the production of detectable levels of phytoalexins. This does not disprove the theory, since no one has yet determined the time of appearance or concentration of phytoalexins in the microenvironment of the individual host cell. Conversely, the proof of the theory would rest in determining that phytoalexins are present in sufficient quantities early enough in the infected, incompatible cell to inhibit fungal invasion and in identifying the mechanism of inhibition. Hohl and Stössel speculated that phytoalexins inhibit fungal glucanases necessary for both fungal growth and disruption of β-1,3-glucan barriers in the host cell (58).

Ultrastructural Studies

The compatible and incompatible interactions between *P. infestans* and potato host cells have been studied by scanning electron microscopy and transmission electron microscopy to yield information complementary to biochemical studies.

Using a scanning electron microscope, Jones *et al*. (59) noted sharp boundaries between obviously infected cells near the surface and uninfected cells below the surface in the compatible interaction of P. infestans with potato tuber tissue. In this interaction, hyphae penetrated intercellularly deep into the tissue where intact, live potato cells remained to support the colony with nutrients. A sharp boundary between necrotic cells and live cells was not observed in the incompatible interaction.

In electron microscopy studies of the infection of potato leaf cells, Shimony and Friend (60) observed penetration of host cells by P. infestans hyphae in both compatible and incompatible interactions. Although the rate of initial penetration did not differ, they reported visible ultrastructural differences between the compatible and incompatible interactions 7 hr after inoculation (with penetration occurring as late as 4-1/2 to 6-1/2 hr after inoculation). Between 9 and 12 hr, host cells surrounding the site of infection on the incompatible variety appeared to be dead and fungal hyphae were contained within the necrotic area; at 24 hr, severely damaged hyphae were observed; after 48 hr, there were no detectable living cells of either fungus or host in the small lesion. An observed feature of both the compatible and incompatible interactions was the formation of a sheath or encapsulation surrounding the intracellular hyphae. In a parallel ultrastructural study of the incompatible interaction between lettuce (Lactuca sativa) leaf tissue and a fungal parasite (Bremia lactucae), Maclean et al. (61) also observed that death of penetrated host cells preceded the cessation of growth and death of fungal cells by several hours.

Hohl and Stössel (58) also reported that fungal hyphae penetrated potato tuber host cells in both the compatible and incompatible interactions. The type of encapsulation described by Shimony and Friend (60) was observed in both incompatible and compatible interactions and was referred to as an extrahaustorial matrix. Hohl and Stössel, however, reported that the typical haustorium found in the incompatible tuber host cell differed from that of the compatible host cell by the presence of an additional entity, wall appositions, which surrounded and encased the haustorium and the extrahaustorial matrix. The encasement of haustoria has also been reported for the incompatible, but not the compatible, interaction of Uromyces phaseoli var. vignae with host cells of the cowpea (62). In two additional studies of a Phytophthora pathogen in compatible and incompatible interactions, the presence of wall appositions was not reported (63, 64).

In studies of leaf tissue of the same two potato varieties employed in tuber tissue studies by Hohl and Stössel (58), Hohl and Suter (65) found that haustoria in both the compatible and incompatible interactions were similar and were always surrounded by an extrahaustorial matrix and sometimes also by wall appositions. As a possible explanation for the marked ultrastructural differences in haustoria observed between the two types of tubers but not in the leaf tissues, the authors pointed out that the leaflets of the two varieties differ less in resistance than the corresponding tubers. Various types of fungal invasion of potato leaf cells ranging from hyphae which penetrated cells and then emerged at another point to fully developed haustoria were observed by Hohl and Suter. Various degrees of haustorial encasement were also observed as shown in Fig. 3, and the authors suggested that host cell resistance may be related to the degree of encasement that takes place. They further noted that hyphae enter the leaf primarily through stomata but are also able to penetrate epidermal cells or hair cells. In the incompatible interaction, the parasite did not penetrate deeply into the host tissue and rarely developed sufficiently to produce sporangia.

While ultrastructural studies of compatible and incompatible interactions of *P. infestans* with the potato host are not in complete agreement, common findings have emerged: cell penetration occurs in both circumstances; in the incompatible interaction, penetration occurs to a depth of very few cells followed by death of the host cells and later death of the fungal cells.

Conclusion

Certain aspects of the incompatible interaction between *P. infestans* and the potato host seem to be reasonably well-documented. A number of compounds with demonstrated fungitoxicity are produced when tuber tissue is invaded, host cells become necrotic and fungal cells later die. Identical symptoms in tuber host cells can be initiated by the application of cell-free preparations of the fungus. However, the observed symptoms of incompatibility may be far removed from the primary event that triggers the sequence of known events. The fact that fungitoxic compounds are evolved is not complete proof that they are, indeed, the cause of incompatibility in the complex cell-cell interaction. Gene-for-gene interactions between host and parasite can be systematically tabulated, but only in the empirical terms of compatible or incompatible. The phenotypic mechanisms involved in interactions at the cell and molecular levels remain to be described. Elucidation of the molecular basis for the specificity exhibited in gene-for-gene, host-pathogen interactions is within reach and is an important objective.

Although valuable information has been gained through studies with tuber tissue, it appears that additional studies with foliar tissue should be emphasized. Phenomena observed in interactions of *P. infestans* with sliced or otherwise disrupted tuber tissue cannot be assumed to occur in foliar tissue. Since resistance to *P. infestans* must be expressed in the foliar portion of the potato plant in order to break the cycle of infection shown in Fig. 1, it is of primary importance to describe and understand resistance in terms of the leaflets rather than the tubers.

Great emphasis has been placed on understanding vertical resistance in the potato, while horizontal resistance has received comparatively little study. Since vertical resistance is temporary and effective only until a new race of *P. infestans* capable of

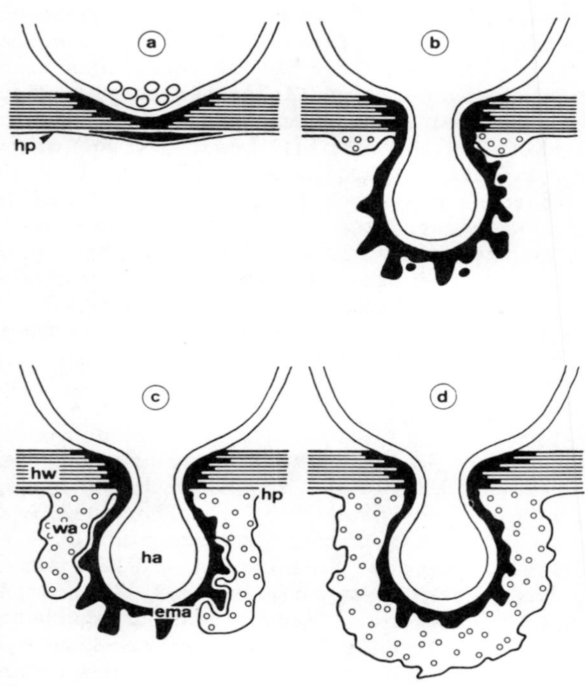

Canadian Journal of Botany

Figure 3. Diagram of haustorial development of Phytophthora infestans *on potato leaves. (a) Early penetration of host wall with incipient extrahaustorial matrix formation between host wall and host plasmalemma (hp). (b) Haustorium ensheathed by extrahaustorial matrix and collared by wall appositions at its neck. (c) Progressive encasement of haustorium (ha) by wall appositions (wa). The host plasma membrane (hp) may double-over on itself at the interface of the extrahaustorial matrix (ema) and the wall apposition. (d) Fully encased haustorium with compressed extrahaustorial matrix. Haustorial development may terminate at (b), (c), or (d), probably reflecting rising degrees of host cell resistance (65).*

bypassing the resistance mechanism evolves, additional studies aimed at identifying the physical and chemical features of the potato that contribute to horizontal resistance seem in order.

Despite the extensive efforts that have been devoted to breeding potato varieties resistant to *P. infestans*, potato growers must rely heavily on the use of fungicides to control late blight. Environmental and economic considerations discourage the continued use of conventional fungicides and encourage further research to understand the chemical basis for potato plant resistance to *P. infestans*. Robinson (*1*), critical of the unholistic approach to disease resistance, pointed out that the disciplines involved have concentrated too heavily upon their particular interests. Effective teams that include plant breeders, plant pathologists, microscopists, and chemists will probably have the greatest impact in the further conceptualization of resistance and the practical application of the concepts to the development of varieties with improved resistance.

Literature Cited

1. Robinson, R. A. "Plant Pathosystems," pp 15-54, Springer-Verlag, Berlin, Heidelberg, New York, 1976.
2. Burton, W. G. "The Potato," pp 97-107, 248-253, H. Veenman and Zonen, N. V., Wageningen, Holland, 1966.
3. Plank, J. E., van der. "Principles of Plant Infection," p 216, Academic Press, New York-London, 1975.
4. Plank, J. E., van der. "Plant Diseases. Epidemics and Control," p 349, Academic Press, New York-London, 1963.
5. Day, P. R. "Genetics of Host-Parasite Interaction," p 238, W. H. Freeman and Co., San Francisco, 1974.
6. Albersheim, P., and Anderson-Prouty, A. J., Ann. Rev. Plant Physiol. (1975) *26*, 31-52.
7. Black, W., Mastenbroek, C., Mills, W. R., and Peterson, L. C., Euphytica (1953) *2*, 173-178.
8. Kuć, J., Currier, W. "Mycotoxins and Other Fungal Related Food Problems," Advances in Chemistry Series 149, pp 356-368, American Chemical Society, Washington, D.C., 1976.
9. Sakuma, T., and Tomiyama, K., Phytopathol. Soc., Jap. (1967) *33*, 48-58.
10. Kuć, J., Henze, R. E., and Ullstrup, A. J., J. Amer. Chem. Soc. (1956) *78*, 3123-3125.
11. McKee, R., Ann. Appl. Biol. (1955) *43*, 147-148.
12. Locci, R., and Kuć, J., Phytopathology (1967) 57, 1272-1273.
13. Korableva, N. P., Morozova, E. V., and Metlitskii, L. V., Dokl. Akad. Nauk SSSR (1973) *212*, 1000-1002.
14. Clarke, D. D., Phytochemistry (1969) *8*, 7.
15. Clarke, D. D. and Baines, P. S., Physiol. Plant Pathol. (1976) *9*, 199-203.
16. Zacharius, R. M., Osman, S. F., Heisler, E. G., and Kissinger, J. C., Phytopathology (1976) *66*, 964-966.

17. Tomiyama, K., Sakuma, T., Ishizaka, N., Sato, N., Katsui, N., Takasugi, M., and Masamune, T., Phytopathology (1968) 58, 115-116.
18. Katsui, N., Murai, A., Takasugi, M., Imaizumi, K., and Masamune, T., Chem. Commun. (Sect. D) (1968), 43-44.
19. Katsui, N., Matsunaga, A., Imaizumi, K., and Masamune, T., Tetrahedron Lett. (1971) No. 2, 83-86.
20. Katsui, N., Matsunaga, A., Imaizumi, K., and Masamune, T., Bull. Chem. Soc. Jap. (1973) 45, 2871-2877.
21. Varns, J., Kuć, J., and Williams, E. B., Phytopathology (1971) 61, 174-177.
22. Hughes, D. L., and Coxon, D. T., J. Chem. Soc., D. (1974), 822-823.
23. Metlitskii, L. V., Ozeretskovskaya, O. L., Chalova, L. I., Vasyukova, N. I., and Davydova, M. A., Mikol. Fitopatol. (1971) 5, 263-271.
24. Stoessl, A., Strothers, J. B., and Ward, E. W. B., J. Chem. Soc., D. (1974), 709-710.
25. Katsui, N., Matsunaga, A., and Masamune, T., Tetrahedron Lett. (1974) No. 51-52, 4483-4486.
26. Kalan, E. B., and Osman, S. F., Phytochemistry (1976) 15, 775-776.
27. Coxon, D. T., Price, K. R., Howard, B., Osman, S. F., Kalan, E. B., and Zacharius, R. M., Tetrahedron Lett. (1974) No. 34, 2921-2924.
28. Metlitskii, L. V., Ozeretskovskaya, O. L., Vasyukova, N. I., Davydova, M. A., Savel'eva, O. H., and D'yakov, Yu, T., Mikol. Fitopatol. (1974) 8, 42-49.
29. Zacharius, R. M., Osman, S. F., Kalan, E. B., and Thomas, S. R., Proc. Am. Phytopath. Soc. (1974) 1, 63.
30. Lyon, G. D., Physiol. Plant Pathol. (1972) 2, 411-416.
31. Lyon, G. D., Lund, B. M., Bayliss, C. E., and Wyatt, G. M., Physiol. Plant Pathol. (1975) 6, 43-50.
32. Beczner, J., and Lund, B. M., Acta Phytopathologica Academiae Scientiarum Hungaricae (1975) 10, 269-274.
33. Metlitskii, L. V., D'yakov, Yu. T., Ozeretskovskaya, O. L., Yurganova, L. A., Chalova, L. I., and Vasyukova, N. I., Izvestiya Akad. Nauk SSSR, Ser. Biol. (1971), 399-407.
34. Zacharius, R. M., Kalan, E. B., Osman, S. F., and Herb, S. F., Physiol. Plant Pathol. (1975) 6, 301-305.
35. Tomiyama, K., and Stahmann, M. A., Plant Physiol. (1964) 39, 483-490.
36. Yamamoto, M., and Konno, K., Plant and Cell Physiol. (1976) 17, 843-846.
37. Varns, J. L., Currier, W. W., and Kuć, J., Phytopathology (1971) 61, 968-971.
38. Ayers, A. R., Ebel, J., Valent, B., and Albersheim, P., Plant Physiol. (1976) 57, 760-765.
39. Kota, D. A., Stelzig, D. A., Proc. Am. Phytopathol. Soc. (1977) 4, in press.

40. Strobel, G. A., Scientific American (1975) 232, 81-88.
41. Ward, E. W. B., Stoessl, A., Phytopathology (1976) 66, 940-941.
42. Metlitskii, L. V., D'yakov, Yu. T., and Ozeretskovskaya, O. L., Biol. Abs. (1974) 58, No. 5, 2978.
43. Keen, N. T., Science (1975) 187, 74-75.
44. Müller, K., and Borger, H., Arb. Biol. Reichsanst. Land Forstwirt, Berlin (1940) 23, 189-231.
45. Müller, K., and Behr, L., Nature (1949) 163, 498-499.
46. Beczner, J., and Érsek, T., Acta Phytopathologica Academiae Scientiarum Hungaricae (1976) 11, 59-64.
47. Ward, E. W. B., Unwin, C. H., and Stoessl, A., Can. J. Botany (1974) 52, 2481-2488.
48. Varns, J. L. "Biochemical Response and its Control in the Irish Potato (*Solanum tuberosum*) -*Phytophthora infestans* Interactions," Ph.D. Thesis, Purdue University, Lafayette, Indiana, 1970.
49. Metlitskii, L. V., and Ozeretskovskaya, O. L., Fitoalexini. Akad. Nauk, SSSR, Nauka, Moskva, pp 176, 1973.
50. Tomiyama, K., Ishizaka, N., Sato, N., Masamune, T., and Katsui, N. "Biochemical Regulation in Diseased Plants or Injury," pp 287-292, Phytopath. Soc., Japan, Tokyo, 1968.
51. Ishizaka, N., Tomiyama, K., Katsui, N., Murai, A., and Masamune, T., Plant Cell Physiol. (1969) 10, 183-192.
52. Keen, N. T., Physiol. Plant Pathol. (1971) 1, 265-275.
53. Rahe, J. E., Kuć, J., Chuang, C. M., and Williams, E. B., Neth. J. Plant Pathol. (1969) 75, 58-71.
54. Sato, N., Kitazawa, K., and Tomiyama, K., Physiol. Plant Pathol. (1971) 1, 289-295.
55. Skipp, R. A., Physiol. Plant Pathol. (1972) 2, 357-374.
56. Bailey, J. A., and Deverall, B. J., Physiol. Plant Pathol. (1971) 1, 435-439.
57. Frank, J. A., and Paxton, J. D., Phytopathology (1970) 60, 315-318.
58. Hohl, H. R., and Stössel, P., Can. J. Bot. (1976) 54, 900-912.
59. Jones, S. B., Carroll, R. J., and Kalan, E. B. "Scanning Electron Microscopy, Part 2, Proceedings of the Workshop on Scanning Electron Microscopy and the Plant Sciences," pp 397-404, IIT Research Institute, Chicago, Illinois, 1974.
60. Shimony, C., and Friend, J., New Phytol. (1975) 74, 59-65.
61. Maclean, D. J., Sargent, J. A., Tommerup, I. C., and Ingram, D. S., Nature (1974) 249, 186-187.
62. Heath, M. C., and Heath, I. B., Physiol. Plant Pathol. (1971) 1, 277-287.
63. Klarman, W. L., and Corbett, M. K., Phytopathology (1974) 64, 971-975.
64. Hanchey, P., and Wheeler, H., Phytopathology (1971) 61, 33-39.
65. Hohl, H. R., and Suter, Elisabeth, Can. J. Bot. (1976) 54, 1956-1970.

5

Biosynthetic Relationships of Sesquiterpenoidal Stress Compounds from the Solanaceae

ALBERT STOESSL and E. W. B. WARD

Agriculture Canada, Research Institute, University Sub Post Office, London, Ontario, Canada N6A 5B7

J. B. STOTHERS

Chemistry Department, University of Western Ontario, London, Ontario, Canada N6A 3K7

Rishitin (I) was reported as a phytoalexin of potatoes (1) and tomatoes (2) in 1968 and was the first sesquiterpenoidal stress compound to be isolated and characterized from the Solanaceae. In the 9 years since, some 20 other, structurally more or less closely related sesquiterpenes have been described from potato, aubergine, thornapple, sweet pepper, and tobacco species (Tables I - V) (3 - 19). All of these compounds are produced essentially only under stress conditions, that is, after infection of the plant with fungi, bacteria, or viruses, or through wounding, exposure to UV, or treatment with deleterious substances. Not all, however, are necessarily phytoalexins because some lack the antifungal activity which is requisite for that function. In addition, several other compounds (Table VI) (20 - 22) which are structurally and biogenetically closely related to the stress metabolites, have been obtained in very low amounts from cured tobacco leaves. Because of this special circumstance, it is largely unclear whether these are stress compounds which were formed through undetected infection, or through other stress processes in the ageing or curing of the leaf, or whether they may have been present as normal constituents of the healthy tissue. In any event, their structures render them relevant to the present discussion.

All bicyclic compounds in Tables (I - VI) are either eudesmanes or can be regarded as derived from eudesmanes by plausible rearrangements. Except for the single case of rishitinol (II), the rearrangements can be represented formally as proceeding through intermediates with C-4 and C-5 as cationic migration termini. Such a formal intermediate (XXVII) and its relationship with some of the metabolites are set out in Scheme 1. It is also notable that the bicyclic compounds, again with the exception of rishitinol, carry oxygen atoms on one or more of carbons 1 to 4 but none on carbons 6 to 9. Rishitinol is oxygenated at C-8, a position which corresponds to a carbon atom which is also oxygenated in the acyclic nerolidol derivatives XIV and XV from eggplant. These close structural relationships must reflect largely common biogenetic origins and pathways, and also, since they are

Table I. Solanum tuberosum stress metabolites

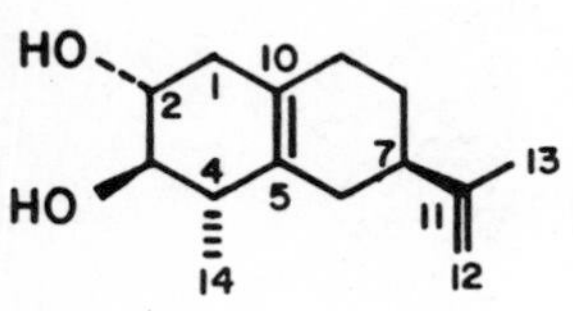

rishitin
(I) (1,2)

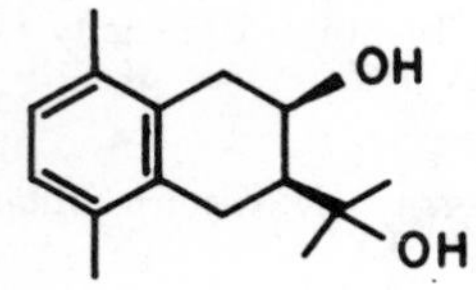

rishitinol
(II) (3)

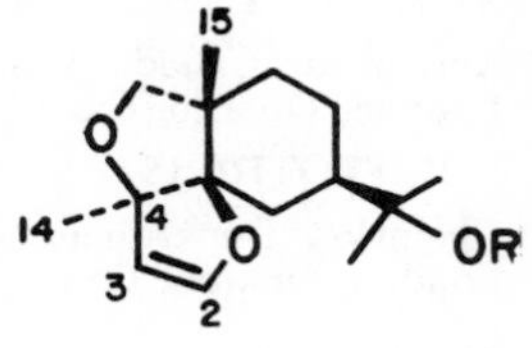

phytuberin
(III, R=Ac) (4,5)

phytuberol
(IV, R=H) (6,7)

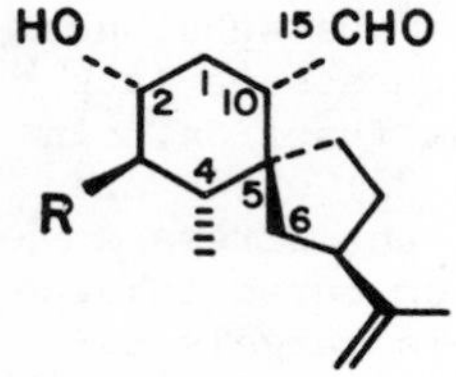

lubimin
(V, R=H) (8,9)

hydroxylubimin
(VI, R=OH) (10)

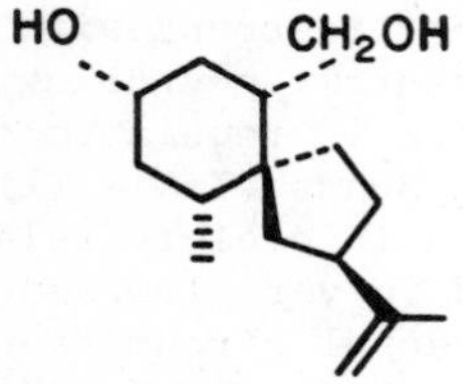

15-dihydrolubimin
(VII) (11)

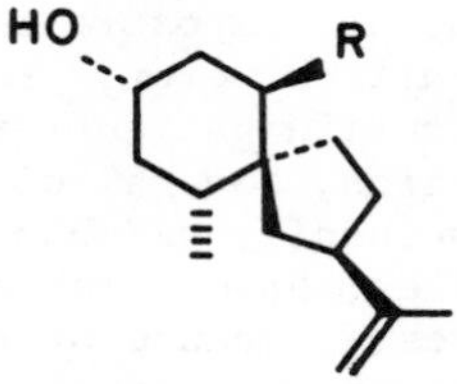

10-epilubimin
(VIII, R=CHO) (11)

15-dihydro-10-epilubimin
(IX, R=CH_2OH) (11)

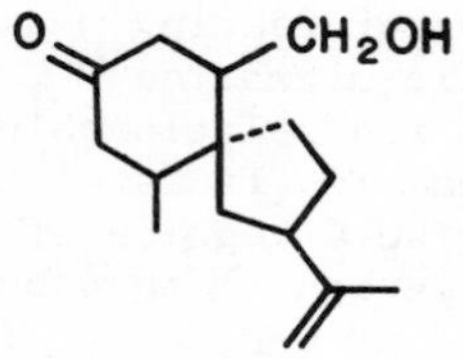

isolubimin
(X) (12)

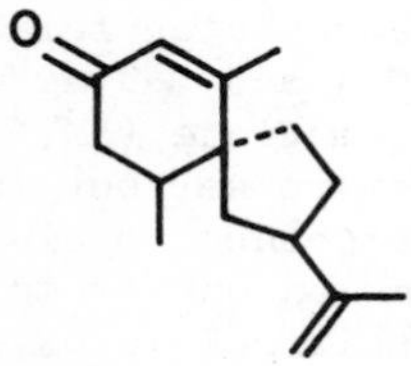

solavetivone
(XI) (13)

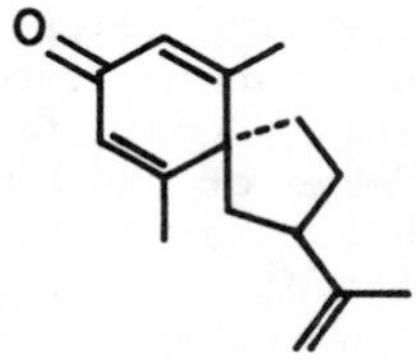

anhydro-β-rotunol
(XII) (13)

Table II. Solanum melongena stress metabolites (9)

lubimin
(V, R=H)

aubergenone
(XIII)

9-oxonerolidol
(XIV)

9-hydroxynerolidol
(XV)

(XVI)

Table III. Datura stramonium stress metabolites (10)

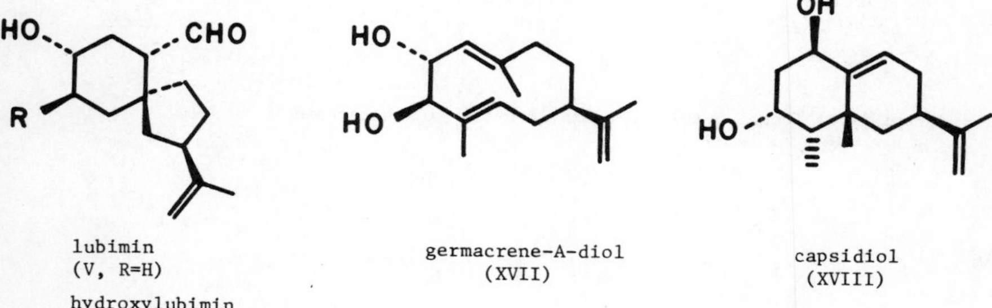

lubimin
(V, R=H)
hydroxylubimin
(VI, R=OH)

germacrene-A-diol
(XVII)

capsidiol
(XVIII)

Table IV. Capsicum frutescens stress metabolites

capsidiol
(XVIII) (14,15)

13-hydroxycapsidiol
(XIX) (16)

Table V. Nicotiana stress metabolites

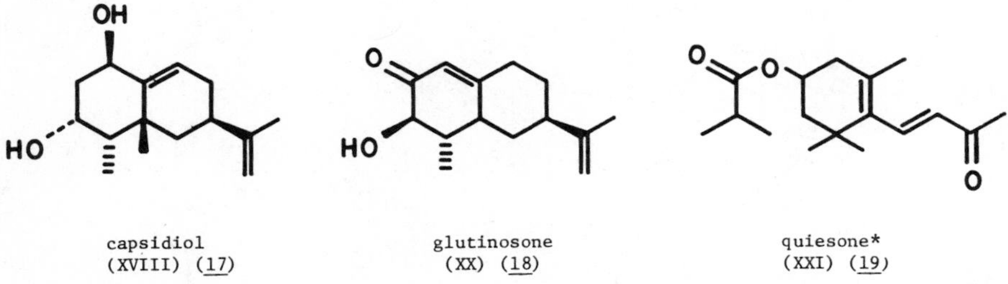

capsidiol
(XVIII) (17)

glutinosone
(XX) (18)

quiesone*
(XXI) (19)

*possibly of fungal origin; biogenetic derivation uncertain

Table VI. Nicotiana sesquiterpenes not proven as stress compounds

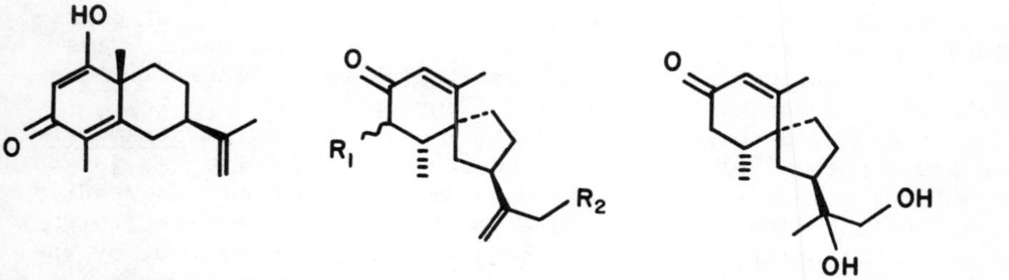
HO
O
O
R1
R2
O
OH
OH

1-keto-α-cyperone (XXI) (20)

solavetivone (XI, R_1=R_2=H) (21)
3-α-hydroxysolavetivone* (XXIII, R_1=αH, R_2=H
3-β-hydroxysolavetivone* (XXIV, R_1=βH, R_2=H
13-hydroxysolavetivone* (XXV, R_1=H, R_2=OH)

11,13-dihydroxysolavetivone* (XXVI)

*occur and isolated as glucosides (22)

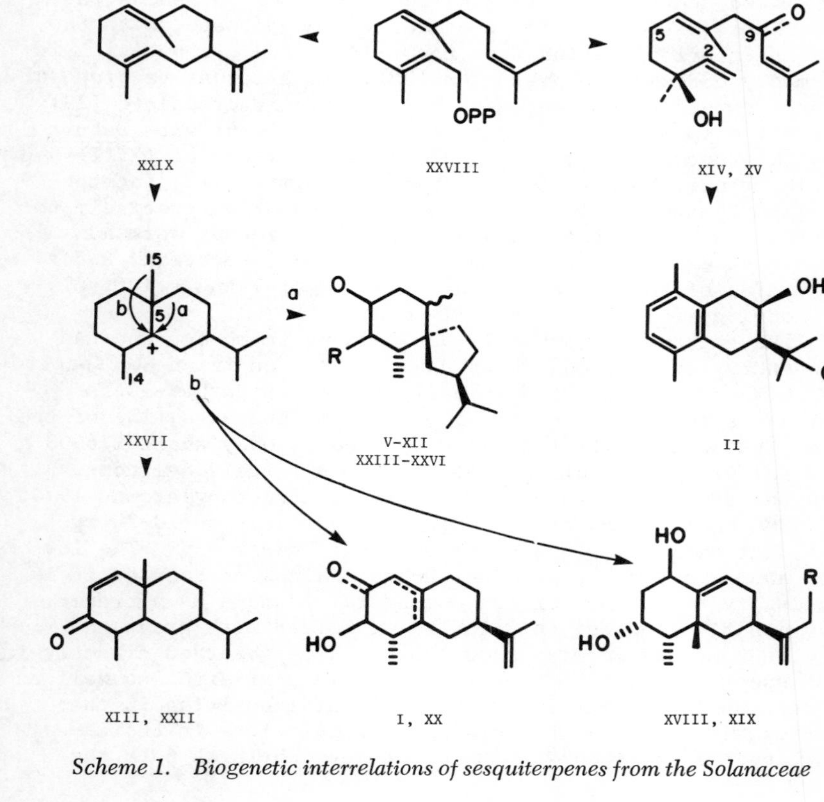
OPP
O
5
9
2
OH
XXIX
XXVIII
XIV, XV
15
b
5
a
+
14
b
a
O
R
OH
OH
XXVII
V-XII
XXIII-XXVI
II
HO
R
O
O
HO
HO
XIII, XXII
I, XX
XVIII, XIX

Scheme 1. Biogenetic interrelations of sesquiterpenes from the Solanaceae

unique to the Solanaceae, a considerable degree of genetic specialization.

Several more or less equivalent schemes which set out possible biogenetic routes in greater though speculative detail were discussed previously (23). The examples shown in Scheme 2 illustrate the stereochemical features of such proposals. A key element is the germacrenediol (XVII, R=OH) from Datura stramonium and its hypothetical monohydroxy analogue (XVII, R=H). As judged by molecular models, these compounds are flexible and can readily adopt conformations which would allow them to be converted to the bicyclic metabolites, both of Datura and other Solanaceae, by the pathways indicated. The Scheme is only illustrative and the intermediacy of other, differently substituted germacrenes is not excluded and is probable for at least some of the Solanaceae metabolites. In general, however, a function of germacrenes as biogenetic precursors of bicyclic sesquiterpenes has long been postulated and is widely accepted (24)[1].

Many of the features of these suggested pathways should be quite readily susceptible to experimental test and some progress in this direction has been made. Thus, Brooks and associates have utilized acetate, doubly-labelled with ^{13}C, to demonstrate that capsidiol is formed from acetate, via mevalonate, with the postulated methyl migration from the C-10 to the C-5 position in a presumably eudesmanoid precursor (28). An alternative proposal invoking a two-fold rearrangement via a spiro-intermediate (29) was thereby eliminated. In an analogous experiment with Datura stramonium, we could show (30) that the germacrenediol (XVII) and the vetispiranes, lubimin (V) and hydroxylubimin (VI), incorporated 1,2-$^{13}C_2$-acetate in a pattern which conformed precisely to the postulated biogenetic pathways. Our more recent work has also relied almost exclusively on the use of 1,2-$^{13}C_2$-acetate, and a brief outline of the essentials of this powerful and relatively new methodology (31 - 34) may be in order.

Unlike carbon-12, carbon-13 has a magnetic moment and can undergo energy transitions which can be observed in an NMR spectrometer. However, because the magnetic moment of carbon-13 is smaller and also because of the low natural abundance (1%) of the isotope, the sensitivity for its detection is only about 1/6000 of that for protons. This problem has been largely overcome, through the advent of Fourier transform spectroscopy around 1970, and ^{13}C-NMR spectra can now be obtained routinely on 10-20 mg samples, but smaller amounts can suffice if necessary. The low natural abundance of ^{13}C has emerged as a blessing because it is the necessary condition for the use of the isotope as a tracer. In biosynthetic studies, this can be exploited with great advantage because any incorporated label can be observed directly in the NMR spectrum which will define the molecular sites as well as the extent of incorporation. Low levels of impurities in the product usually will neither mislead nor interfere in the determination. Thus, the ^{13}C-NMR methodology obviates both the

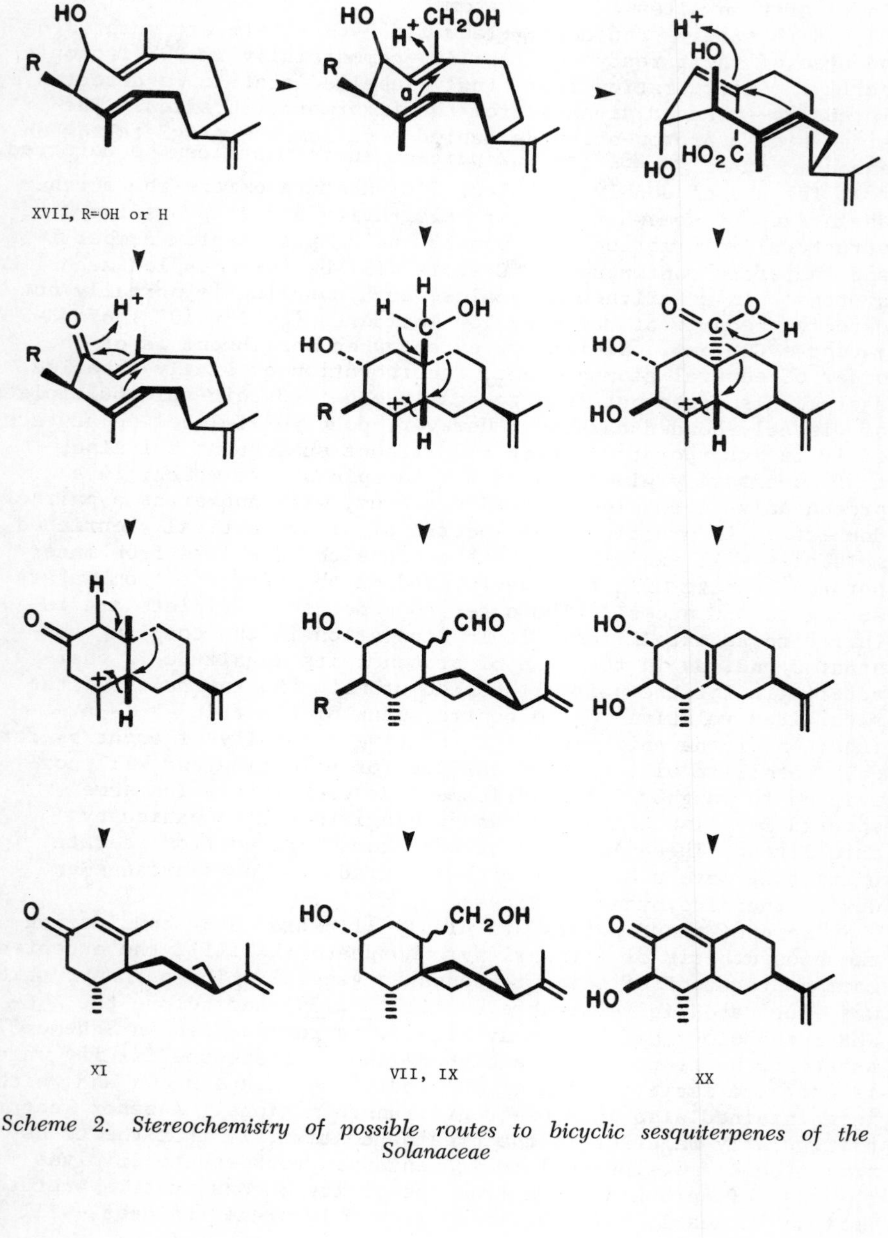

XVII, R=OH or H

XI

VII, IX

XX

Scheme 2. Stereochemistry of possible routes to bicyclic sesquiterpenes of the Solanaceae

rigorous purification requirements of radioactive tracer methods and particularly, also the often difficult, material- and time-consuming degradative procedures which are needed to locate the incorporation sites of radio-atoms.

Both singly- and doubly-labelled ^{13}C-acetate are much used because of their ready availability commercially in 90% isotopic purity. Incorporation from singly-labelled acetate is detected by intensity-enhanced signals from the incorporation sites. The signals of a proton-noise decoupled spectrum are singlets and an enhancement of ca 20% (ca 0.2 percent enrichment) can be detected.

The use of doubly labelled, $^{13}C_2$-acetate offers the further advantages of somewhat greater sensitivity and of providing more structural information. Carbon-13 has a spin-quantum number $I=\frac{1}{2}$ and therefore continguous ^{13}C atoms will be spin-coupled much like protons. In unenriched molecules, such coupling is normally not detected because of the very low incidence (ca 1×10^{-4}) of adjacent ^{13}C atoms. Similarly, except when enrichment is of the order of several atom percent, incorporation of singly-labelled acetate also does not lead to easily observed spin-spin multiplets of directly bonded carbons. However, doubly-^{13}C-labelled acetate, if it is incorporated intact and without subsequent scission, must necessarily give rise to a spin-spin system which, in a proton-noise decoupled ^{13}C-NMR spectrum, will appear as a pair of doublets. In practice, the spectra of biosynthetically enriched molecules will exhibit triplets because the doublets from incorporated acetate will be superimposed on the singlets from natural abundance ^{13}C atoms[2]. The outer components of triplets are referred to as satellites. Their separation is the coupling constant J and, as in the case of protons, its magnitude is characteristic for the pair of coupled atoms. The intensity of the satellites relative to the central peak of the triplet is a function of the enrichment. A relative intensity of about 5% for each satellite will readily suffice for detection and will correspond to roughly 0.14% enrichment at each site. The same spectra may also contain enhanced singlets unaccompanied by satellites. These indicate carbon atoms derived from acetate units that were cleaved by either degradation or rearrangement during the biosynthetic process.

Some of these points are nicely illustrated by considering the biosynthesis of farnesyl pyrophosphate (XXVIII), the accepted common precursor of sesquiterpenes in general. The pathway, which was elucidated in remarkable detail (35, 36) and before the ^{13}C-NMR methodology had become available, is represented in Scheme 3. A noteworthy feature is the high degree of stereospecificity which is characteristic of each of the individual steps shown and which is maintained also in subsequent transformations. Another feature, particularly relevant in the present context, is that the transformation of mevalonate into isopentenylpyrophosphate involves the loss of carbon dioxide from one of its 3 constitutive acetate units. Farnesol, being built up from 3 isoprene residues, will therefore comprise 6 intact acetate residues, shown in the scheme by heavy bonds, and 3 lone carbon atoms derived from C-2 of mevalonate and indicated by heavy dots. In the ^{13}C-NMR spectrum of farnesol biosynthesized from doubly-labelled acetate, the atoms linked by heavy bonds therefore would exhibit triplet while the lone carbons will show singlet absorption. This pattern will be maintained if the farnesol undergoes subsequent substitution and cyclization reactions without rearrangement or carbon loss, that is, if it is converted into sesquiterpenes which obey Ruzicka's isoprene rule. This is exactly what is observed, for instance in the biosynthetic study (30) of the germacrenediol (XVII) from *D. stramonium*. The ^{13}C-NMR spectrum of the enriched compound is shown in Fig. 1 and exhibits all the expected features.

Doubly-labelled acetate is particularly well suited for studying the biosynthesis of sesquiterpenes which do not obey the isoprene rule. This is illustrated by the already cited study of capsidiol by Brooks and coworkers (28) and by the recent work with the potato sesquiterpenes (37) which is now to be described in some detail.

To induce the biosynthesis of the sesquiterpenes, freshly cut potato slices were inoculated with spore suspensions of *Monilinia fructicola* in one experiment, and of *Glomerella cingulata* in another. The inoculation sites consisted of hemispherical wells of ca 1 cm. radius cut into the slices with a melon baller. Each received 2 ml of spore suspension and ^{13}C-sodium acetate, of about 90% isotopic purity at both C-1 and C-2, was added a day later, at the rate of 0.5 mg in 0.2 ml water per well. The liquid contents of the wells were collected after another day and extracted with ether. The product so obtained was chromatographed over a large column (1 g/mg extract) of silica as used for TLC (but not activated by heating). Ether-light petrol 1:1 or methanol-chloroform 5:95 are suitable solvents. The materials obtained in this basic fractionation were then further purified as necessary by one or more column or preparative TLC separations. In this manner, the interaction of the potatoes with *M. fructicola* yielded the compounds listed in Table VII, in the amounts indicated. In their ^{13}C-NMR spectra, all the isolates exhibited the expected labelling patterns. For example, rishitin (I), the main component of the product, gave rise to 5 pairs of triplets and 4 singlets, entirely consistent with the oxidative loss of C-15 in its formation (Fig. 2). In this spectrum, two other features can be observed which were seen also in the spectra of all the metabolites which we examined. Firstly, intensities of the satellites relative to each central peak are all of the same order of magnitude, implying that the enrichment level is roughly the same at all sites. This is consistent with rapid *de novo* synthesis without recourse to already present metabolic pools. Secondly, although an allylic shift equilibration of the side-chain double bond between the 11, 12 and 11, 13 positions can be envisaged as easily occurring throughout the biosynthetic sequence,

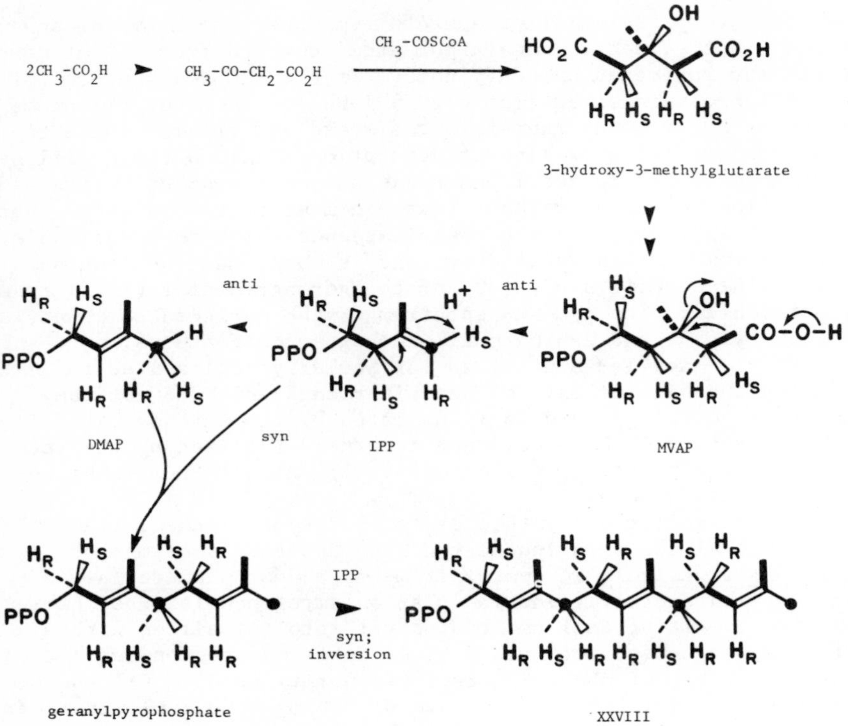

Scheme 3. Biosynthesis of farnesol

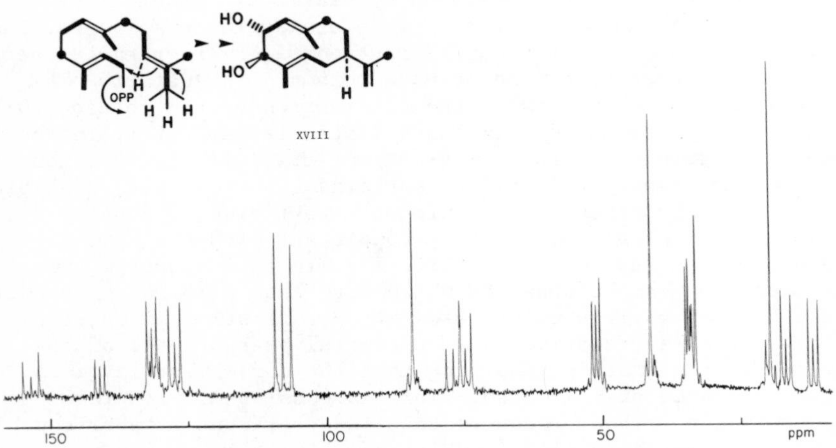

Figure 1. ^{13}C NMR spectrum of 2,3-dihydroxygermacrene (XVII) biosynthesized from 1,2-^{13}C-acetate

Table VII ^{13}C-Labelled Compounds from the Potato - Monilinia fructicola interaction.

Compound	yield (mg)*
rishitin (I)	69
lubimin (V)	15
epilubimin (VIII)	3
hydroxylubimin (VI)	8
dihydroepilubimin (IX)	4
phytuberin (III)	<1

* after final purification; from 120 lbs potatoes

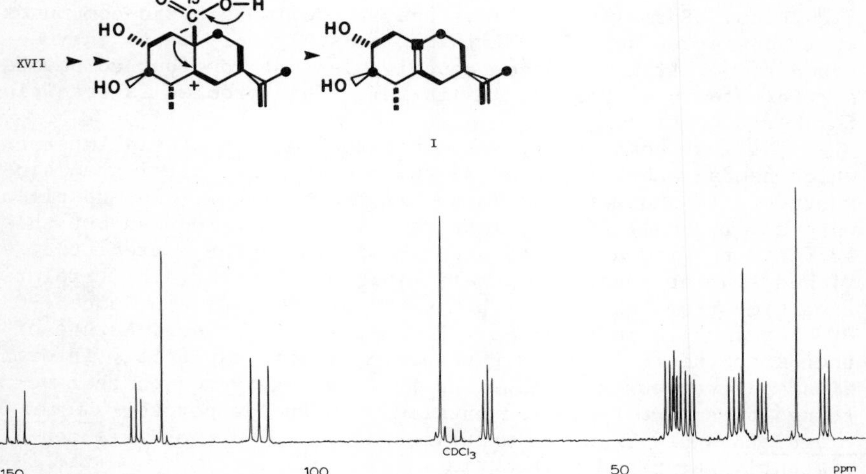

Figure 2. ^{13}C NMR spectrum of rishitin biosynthesized from 1,2-^{13}C-acetate

the spectra show that such scrambling does not, in fact, occur. Thus, the only carbon detectable as coupled to C-11 is olefinic and, similarly, the signal from the isopropenyl methyl group is of singlet character only.

The spectrum of lubimin in Fig. 3 illustrates the case of a metabolite which exhibits the normal triplets and 3 singlets although a molecular arrangement did intervene in its biosynthesis. Again, this is the expected result because the bond broken in the rearrangement was between two acetate residues and not within one. Entirely analogous spectra were obtained also from hydroxylubimin (VI) and from 10-epi- and 15-dihydro-10-epilubimin (VII and IX). The last two compounds have not been previously reported. Their structures followed readily from their ^{1}H- and ^{13}C-NMR spectra and in passing, it may be noted that the ^{13}C- coupling patterns were themselves useful in the structure elucidation. The biogenetic origin of the two compounds has not been clarified but possibly, epilubimin (VIII) is formed via the enolisation of the aldehyde group of lubimin (V). It may then give rise to the dihydro compound IX by reduction. However, the alternative pathway, from a germacrene to dihydroepilubimin (IX) and hence to lubimin (V) *via* epilubimin (VIII), appears equally attractive.

The mother liquors from which dihydroepilubimin (IX) had separated also contained a little 15-dihydrolubimin (VII). The amount was too small to allow its isolation but its presence was indicated by chromatography and established by the ^{1}H-NMR spectrum of the mother liquors which was the composite of the spectra of pure VII and IX. Again, it is not clear whether dihydrolubimin (VII) precedes lubimin (V) biogenetically or is formed from it by reduction. Formally, at least one of the two dihydro-compounds must be a precursor of lubimin but possibly only in an enzyme-bound form. It is, indeed, conceivable that the four compounds participate in a dynamic equilibrium. This problem is receiving further attention.

The last compound from the potato - *M. fructicola* interaction which needs to be discussed is phytuberin. Its predicated biosynthesis is shown in Scheme 4. In the *M. fructicola* experiment, only a very small amount, less than 1 mg, was produced but this sufficed to give a ^{13}C-NMR spectrum in which the 5 predicted, enriched singlet absorptions were visible. However, the triplets resulting from labelled carbon pairs and natural abundance ^{13}C were too weak to be observed. Because of the unusual nature of the biosynthetic route to phytuberin, a more explicitly informative spectrum was desirable and it is for this reason that we turned to an incorporation experiment with the potato - *Glomerella cingulata* system. Other experiments had shown that this consistently furnished phytuberin in good yields.

The metabolites which were isolated are listed in Table VIII. Phytuberin (III) was obtained in gratifying amount and gave the spectrum shown in Fig. 4. The spectrum is entirely consistent with the postulated rearrangement of the carbon skeleton of a

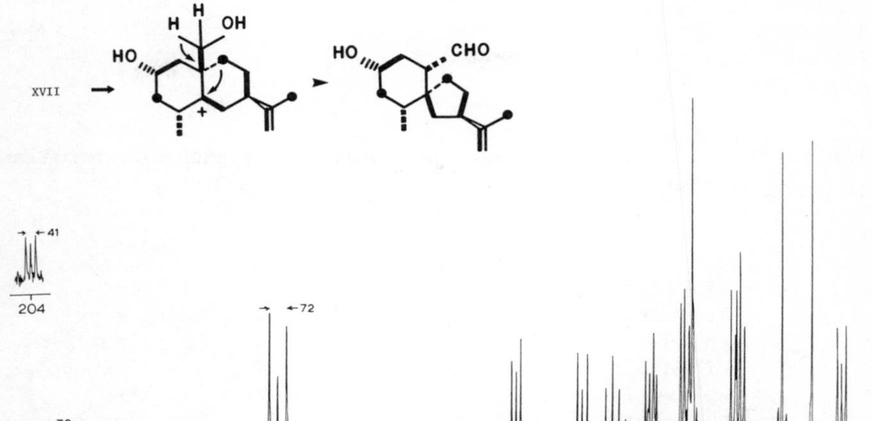

Figure 3. ^{13}C NMR spectrum of lubimin biosynthesized from 1,2-^{13}C-acetate

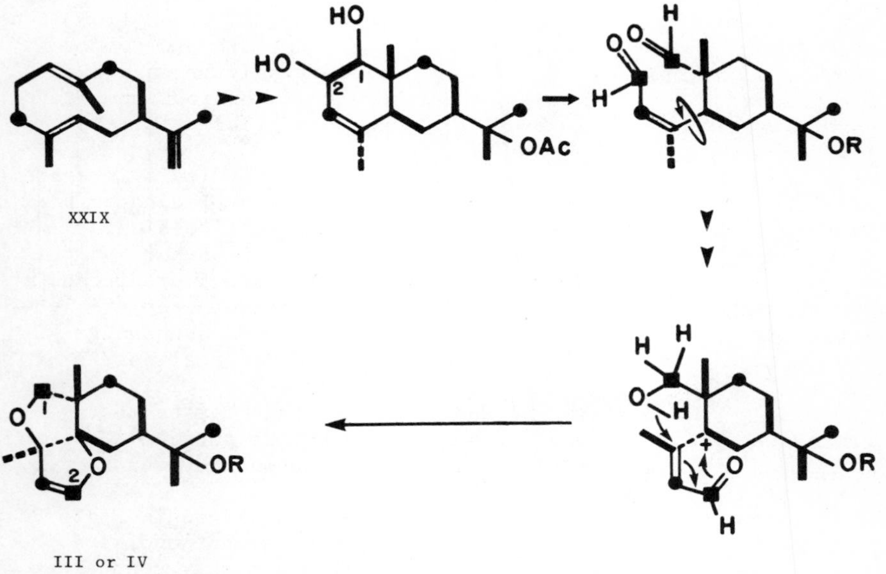

XXIX

III or IV

Scheme 4. Postulated biosynthesis of phytuberin

Table VIII ^{13}C-Labelled Compounds from the Potato - <u>Glomerella cingulata</u> interaction.

Compound	yield (mg)*
rishitin (I)	53
lubimin (V) epilubimin (VIII)	27
dihydrolubimin (VII)	79
solavetivone (XI)	19
phytuberin (III)	39
phytuberol (IV)	11

* after final purification; from 64 lbs potatoes

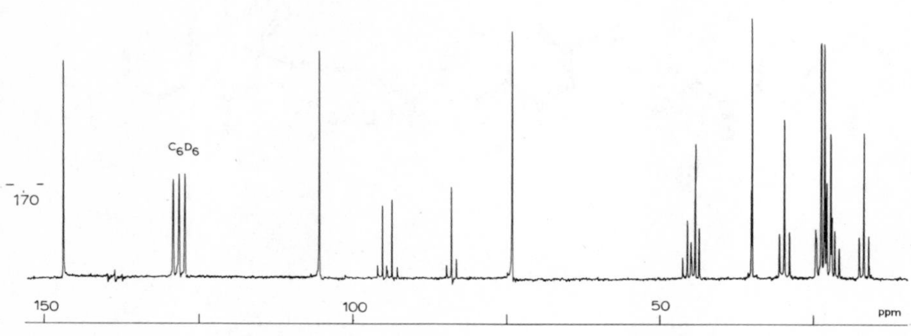

Figure 4. ^{13}C *NMR spectrum of enriched phytuberin*

eudesmane precursor, and in particular, with the scission of a C-1 - C-2 bond and reattachment of C-2 to C-5 *via* oxygen. Entirely analogous results were obtained from phytuberol (desacetylphytuberin; IV) which was isolated in the same experiment. The spectra do not provide evidence other than that pertaining to the making and breaking of skeletal bonds and many of the details of this intriguing biosynthesis remain to be studied. Experiments to this end are in hand.

Finally, a few words about the other metabolites isolated from the *G. cingulata* interaction. The spectrum obtained from rishitin (I) did not differ in any essential respect from that discussed earlier. Lubimin (V) was isolated only in admixture with small amounts of epilubimin (VIII) and of a third, as yet unidentified component, but the spectrum of the mixture incorporated all the features observed earlier in the enriched pure samples of V and VIII. The spectrum of solavetivone (XI), a compound which was not detected in the *M. fructicola* interaction, showed, as expected, the same coupling pattern as that shown earlier for lubimin and this pattern could also be seen in the spectrum of dihydrolubimin (VII). The high yield in which VII was isolated from the potato *G. cingulata* interaction contrasts strongly with its trace occurence in the *M. fructicola* experiment and is, without doubt, to a large extent a consequence of its formation from V by the fungus. The reduction of V to VII is carried out readily by *G. cingulata* and several other fungi in pure culture (*38*). However, this conversion does not appear to take place in *M. fructicola* cultures. For this reason, the occurrence of traces of VII in the potato - *M. fructicola* interaction suggests that VII is formed by potato enzymes as discussed earlier. This problem, too, should be amenable to solution by further tracer studies.

Footnotes

[1]An example of the conformational mobility of germacrenes was seen by Yamamura and coworkers in the cyclization of acoragermacrone by chemical means to both *cis*- and *trans*-decalins (*25*). However, in laboratory cyclizations, 1(10):4-germacrenes usually exhibit strong conformational preferences which lead to *trans*-decalins (*26*) and, as was pointed out (*26*), Yamamura's results are not unambiguous. Nevertheless, when enzyme-bound in biosynthetic reactions, the compounds may well adopt conformations that are not favoured in the laboratory. Certainly *cis*-decalins, inclduing *cis*-eudesmanes, occur in nature. The problem is briefly but lucidly discussed in a recent monograph (*27*).

[2]Commercially available 1,2-$^{13}C_2$-acetate always contains an appreciable amount of singly labelled (both 1- and 2-) acetate, which will be incorporated at the same rate and will also contribute to the singlet absorption.

Literature Cited

(*1*) Tomiyama, K., Sakuma, T., Ishizaka, N., Sato, N., Katsui, N., Takasugi, M., and Masamune, T. Phytopathology (1968) (*58*),

115.
(2) Sato, N., Tomiyama, K., Katsui, N., and Masamune, T. Ann. Phytopathol. Soc. Japan (1968) 34, 344.
(3) Katsui, N., Matsunaga, A., Imaizumi, K., Masamune, T., and Tomiyama, K. Tetrahedron Letters (1971) 83.
(4) Varns, J. L., Kuć, J., and Williams, E. B. Phytopathology (1971) 61, 174.
(5) Coxon, D. T., Price, K. R., Howard, B., and Curtis, R. F. J. Chem. Soc. Perkin I (1977) 53.
(6) Currier, W. W. Diss. Abstr. Intern. B. (1975) 36, 685.
(7) Price, K. R., Howard, B., and Coxon, D. T. Physiol. Plant Path. (1976) 9, 189.
(8) Ozeretskovskaya, O. L., Vasyukova, N. I., and Metlitskii, L. V. Dokl. Bot. Sci. (1969) 187-189, 158.
(9) Stoessl, A., Stothers, J. B., and Ward, E. W. B. Can. J. Chem. (1975) 53, 3351.
(10) Stoessl, A., Stothers, J. B., and Ward, E. W. B. J. Chem. Soc. Chem. Commun. (1975) 431.
(11) Stoessl, A., Ward, E. W. B., and Stothers, J. B. in preparation.
(12) Kalan, E. B. and Osman, S. F. Phytochemistry (1976) 15, 775.
(13) Coxon, D. T., Price, K. R., Howard, B., Osman, S. F., Kalan, E. B., and Zacharius, R. M. Tetrahedron Letters (1974) 2921.
(14) Stoessl, A., Unwin, C. H., and Ward, E. W. B. Phytopathol. Z. (1972) 74, 141.
(15) Birnbaum, G. I., Stoessl, A., Grover, S. H., and Stothers, J. B. Can. J. Chem. (1974) 52, 993.
(16) Ward, E. W. B., Stoessl, A., and Stothers, J. B. Phytochemistry, in press, 1977.
(17) Bailey, J. A., Burden, R. S., and Vincent, G. G. Phytochemistry (1975) 14, 597.
(18) Burden, R. S., Bailey, J. A., and Vincent, G. G. Phytochemistry (1975) 14, 221.
(19) Leppik, R. A., Hollomon, D. W., and Bottomley, W. Phytochemistry (1972) 11, 2055.
(20) Roberts, D. L. Phytochemistry (1972) 11, 2077.
(21) Fujimori, T., Kasuga, R., Kaneko, H., and Noguchi, M. Phytochemistry (1977) 16, 392.
(22) Anderson, R. C., Gunn, D. M., Murray-Rust, J., Murray-Rust, P., and Roberts, J. S. J. Chem. Soc. Chem. Commun. (1977) 27.
(23) Stoessl, A., Stothers, J. B., and Ward, E. W. B. Phytochemistry (1976) 15, 855.
(24) Parker, W., Roberts, J. S., and Ramage, R. Quart. Rev. (1967) 21, 331.
(25) Iguchi, M., Niwa, M., and Yamamura, S. Tetrahedron Lett. (1973) 1687, 4367, and references there cited.
(26) Sutherland, J. K. Tetrahedron (1974) 30, 1651.
(27) Coates, R. M. Fortschr. Chem. Org. Naturstoffe (1976) 33,

73.
(28) Baker, F. C., Brooks, C. J. W., and Hutchinson, S. A. J. Chem. Soc. Chem. Commun. (1975) 293.
(29) Dunham, D. J. and Lawton, R. G. J. Amer. Chem. Soc. (1971) 93, 2075.
(30) Birnbaum, G. I., Huber, C. P., Post, M. L., Stothers, J. B., Robinson, J. R., Stoessl, A., and Ward, E. W. B. J. Chem. Soc. Chem. Commun. (1976) 330.
(31) Stothers, J. B. "Carbon-13 NMR Spectroscopy" Academic Press, New York, 1972.
(32) Grutzner, J. B. Lloydia (1972) 35, 375.
(33) Séquin, V. and Scott, A. I. Science (1974) 186, 101.
(34) McInnes, A. G. and Wright, J. L. C. Accts. Chem. Res. (1975) 8, 313.
(35) Cornforth, J. W. Angew. Chem. Internat. Edn. (1968) 7, 903.
(36) Cordell, J. A. Chem. Rev. (1976) 76, 425.
(37) Stoessl, A., Ward, E. W. B., and Stothers, J. B. Tetrahedron Lett. (1976) 3271.
(38) Ward, E. W. B. and Stoessl, A. Phytopathology (1977) 67, 468.

6

Activated Coordinated Chemical Defense against Disease in Plants

JOSEPH KUĆ and FRANK L. CARUSO

Department of Plant Pathology, University of Kentucky, Lexington, KY 40506

It has been repeatedly demonstrated that plants are protected against disease by using procedures that are essentially identical to those used to immunize animals (1-9). Plants inoculated or treated with cultivar nonpathogenic races of pathogens, avirulent forms of pathogens, non-pathogens, heat-attenuated pathogens and high molecular weight products of infectious agents are protected against disease caused by subsequent infection by pathogens (1, 2, 4-6, 10-23). Recovery from disease or limited disease caused by a pathogen also systemically protect plants against disease caused by subsequent infection by the same pathogen or occasionally unrelated pathogens (4, 5, 17, 24-26). It is also evident that infectious agents elicit the accumulation of antibiotic chemicals (phytoalexins or stress metabolites) around sites of infection, and in resistant interactions these substances often rapidly attain concentrations that inhibit development of many infectious agents (1, 18, 27-32). Clearly, the elicitation of disease resistance in a animals and plants is not based upon the introduction of new genetic information for resistance into all cells of the host. The basic metabolic mechanisms are present and the success and versatility of the immune response in animals is based upon its rapid expression. It is not clear, however, that the immune response in plants depends upon the activation of a mechanism in the host which is unique for its role in defense against disease, and, which in turn, is elicited by structural components or metabolites unique to the infectious agents.

Though the literature reporting induced resistance in plants dates back at least 100 years (33, 34), the reports of the successful control of plant disease in the field using the technique are very recent. This is puzzling since immunization rapidly became and still froms the basis of preventative medicine against infectious disease in animals.

The review by Chester (34), though critical of experimental procedures in many reports he cited, did include reports that complete or partial immunity to disease occurred in some plants after recovery from some diseases. Hypovirulent strains of tobacco mosaic virus have been employed in the Netherlands (35), Japan (36) and the U.S.S.R. (personal communication) to control the disease caused by virulent strains of the virus on tomato. A mechanism to explain this protection has not been established. Chestnut tree blight, incited by Endothia parasitica, has virtually made the American Chestnut extinct. Recently, limited biological control of the disease has been attained by inoculating plants with a hypovirulent strain of the fungus (37, 38). The mechanism for the protection is not understood. It is possible that hypovirulence is determined by a transmissable factor, possibly a virus. Thus, virulent strains of the fungus may become hypovirulent. It is uncertain, however, whether the suppression of virulence itself is solely responsible for the protection or whether the suppressed hypovirulent strains of the pathogen also elicit a defense reaction in the chestnut soon enough and with sufficient magnitude to control development of the fungus. Plants susceptible to Agrobacterium tumefaciens, the incitant of crown gall, are protected from the disease by prior inoculation with the nonpathogen A. radiobacter (39, 40). This has proven a practical means for protection against the disease. The mechanism for protection is uncertain. Evidence exists for a chemical antagonism due to a bacteriocin produced by A. radiobacter as well as competitive blocking of infectible binding sites by A. radiobacter (39, 41). Perhaps more than one mechanism explains the protection.

Two plant-pathogen interactions (green bean - Colletotrichum lindemuthianum and cucurbits - Colletotrichum lagenarium) have received considerable attention in our laboratory. In both the activation of chemical defense appears part of the plant's resistance mechanism to disease, and, in both, protection against disease can be elicited by methods which resemble immunization in animals.

Protection of green bean (Phaseolus vulgaris L.) against Colletotrichum lindemuthianum

Protection of green bean against bean anthracnose, caused by C. lindemuthianum, was first demonstrated by infecting the hypocotyls of bean cultivars, resistant to some but not all races of the fungus (Figure 1) with cultivar nonpathogenic races of the organism. The inoculated plants were then locally (inducer inoculum and challenge inoculum applied to the same site) and systemically (inducer inoculum and challenge inoculum applied to different sites) protected from disease caused by subsequent infection with cultivar pathogenic races of the fungus (12, 42, 43). Symptoms on protected plants were microscopically and macroscopically indistinguishable from the normal resistance reactions produced in response to inoculation of a cultivar with a cultivar nonpathogenic race (12). It is apparent, therefore, that if a cultivar has resistance to a race of a pathogen, it could be made resistant to all races of the pathogen.

A large number of bean cultivars are, however, susceptible to all known races of C. lindemuthianum (44, 45), and the question remained whether these were susceptible because of an absolute lack of resistance mechanisms, or the inability to elicit resistance soon enough and with sufficient magnitude to contain an infectious agent. Evidence existed to support the latter possibility (46).

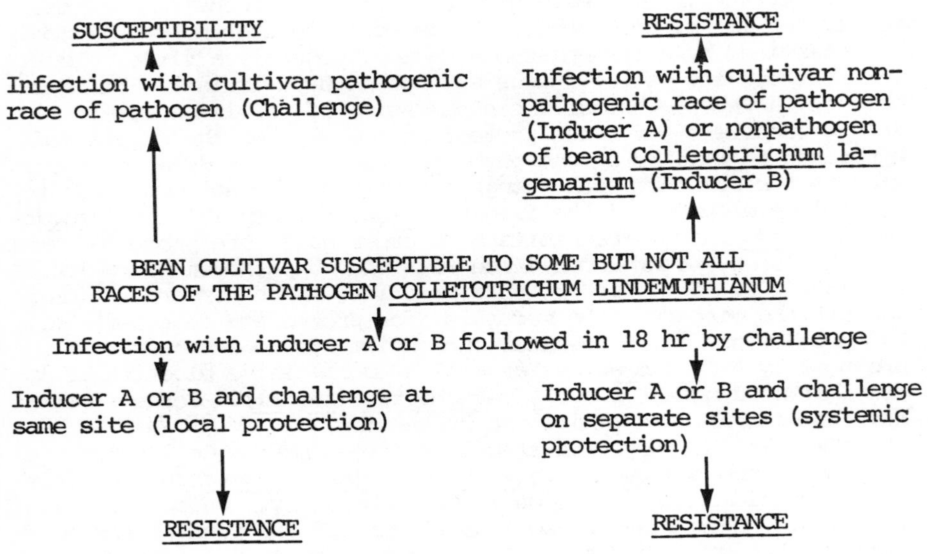

Figure 1. Elicitation of resistance in a cultivar of bean susceptible to some but not all races of Colletotrichum lindemuthianum

Hypocotyl tissue of even completely susceptible cultivars is resistant close to and within the root zone (nonexpanding mature tissue). An effective resistance mechanism is functional in this tissue and the classification of a cultivar as susceptible to C. lindemuthianum is based only on the reaction of hypocotyl tissue distant from the root zone. Further experiments in our laboratory demonstrated that cultivars of bean susceptible to all races of C. lindemuthianum (Figure 2) were locally and systemically protected against the disease by infection with C. lagenarium, a pathogen of cucurbits, prior to infection with the pathogen of bean (13). C. lagenarium was equally effective in inducing systemic protection in cultivars susceptible to all races of C. lindemuthianum as in cultivars resistant to one or more races of the pathogen (Figure 1). Protection induced by races of C. lagenarium was generally not elicited as rapidly nor as extensively as protection elicited by cultivar nonpathogenic races of *C. lindemuthianum*. However, like protection elicited by *C. lindemuthianum*, systemic protection elicited by *C. lagenarium* became effective earliest in the most mature region of the hypocotyl. The challenge fungus developed in protected tissue to the same extent as in tissue protected by cultivar nonpathogenic races of *C. lindemuthianum*. Protection elicited by *C. lagenarium* protected against all races of *C. lindemuthianum* tested as did protection elicited by cultivar nonpathogenic races of *C. lindemuthianum*. These observations suggest that the same mechanisms may be involved in protection elicited by both fungi. Additional evidence for the ability to activate a resistance mechanism in cultivars susceptible to all races of *C. lindemuthianum* was obtained by heat-attenuating cultivar pathogenic races of the pathogen (Figure 2) in host tissue prior to the expression of symptoms (47). Such plants were protected from disease caused by infection with the same or other cultivar pathogenic races of the fungus. This experiment also suggested that fungal components or metabolites, even in cultivar pathogenic races, can elicit resistance. It is evident, therefore, that resistance mechanisms exist and can be activated in bean cultivars resistant to some and susceptible to other races of *C. lindemuthianum* as well as in cultivars susceptible to all races of the pathogen. It is also evident that even cultivar pathogenic races have the chemical potential to elicit resistance.

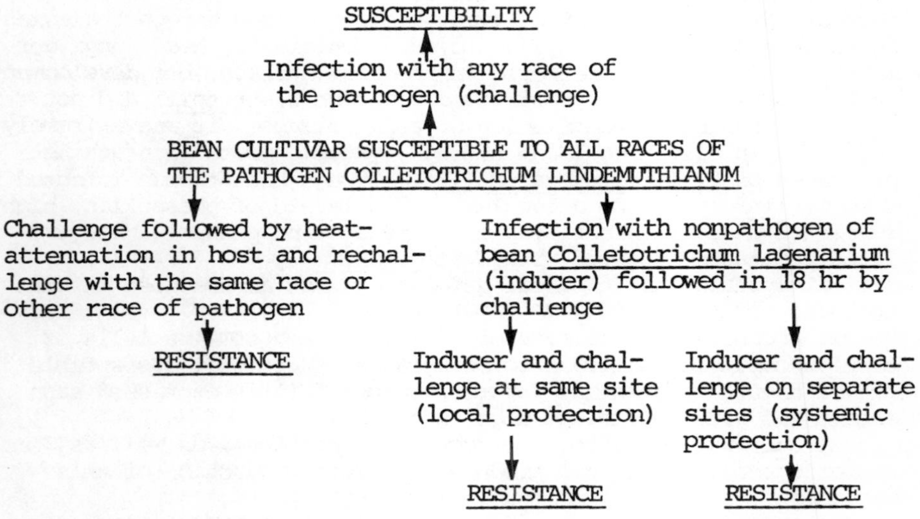

Figure 2. Elicitation of resistance in a cultivar of bean susceptible to all races of Colletotrichum lindemuthianum

Systemic protection elicited by cultivar nonpathogenic races of *C. lindemuthianum* and by *C. lagenarium* may be mechanistically similar to the resistance of mature tissue (14, 15). The nature and extent of fungal development in these tissues is similar and heat treatment prior to infection with the inducer fungus markedly reduced systemic protection and mature tissue resistance. Heat treatment did not reduce the effectiveness of race-specific resistance to *C. lindemuthianum*, resistance to *C. lagenarium* or local protection elicited by these fungi.

Though phytoalexin accumulation may explain race-specific resistance of bean cultivars to *C. lindemuthianum* as well as induced local protection by *C. lindemuthianum* or *C. lagenarium*, it does not itself explain systemic protection (16). Phaseollin and other phytoalexins were not detected in systemically protected unchallenged tissue, and became apparent in systemically protected challenged tissue only when restricted collapse and browning of host cells became evident. It is possible, therefore, to separate two facets of the induced resistance phenomenon: (1) the chemical agents including phaseollin and other isoflavonoid phytoalexins which accumulate around the site of infection and which contribute to the inhibition of fungal development; (2) the signal which commits cells removed from the site of an inducing inoculation (systemic protection) to resistance. Further evidence indicates resistance in bean cannot be attributed solely to the accumulation of phytoalexins (16). *C. gossypii*, *C. phomoides* and *C. trifolii*, nonpathogens of bean, did not cause hypersensiteve browning of host cells and did not elicit the accumulation of phaseollin, phaseollidin, phaseollinisoflavan, kievitone or other phytoalexins detectible by the bioassays employed. Conidia of these fungi germinated and formed appressoria on the host surface, but development in the host was rapidly restricted. Though *C. trifolii* did not elicit phytoalexin accumulation or cell collapse, it was extremely effective in eliciting local but not systemic protection against all races of *C. lindemuthianum* (13). Though bean tissue infected with *C. lindemuthianum* accumulated high levels of phaseollin, high levels of phaseollinisoflavan and little or no phaseollin accumulated in tissue infected with *C. lagenarium*.

It appears that chemical defense against anthracnose in bean includes the following considerations:

(1) production or release of a signal which commits cells of even susceptible cultivars to resistance; therefore, susceptible plants have the potential for resistance if it is expressed soon enough and with sufficient magnitude.

(2) cultivar nonpathogenic races of a pathogen, as well as some nonpathogens of bean, can produce or elicit production of this signal.

(3) systemically protected unchallenged tissue does not contain isoflavonoid phytoalexins.

(4) phytoalexins are not detected in all reactions between nonpathogens and bean.

(5) the fungus influences which phytoalexins accumulate.
(6) induced protection is not evident before spore germination, appressorium formation and penetration into the host.

Protection of cucurbits against Colletotrichum lagenarium - aspects of the phenomenon

Recent work in our laboratory using the interaction of cucurbits with the pathogen C. lagenarium has supported the work with the interaction of bean and the pathogen C. lindemuthianum. Infection of a cotyledon or first true leaf (leaf one) of cucumber with C. lagenarium systemically protected tissue above (developed or not yet developed) against disease caused by the pathogen (4). Physical damage or chemical injury did not elicit protection. Susceptibility in this interaction is characterized by the formation of large but defined lesions. Since many such lesions may form on leaves, stems and fruit of cucurbits, the growth of plants and their productivity, as well as the quality of fruit, are adversely affected. Nevertheless, the eventual restriction of lesion development suggests the presence of a mechanism for resistance in susceptible plants. Whether the mechanism which restricts the development of lesions is identical to that which is systemically induced by the pathogen is unknown, but similarities are evident and will be discussed later.

Systemically induced protection against the pathogen by the pathogen is evident as a delay in symptom expression and a reduction in the number and size of lesions (5). Infection of leaf one when the second true leaf was one fourth to one third expanded, systemically protected plants for 4-5 wk. A second or booster inoculation 3 wk after the first inoculation extended the time of protection into the fruiting period. Protection was elicited by and effective against six races of the fungus and was evident not only with more than 20 susceptible cultivars but also with two cultivars which express some reisitance to the pathogen. Resistance in these two cultivars is expressed, as is systemic induced resistance, by a delay in symptom appearance and a reduction in the number and size of lesions. A single lesion on leaf one produced significant protection (Figure 3). Protection was evident on the second leaf 72-96 hr after inoculating leaf one. Excising leaf one 72-96 hr. after inoculating leaf one did not reduce protection of leaf two. Leaf two was protected if excised 96-120 hr. after leaf one was inoculated. It is evident that it is not necessary for the inducer to be present once protection has been initiated and protection continues to be expressed in all developing leaves for 4-5 wk. It is also evident that the protected leaf need not be attached to the plant to maintain protection.

The pattern of protection in watermelon and muskmelon resembles that in cucumber (48, 49). Infection of the cotyledons or first true leaf of four cultivars of watermelon and four cultivars of muskmelon with C. lagenarium systemically protected the plants from disease caused by subsequent infection with the pathogen. Protection was noted as a reduction in the number and size of lesions, and plants remained protected 4 wk. after the protecting inoculation. Race 1, 2 and 3 of the fungus elicited protection and a single lesion on leaf one elicited significant protection. In three separate field trials, cucumber plants were protected against a challenge inoculation with C. lagenarium by a prior inoculation with the pathogen (50). Protection of watermelon was evident in two trials and indications were that muskmelon could also be protected. Lesions on protected plants were reduced in number and size as compared to lesions on unprotected plants (Figure 4).

Though chemical or physical injury without the presence of an infectious agent did not elicit protection, protection against C. lagenarium was also elicited by the infection of cotyledons with tobacco necrosis virus (TNV) (17) and the infection of leaf one with the bacterial incitant of angular leaf spot, Pseudomonas lachrymans (51). Infection with C. lagenarium also protected cucumber against disease caused by P. lachrymans, and plant breeders have observed that cultivars are either resistant to both or susceptible to both pathogens (personal communication). Infection with C. lagenarium did not protect plants against TNV and TNV did not protect against TNV.

The above data support the hypothesis that even susceptible plants have effective mechanisms for disease resistance if they can be expressed soon enough and with sufficient magnitude. Induced resistance can be systemic and of quite long duration. The phenomenon of systemic induced protection in plants and immunization in animals are similar at least in outward appearance, and the phenomenon suggests a new approach for the practical control of disease in plants.

Mechanisms for induced systemic protection

Investigators in the field may argue about which is the initial or most important mechanism for disease resistance (or susceptibility) in plants. Such arguments may be fruitless in that two or more distinct mechanisms may be operative and the presence of both mechanisms and their coordination may determine the specificity of the interaction. Interactions in which phytoalexins accumulate but the plant is judged susceptible may be due to a lack of or inadequacy in a second mechanism. The second mechanism may be the presence or induced production of agglutinins.

An extremely stimulating and thought-provoking paper by Uritani and colleagues (52) presents data which support this contention. They report that resistance of sweet potato root to nonpathogenic isolates of C. fimbriata is based on at least two mechanisms. In the initial stage of infection, agglutinating factors in host cells may agglutinate spores and germinating hyphae thereby localizing the fungus in the host. In the second stage, both

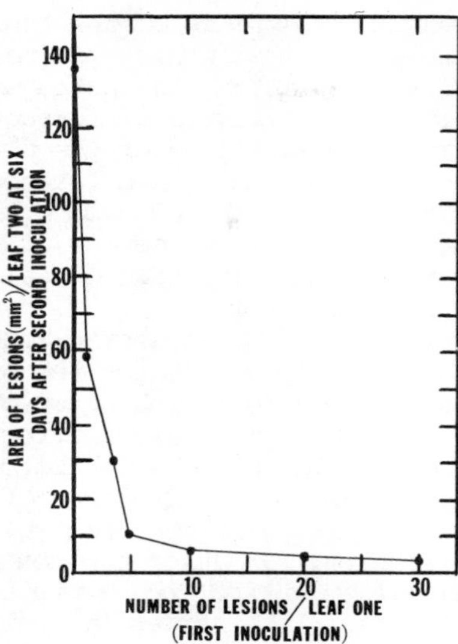

Figure 3. The effect of the number of lesions on leaf one on the protection of leaf two of cucumber against Colletotrichum lagenarium

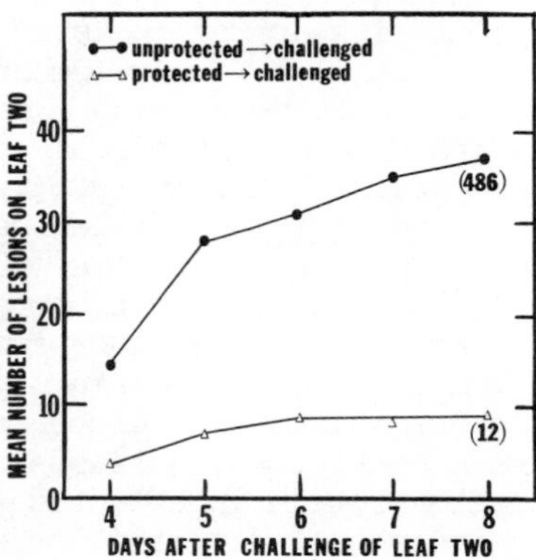

Figure 4. Protection of cucumber against Colletotrichum lagenarium *by* Colletotrichum lagenarium *in the field. The area of lesions per leaf eight days after the inoculation of leaf two is given in parentheses.*

pathogenic and nonpathogenic strains of the fungus elicit the accumulation of furanoterpenoids in host tissue. The implication is that the speed of development of the pathogenic strain exceeds the accumulation of fungitoxic levels of furanoterpenoids due to a lack of agglutination, whereas spread of the nonpathogenic strain is limited by agglutination which allows furanoterpenoids to accumulate to fungitoxic levels in and around the site of interaction between host and fungus. This concept is consistant with our investigations of the systemically induced resistance of cucumber to C. lagenarium by C. lagenarium.

There is no evidence that the systemic accumulation of classical phytoalexins is alone responsible for the systemic protection of cucurbits. A chloroform-soluble inhibitor of the growth of C. lagenarium has been obtained from tissue surrounding lesions on protected and unprotected plants. The inhibitor may be important in restricting lesion development and account for the normally restricted lesions characteristic of cucurbit anthracnose. Spore germination and appressorium formation are not inhibited on protected plants, but development from the appressoria appears reduced in protected plants. In a series of tests, penetration from appressoria was 20 to 40% and from less than 1 to 5% in unprotected and protected plants, respectively. The presence of preformed agglutinins appears closely associated with resistance of cucumber to scab incited by Cladosporium cucumerinum (53). Studies also indicate an agglutinating factor accumulates in unchallenged plants systemically protected by C. lagenarium. It is possible, therefore, that at least three mechanisms contribute to acquired resistance of cucurbits to C. lagenarium. One mechanism restricts penetration into the host, a second agglutinates developing hyphae in penetrated tissue to minimize spread, and the third is the production of classical phytoalexins around the site of infection. It is entirely possible that still a fourth mechanism may be operative. The agglutinins may also react with extracellular hydrolytic enzymes or toxins produced by infectious agents thereby neutralizing their "symptom-causing" activity. The systemic nature of the protection of C. lagenarium by C. lagenarium resembles the systemic elicitation of protease inhibitors in wounded plants (54-57), though injury does not itself elicit protection. Injury of leaves of plants representing several plant families or the introduction of extracts from injured tissues into plants is sufficient to elicit the systemic accumulation of protease inhibitors. The character of the protein in responding tissues markedly changes and as much as 12% of the protein in affected tissues is protease inhibitor with a high content of disulfide cross linkages. We are investigating the relationship between the production of extracellular proteases by the pathogen and the induction of protease and other enzymatic inhibitors in systemically protected plants.

The systemic elicitor of protease accumulation reported by Ryan and his colleagues is still uncharacterized but the implications of this "long distance" communication are profound in relation to disease resistance and practical control of disease. Several possible ingredients of a defense system become apparent: (1) a rapid "long distance" system of communication (2) presence and/or elicitation of protein or glycoprotein enzyme inhibitors and agglutinins (3) presence and/or elicitation of classical low molecular weight phytoalexins. How far apart in basic metabolic mechanisms are the immune systems in plants and animals? Genes for resistance and susceptibility can control the presence or absecce of any one of the mechanisms and the rates of the mechanisms. In a coordinated defense against disease, any one or two mechanisms for resistance may be active, but if a third is inactive, it may become the limiting factor for resistance and the plant is susceptible. Plants with mechanisms for resistance, therefore, can in fact be susceptible. Induced resistance may be based on the marked activation of a mechanism, which is not limiting to defense, to a level that overcomes the limitation of a deficient factor, e.g., phytoalexins accumulate rapidly enough and with sufficient magnitude to eliminate the need of an agglutinin. In other situations, induced resistance may activate a deficient or limiting factor in the coordinated mechanisms. The key to disease resistance in plants is the functioning of multiple mechanisms for resistance and the key concept in understanding their interaction is one of "coordinated defense". Of course the infectious agent also influences the effectiveness or coordination of host defense. The active contribution of the infectious agent makes the interaction extremely complex but at the same time increases many fold the possibilities for specificity.

This is paper 77-10-124 of the Kentucky Agricultural Experiment Station, Lexington, Kentucky 40506. The authors' work reported in this manuscript was supported in part by a grant from the Herman Frasch Foundation and grant 316-15-51 of the Cooperative State Research Service of the United States Department of Agriculture.

Literature Cited

1. Kuc´, J. Annu. Rev. Phytopathology (1972) 10, 207-232.
2. Kuc´, J. World Rev. of Pest Control (1968) 7, 42-55.
3. Hammerschmidt, R., Acres, S., Kuc´, J. Phytopathology (1976) 66, 790-794.
4. Kuć, J., Shockley, G., Kearney, K. Physiol. Plant Pathology (1975) 7, 195-199.
5. Kuć, J., Richmond, S. Phytopathology (1977) 67, 533-536.
6. Matta, A. Annu. Rev. Phytopathology (1971) 9, 387-410.
7. Ross, A. Proceed. Internatl. Conf. Plant Viruses (1966) July, 127-150.
8. Yarwood, C. Proceed. Natl. Acad. Sci. U.S.A. (1954) 40, 374-374-377.
9. Yarwood, C. Phytopathology (1956) 46, 540-544.
10. Bell, A., Presley, J. Phytopathology (1969) 59, 1147-1151.

11. Deverall, B. "Defense Mechanisms in Plants" 110 p., Cambridge Univ. Press, London, 1977.
12. Elliston, J., Kuc´, J., Williams, E. Phytopathology (1971) 61, 1110-1112.
13. Elliston, J., Kuc´, J., Williams, E. Phytopathol. Z. (1976) 86, 117-126.
14. Elliston, J., Kuc´, J., Williams, E. Phytopathol. Z. (1976) 87, 289-303.
15. Eliston, J., Kuc´, J., Williams, E. Phytopathol. Z. (1977) 88, 43-52.
16. Elliston, J., Kuc´, J., Williams, E., Rahe, J. Phytopathol. Z. (1977) 88, 114-130.
17. Jenns, A., Kuc´, J. Physiol. Plant Pathol. (1977) In press.
18. Kuc´, J. "Specificity in Plant Disease" pp. 253-271, Plenum Press, N.Y. 1976.
19. Kuc´, J., Currier, W., Elliston, J., McIntyre, J. "Biochemistry and Cytology of Plant-Parasite Interaction" pp. 168-180, Elsevier, N.Y. 1976.
20. Lovrekovich, L., Farkas, G. Nature (1965) 205, 823-824.
21. Main, C. Phytopathology (1968) 58, 1058-1059.
22. McIntyre, J., Kuc´, J., Williams, E. Physiol. Plant Pathol. (1975) 7, 153-170.
23. Heale, J. B., Sharman, S. Physiol. Plant Pathol. (1977) 10, 51-61.
24. Braun, J., Helton, A. Phytopathology (1971) 61, 685-687.
25. Randall, R., Helton, A. Phytopathology (1976) 66, 206-207.
26. Cruickshank, I., Mandryk, M. J. Australian Inst. Agric. Sci. (1960) 26, 369-372.
27. Kuc´, J. "Encyclopedia of Plant Physiology" Vol. 4, pp. 632-652, Springer-Verlag, Berlin, 1976.
28. Kuc´, J., Currier, W., Shih, M. "Biochemical Aspects of Plant-Parasite Relationships" pp. 225-257, Academic Press, London 1976.
29. Kosuge, T. Annu. Rev. Phytopathology (1969) 7, 195-222.
30. Stoessel, A., Stothers, J., Ward, E. Phytochem. (1976) 15, 855-872.
31. Ingham, J. Botan, Rev. (1972) 38, 343-424.
32. Cruickshank, I., Biggs, D., Perrin, D. J. Indian Botan. Soc. (1971) 50A, 1-11.
33. Bernard, N. Ann. Sci. Natl. Bot. (1909) 1, 1-196.
34. Chester, K. S. Quart. Rev. Biol. (1933) 8, 129-154, 275-324.
35. Rast, A. Agricultural Research Report of the Inst. of Phytopathol. Res., Wageningen, The Netherlands (1975) 834, 1-76.
36. Komochi, S., Goto, T., Oshima, N. J. Hort. Assn. Japan (1966) 35, 269-276.
37. Anagnostakis, S., Jaynes, R. Plant Dis. Reptr. (1973) 57, 225-226.
38. Van Alfen, N., Jaynes, R., Anagnostakis, S. Science (1975) 189, 890-891.
39. New, P., Kerr, A. J. Applied Bacteriol. (1972) 35, 279-287.

40. Kerr, A. J. Applied Bacteriol. (1972) 35, 493-497.
41. Lippincott, B., Lippincott, J. J. Bacteriol. (1969) 97, 620-628.
42. Rahe, J., Kuc´, J., Chuang, C., Williams, E. Phytopathology (1969) 59, 1641-1645.
43. Skipp, R., Deverall, B. Physiol. Plant Pathol. (1973) 3, 299-313.
44. Bannerot, H. Ann. Amel. Plantes (1965) 15, 201-222.
45. Goth, R., Zaumeyer, W. Plant Dis. Reptr. (1965) 49, 815-818.
46. Leach, J. Minn. Agric. Expt. Sta. Techn. Bull. (1923) 14, 1-41.
47. Rahe, J. Phytopathology (1973) 63, 572-577.
48. Caruso, F., Kuc´, J. Phytopathology In Press.
49. Caruso, F., Elliston, J., Kuc´, J. Proc. Amer. Phytopathol. Soc. (1976) 3, 259.
50. Caruso, F., Kuc´, J. Phytopathology In Press.
51. Caruso, F. Proc. Amer. Phytopathol. Soc. (1977) In Press.
52. Uritani, I., Oba, K., Kojima M., Kim, W., Ohuni, I., Suzuki, H. "Biochemistry and Cytology of Plant-Parasite Interaction" pp. 239-252, Elsevier, N.Y. 1976.
53. Hammerschmidt, R., Kuc , J. Proc. Amer. Phytopathol. Soc. (1977) In Press.
54. Peng, J., Black, L. Phytopathology (1976) 66, 958-963.
55. Ryan, C. Annu. Rev. Plant Physiol. (1973) 24, 173-196.
56. Gustafson, G., Ryan, C. J. Biol. Chem. (1976) 251, 7004-7010.
57. Walker-Simmons, M., Ryan, C. Plant Physiol. (1977) 59, 437-439.

7

Biochemical and Ultrastructural Aspects of Southern Corn Leaf Blight Disease

PETER GREGORY, ELIZABETH D. EARLE, and VERNON E. GRACEN

Department of Plant Breeding and Biometry, Cornell University, Ithaca, NY 14853

The study of southern corn leaf blight disease is of major economic and scientific value. The economic importance of this fungal disease was dramatically emphasized in 1970 when it reached nearly epidemic proportions in the U.S.A., resulting in severe losses for corn growers. Scientifically, research on the disease may result in novel and major gains in plant pathology, plant breeding and higher plant genetics.

The main aims of this paper are to critically describe the present state of biochemical and ultrastructural knowledge about the mechanism of the disease, to highlight the most important gaps in our knowledge and to discuss some of the ways in which these gaps can be filled.

General Characteristics of Southern Corn Leaf Blight Disease

The causal organism of southern corn leaf blight disease is the fungus *Helminthosporium maydis* Race T (HmT). This is one of several members of the genus *Helminthosporium* to produce host-specific toxins. HmT and the toxin(s) from HmT (HmT toxin) only affect corn which contains the Texas (T) source of male sterile cytoplasm. Plants containing the non-male sterile (N) or the C or S types of male sterile cytoplasms are resistant to the fungus and insensitive to HmT toxin (*1*,*2*). Nuclear genes which restore T cytoplasm to male fertility (Rf genes) may slightly modify the disease reaction and toxin response of corn varieties containing the T cytoplasm (*3*,*4*). Nuclear genes other than Rf genes can affect the HmT disease reaction (*5*,*6*,*7*) with some inbred lines exhibiting a higher level of resistance in T cytoplasm than others. In all cases, however, each inbred is more susceptible in T than in N cytoplasm. Toxin sensitivity of different inbred lines in T cytoplasm also varies but there is no apparent correlation between nuclear control of toxin sensitivity and the degree of disease resistance (*8*).

It is of considerable interest that another race of *H. maydis*, Race O, produces a toxin but shows no specificity for T cytoplasm (*9*). *H. maydis* Race T might essentially represent a Race O isolate that has accumulated additional genes for cytoplasm-specific toxin production (9,10). The fact that HmT toxin can specifically damage T cytoplasm in the absence of the fungus (11) gives southern corn leaf blight disease great potential as a model system in plant pathology.

Effects of HmT Toxin on T Cytoplasm

Treatment with HmT toxin produces morphological, physiological, ultrastructural and biochemical changes in T cytoplasm material. Table I lists these effects, seen in systems ranging in complexity from whole plants to subcellular fractions. The effects listed are specific for T cytoplasm; N cytoplasm material is either unaltered by any toxin concentration tested or is affected less than T cytoplasm at a given concentration. Non-specific effects (e.g. carbohydrate leakage (12)) are not included in Table I. It can be seen in Table I that the timing, sensitivity and degree of specificity of the different effects vary considerably.

The relative importance of the reported effects is difficult to evaluate. Some effects, such as inhibition of root growth and changes in isolated mitochondria, have been observed by many workers; others have been documented only by work from a single laboratory or even by a single publication. Comparison of work in different laboratories is difficult because no standard procedures for preparation and assay of HmT toxin exist. Many different schemes for toxin production and purification are in use (Table II). Toxin activity in the initial toxin source is influenced by culture conditions and the fungal isolate used, but the initial toxin activity is rarely reported. The effectiveness of the purification procedures, the level of purity attained and the amount of nonspecific material remaining are usually difficult to assess. The use of different corn lines in different studies or even as the source of N and T cytoplasm in a given experiment (17) causes further complications.

In spite of problems in interpretation and comparison, studies of toxin effects on T cytoplasm serve several purposes. 1) They provide convenient criteria for identifying T cytoplasm and T cytoplasm components in many different systems, at the field level, in seed or pollen populations or in cell cultures. Usually, only T cytoplasm material is sensitive to toxin. Other male sterile and non-male sterile cytoplasms are resistant. Procedures that alter the usual response of N and T cytoplasm to toxin may yield material which will further understanding of the difference between the two cytoplasms. Because callus from T cytoplasm seeds usually fails to grow on nutrient medium containing toxin, the toxin resistant callus occasionally seen in T cytoplasm cultures is easy to isolate for further analysis (25, 29). Biochemical and genetic analysis of such toxin-resistant 'T' callus is now in progress in our laboratory and in several other laboratories. Similarly failure of T cytoplasm protoplasts to

Table I. Effects of H. maydis Race T toxin on T cytoplasm material

Toxin application (if not in ambient solution	Effect	Timing[a]	References
	PLANTS		
sprayed	lesion formation	3-6 days	(13)
injected into stem	chlorosis, necrosis of leaves	2-4 days	(13,46)
pipetted onto leaf whorl, immature tissue, punctured with needle	chlorotic streaks	2-5 days	(10,14,15)
	SEEDLINGS		
	inhibition of root growth	24-48 hours	(1,15,16, 17,18)
		30 min	(17)
	inhibition of uranyl uptake into vacuoles of root cells	21-28 hours	(19)
	stimulation of electrolyte leakage	1-4 hours	(17)
	ultrastructural damage to mitochondria in root cells	15-120 min	(20)
	failure of root mitochondria to respond to deoxyglucose treatment	90 min	(21)
	depolarization of membrane potential difference of root cells	2-5 min	(22)
	DETACHED LEAVES-LEAF DISCS		
leaf floated on toxin	water soaking	3-4 days	(1)
injected into leaf	lesion formation	3 days	(23)
applied to puncture	lesion formation	12 hours	(24)
		48 hours	(15)
cut end in toxin	chlorophyll retention in dark	3 days	(18)

Table I. (cont.)

Toxin application (if not in ambient solution)	Effect	Timing[a]	References
leaf piece floated on toxin	chlorophyll retention in dark	3 days	(25)
leaf discs	inhibition of dark CO_2 fixation	9-10 hours	(15,18)
leaf discs	inhibition of light CO_2 fixation	9 hours	(18)
cut end in toxin	inhibition of light CO_2 fixation	15-60 min	(26)
cut end in toxin	stimulation of electrolyte leakage	7-23 hours	(27)
		3 hours	(19)
vacuum infiltration of leaf pieces	stimulation of electrolyte leakage	2 hours	(12)
epidermal peels floated on toxin	inhibition of light-stimulated K^+ uptake by guard cells	3 hours	(26)
cut end in toxin	inhibition of transpiration	3 hours	(19)
		5-60 min	(26)
ISOLATED ROOTS			
	stimulation of electrolyte leakage	2 hours (10 min in 1 expt)	(12)
	inhibition of ^{86}Rb uptake	90 min	(7)
toxin present during aeration of roots	inhibition of development of augmented ^{86}Rb uptake	2 hours	(28,47)
CALLUS			
in agar medium	inhibition of growth	28 days	(25,29)
POLLEN (from T cytoplasm plants with TRf genes)			
in agar medium	inhibition of germination and pollen tube growth	2 hours	(30)

Table I. (cont.)

Toxin application (if not in ambient solution)	Effect	Timing[a]	References
	PROTOPLASTS		
	inhibition of volume increase	1 day	(31)
	collapse	1-3 days	(31,32)
	ultrastructural damage to mitochondria	5-60 min	(20,33)
ISOLATED MITOCHONDRIA[b]			
	uncoupling of oxidative phosphorylation		(34-39)
	stimulation of NADH oxidation		(15,23,34 35,37,38 40,41)
	inhibition of malate-pyruvate oxidation		(15,34,35 37,38,40 41)
	inhibition of α-ketoglutarate oxidation		(37,41)
	partial inhibition of succinate oxidation (in sucrose or KCl assay media)		(15,34,35 38)
	stimulation of succinate oxidation (in mannitol or sucrose assay media)		(41,42)
	stimulation of ATPase activity		(38,41)
	activation of cytochrome oxidase		(41)
	activation of succinate cytochrome C reductase		(41)
	inhibition of ^{32}P accumulation		(17,38)
	increase in light transmittance (swelling)		(17,23,34 43)

Table I. (cont.)

Toxin application (if not in ambient solution)	Effect	Timing[a]	References
	ultrastructural changes (swelling, disruption of inner membranes)		(33,43)
	MICROSOMES		
	inhibition of K^+-stimulated ATPase	30-60 min	(44,45)

[a] approximate time from exposure to toxin until effects are noted and/or measured

[b] effects seen within seconds or a few minutes

Table II. Toxin Preparations Used in Physiological, Biochemical and Ultrastructural Studies

Initial Toxin Source	Processing (abridged)	References
infected leaves	boiling water extraction	(30)
	methanol extraction, dialysis	(11)
H. maydis race T mycelium + agar culture medium	methanol extraction	(14)
	methanol, ethyl acetate extraction	(12)
	methanol, ethyl acetate extraction, chromatography (chloroform-methanol), ethyl acetate elution	(37,39,40,41,44,45)
H. maydis race T mycelium + culture filtrate	chloroform extraction of ground mycelium and culture filtrate	(18)
H. maydis race T culture filtrate	none	(3,11,13,16,46)
	dialysis	(1,34)
	desalting, ultrafiltration	(17,23,26,43,70)
	desalting, ultrafiltration, partial Sephadex purification	(17)
	freeze drying, methanol extraction	(19)
	ethyl acetate extraction	(29)
	methanol-water, chloroform extraction	(8,15,38)
	methanol-water, butanol extraction	(28,47)
	chloroform extraction	(7,20,31,32,33)

survive toxin treatment may permit detection of 'T' protoplasts whose response to toxin has been altered by genetic manipulations like protoplast fusion or organelle transfer. 2) They provide a variety of assays for HmT toxin activity. Sensitive bioassays are essential both for rigorous studies of toxin action and for toxin purification procedures. Evaluation of a single toxin preparation in assays involving succinate, NADH or malate-dichlorophenol indophenol (DCPIP) respiration by isolated mitochondria and dark CO_2 fixation by leaf discs showed a quantitative inhibition by the toxin (15). Moreover, these assays are 5-10X as sensitive to toxin as the seedling root growth assay and 20-100X as sensitive as leaf lesion assays (15). Inhibition of protoplast survival is also 5-10X as sensitive to toxin as seedling root growth (32). Electrolyte leakage is one of the least sensitive bioassays (18, 19). 3) They provide information about the mode of toxin action in T cytoplasm cells. The slower effects listed in Table I are probably secondary effects, far removed from the initial interaction of toxin with T cytoplasm. Primary effects should be rapid, highly specific for T cytoplasm and at least as sensitive to toxin as later effects. It should also be possible to relate the primary effects to the observed cytoplasmic inheritance of toxin sensitivity. Both mitochondria and plasma membranes have been implicated as primary sites of early toxin action.

Evidence for the Plasma Membrane as a Site of Toxin Action

Some physiological studies suggest that plasma membranes and mitochondrial membranes are affected by HmT toxin. Evidence for plasma membrane effects is based on observations that after toxin treatment electrolyte leakage occurs (12,17,19,27), K^+ or ^{86}Rb uptake is inhibited (7,28,47), light-stimulated K^+ movement into guard cells is inhibited (26), partial depolarization of the plasma membrane potential difference occurs (22), and K^+ stimulated ATPase from corn root microsomes is inhibited (44,45). This latter point is significant because microsomal preparations from oat roots contain fragments of plasma membranes (48).

Because some of these effects occur within minutes after toxin treatment (12,22) and because of the toxin effect on microsomal K^+ stimulated ATPase, it has been suggested that HmT toxin acts directly on plasma membranes. However the importance of some of the observations implicating plasma membranes is questionable because of the weak specificity for T cytoplasm (28,47), the preliminary nature of the report (22), the inability of other workers to repeat the microsomal ATPase results (49), and the high levels of crude toxin often needed. There is the further problem of interpreting cytoplasmic inheritance of differential membrane sensitivity.

When a highly active, T cytoplasm-specific chloroform extractable toxin is used, no ultrastructural damage to plasma membranes of T protoplasts is seen at times when mitochondrial disruption is apparent (20,33). Although crude HmT culture filtrate caused plasma membrane damage in N and T corn (50), the latter ultrastructural effects could be due to the presence of contaminants in HmT crude filtrate. It seems possible that some of the observed plasma membrane effects may result from prior mitochondrial inactivation. We cannot eliminate the possibility that rapid and specific effects on plasma membranes occur but are below the level of ultrastructural resolution, although we also have preliminary freeze-etch data that fails to detect plasma membrane damage with chloroform-extracted toxin (50). At present, it appears that specificity of HmT toxin for T cytoplasm resides in its interaction with mitochondria. Identification of the differences between N and T mitochondria which limit the effect of toxin to T mitochondria and the mechanism of action of HmT toxin on T mitochondria is of considerable interest.

Evidence for Mitochondria as a Site of Toxin Action

The cytoplasmic inheritance of toxin sensitivity suggests that an altered mitochondrial or chloroplast genome is involved. A chloroplast site seems unlikely both because no effect of toxin on enzyme activities in isolated chloroplast lamellae has been observed (26) and because non-green tissues like roots and callus are very sensitive to toxin.

Several lines of evidence point to the mitochondria as an important and possibly primary site of HmT toxin action.

Effects of HmT Toxin on Isolated Mitochondria. Evidence for an effect of HmT toxin on mitochondrial physiology comes from several reports of toxin-induced alteration in respiration rates with several substrates and uncoupling of oxidative phosphorylation in mitochondria isolated from corn containing T cytoplasm (Table I, Table III). Evidence for HmT toxin-induced structural changes comes from light transmittance and ultrastructural studies which have shown that isolated T mitochondria swell after toxin treatment (Table I). Toxin treatment of isolated mitochondria also results in damage to cristae and loss of matrix density (Table I). Outer mitochondrial membranes appear to be ultrastructurally unaffected by HmT toxin. These multiple effects of HmT toxin are very rapid (Table I).

Mitochondria isolated from T corn with nuclear genes for fertility restoration (TRf) are also sensitive to toxin (23,37), but a longer time is required for complete inhibition of malate oxidation than when T mitochondria are used. The HmT toxin effects are specific for T (and TRf) mitochondria. Even very high concentrations of toxin have no effects on mitochondria isolated from plants with N cytoplasm (Table I, Table III). However, the removal of the outer membrane of N mitochondria results in sensitivity of the inner membranes to toxin as shown by inhibition of malate-DCPIP respiration (70).

Table III. Effect of HmT Toxin on Respiratory Activities of Root Mitochondria from T-cytoplasm and N-cytoplasm Corn

Substrate		Rate of respiration ($nmol\ O_2/min/mg$ protein)	
		W64A(T)	W64A(N)
NADH	State 3	243	273
	State 4	79	88
	+ Toxin	407*	88
Malate + pyruvate	State 3	80	98
	State 4	20	21
	+ Toxin	0*	21
Succinate	State 3	175	193
	State 4	86	69
	+ Toxin	150*	69

* Addition of ADP or the uncoupler 2,4-dinitrophenol (40 μM) did not stimulate the rate of oxygen uptake by T mitochondria after treatment with HmT toxin. N mitochondria remained well-coupled in the presence of HmT toxin.

Oxygen uptake was measured with a Clark-type oxygen electrode (Yellow Springs Instrument Company) in 3 ml of a medium containing 0.4 M sucrose, 10 mM KCl, 2.5 mM $MgCl_2$, 4 mM KH_2PO_4, 20 mM HEPES, pH 7.4, 1 mg/ml bovine serum albumin and 0.4 mg of mitochondria isolated from roots of 3-day-old W64A(T) or W64A(N) seedlings. Other additions, as indicated, were 1.5 μmol NADH, 30 μmol malate plus 30 μmol pyruvate, 30 μmol succinate, and 33 μg HmT toxin (from a preparation which caused 50% inhibition of root growth of W64A(T) seedlings at a concentration of 0.65 μg/ml). State 3 rates were measured after addition of 150 to 300 nmol of ADP, and State 4 rates after subsequent exhaustion of the added ADP. HmT toxin was added during State 4 respiration.

Effects of HmT Toxin on Mitochondria *in situ*. Although the evidence for rapid, sensitive and specific effects of HmT toxin on isolated T mitochondria is clear, the argument for mitochondria as the primary site of toxin action in intact cells has been weakened by a frequently cited report that respiration and ATP content of seedling tissue are not specifically affected by toxin in time periods during which some inhibition of root growth can be detected (17). Unfortunately the method for measuring respiration was not given and the tissues assayed were seedling shoots, for which no rapid growth inhibition by toxin was described. Toxin penetration into the intact 2.5 cm detached shoots used may have been a limiting factor in the experiment.

Recent ultrastructural studies (20,33) by our group should resolve the apparent contradiction between toxin effects on isolated mitochondria and mitochondria *in situ*. After toxin treatment of roots and mesophyll protoplasts of T corn, the first ultrastructural effects observed were changes in the mitochondria similar to those seen in toxin-treated isolated mitochondria. Fifteen minutes exposure of roots to toxin resulted in some damage to the mitochondria of cells in the root cap and elongation zone. By two hours, many root mitochondria showed swelling, reduction in cristae and loss of matrix density (Figure 1). HmT toxin had no effect on mitochondria in N roots (Figure 1). Mitochondria within mesophyll protoplasts from T corn were affected even more rapidly and at lower toxin concentrations probably because penetration problems were minimized (33). Some protoplast mitochondria were damaged within 5 minutes and by 30 minutes, almost all of the protoplasts were affected. Protoplasts treated with toxin for 30 minutes and then washed thoroughly and cultured all collapsed within a few days, like protoplasts continuously exposed to toxin. The rapid toxin damage to the protoplasts was apparently irreversible (32). Toxin partially purified by chloroform extraction had no effect on mitochondria in N protoplasts, even when high concentrations were used (33). Plasma membranes and other components of N and T protoplasts looked normal after 60 minutes exposure to chloroform extractable toxin. Physiological and biochemical studies of protoplasts containing mitochondria damaged by brief toxin treatments have not yet been done; however, it seems likely that rapid toxin effects on mitochondrial functions *in vivo* will be detected, perhaps even before ultrastructural damage is apparent.

Reversibility of Toxin Action

There is some evidence which suggests that HmT toxin action is reversible. Arntzen *et al*. (17) in a study on T corn seedlings showed that roots treated with toxin for one hour and then returned to a toxin-free medium grew more slowly than the control for approximately 6 hours, then started to grow faster, and by 10 hours were growing as vigorously as the controls. The degree to

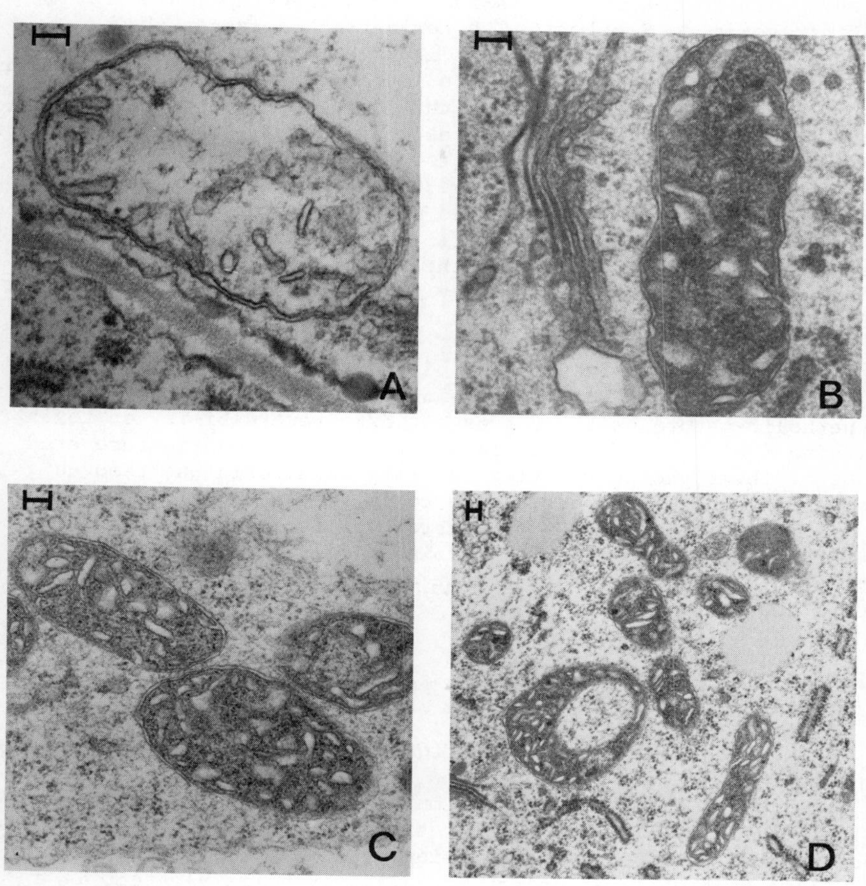

Tissue and Cell

Figure 1. Ultrastructural effects of chloroform-extractable HmT toxin on mitochondria in the region of elongation of susceptible (W64A(T)) and resistant (W64A(N)) corn roots. Experimental conditions are exactly as described in Ref. 20. Treatments were (A) W64A(T), treated for 2 hr with chloroform-extractable toxin; (B) W64A(T), no toxin added; (C) W64A(N), treated for 2 hr with chloroform-extractable toxin; (D) W64A(N), no toxin added. The calibration bar represents 0.1μ in each case.

which the roots grew after the toxin treatment was dependent on toxin concentration (17). Halloin et al (12) treated leaves of T corn with toxin for 3 hours and then washed the leaves for 1 hour in toxin-free medium. The toxin-treated leaves showed much higher rates of electrolyte leakage than the untreated controls. Interpretation of this latter data is difficult because it was not stated whether the increased rate was different from the rates obtained for leaves continuously grown in toxin. Recent experiments in our laboratory (32) showed that washing of T corn protoplasts, following 30 minute treatments with chloroform-extractable toxin, did not prevent protoplast collapse. In the same set of experiments T-corn protoplast collapse was prevented by washing the protoplasts after 15 minute toxin treatments. Two lines of evidence suggest that the effect of HmT toxin on oxidative phosphorylation in isolated T mitochondria is reversible. Firstly, with NADH as the substrate, the degree of toxin-induced inhibition of mitochondrial ATP formation increases with increasing toxin concentration; the hyperbolic nature of the curves suggests reversibility of toxin binding (38). Secondly, toxin-treated T mitochondria recovered most of their phosphorylating capacity when toxin was removed by washing the mitochondria with toxin-free medium (38).

The data on roots, protoplasts, and isolated mitochondria suggest that HmT toxin does not bind firmly to toxin-sensitive sites in T cells, but rather that an equilibrium exists between bound and unbound toxin.

Speculations on the Mechanism of Toxin-Induced Damage to T Mitochondria

The mechanism by which HmT toxin damages T mitochondria is unknown.

Although HmT toxin induces multiple effects in T mitochondria (Table I, Table III) it is theoretically possible that all of the observed effects result from a single, primary effect caused by a single type of toxin molecule. A likely primary effect is a toxin-induced increase in the ion permeability of the inner membranes of T mitochondria. Toxin-induced ion leakage might occur in one of three ways.

Firstly, the toxin could act as an inner mitochondrial membrane disruptant as suggested by Gengenbach et al (35) on the basis of ultrastructural data. Such action would result in extensive structural damage, in addition to causing increased ion permeability, and should lead to leakage of inner membrane matrix components such as malate dehydrogenase. However, recent experiments in our laboratory (42) have shown that toxin treatment does not induce leakage of malate dehydrogenase from T mitochondria (Table IV). One criticism of the latter experiment is that on addition of toxin, malate dehydrogenase could have leaked from the T mitochondrial inner membrane but may have been held within the mitochondria by the outer membrane.

Secondly, the toxin may act as a classical uncoupler and promote the passage of protons through the T mitochondrial inner membrane by acting as a lipophilic weak acid.

A third possibility is that the toxin is an ionophore and induces increased permeability of T mitochondrial inner membranes to one or more types of cation by complexing with the appropriate cation to form a lipid soluble complex which can pass through the inner membrane.

Whichever mode of action is correct, we will be faced with explaining why N mitochondria are completely resistant to HmT toxin. At present we do not know whether the inner or the outer mitochondrial membrane mediates resistance. A possible role of the outer mitochondrial membrane in resistance was suggested by Watrud *et al* (70) because removal of the outer membrane from N mitochondria yielded a toxin-sensitive inner membrane preparation as judged by toxin inhibition of the malate-DCPIP reaction. Interpretation of this latter experiment should be cautious because the technique used to prepare the inner membranes involved osmotic swelling and shrinking of mitochondria in combination with high speed discontinuous sucrose gradient centrifugation. The latter procedure could well have damaged the inner membranes and no physiological data were presented to the contrary (70). Consequently, it is possible that toxin sensitivity of the inner membranes of N mitochondria (70) was due to abnormalities brought about by the inner membrane isolation technique rather than the removal of the outer mitochondrial membrane *per se*.

Much careful work on whole mitochondria and on submitochondrial fractions is needed before we can discover the nature of HmT toxin action on T mitochondria and the lack of toxin action on N mitochondria. This work is a focal point of our present experiments because it is likely to reveal the biochemical basis of southern corn leaf blight disease.

Comparative Biochemistry of Mitochondria from N and T Cytoplasms

The Action of Chemicals (Other than HmT Toxin) on N and T Mitochondria. Several chemicals known to induce one or more responses similar to those of HmT toxin have been applied to N and T mitochondria. Quantitative distinctions between N and T mitochondria have arisen from these experiments. The following chemicals were used:

Valinomycin. Valinomycin (VAL) stimulates NADH oxidation and swelling in mitochondria under certain reaction conditions by act-

Table IV. Effect of HmT Toxin on the Release of Malate Dehydrogenase (MDH) from T Mitochondria

Treatment	MDH activity (nmol NADH oxidized/min)		MDH activity released into supernatant
	Supernatant	Mitochondria	% of total activity
None	190	3,200	6%
	150	2,400	6%
+ HmT Toxin	150	3,600	4%
	190	3,600	5%
+ Triton X-100	3,600	50	99%
	3,500	20	99%

Mitochondria from W64A(T) roots (about 0.5 mg protein) were suspended in 3 ml of 0.4 M sucrose, 10 mM KCl, 2.5 mM $MgCl_2$, 4 mM KH_2PO_4, 20 mM HEPES, pH 7.4 (Medium A) containing 1.5 µmol NADH and 300 nmole ADP. As indicated, HmT toxin (33 µg) or the detergent Triton X-100 (to a concentration of 0.15%) were added. After incubation for 5 min at room temperature the samples were centrifuged at 18,000 xg_{max} for 18 min at 25°. Each pellet was suspended in 0.5 ml of Medium A, and solubilized with Triton X-100 (0.15%) to release the total enzyme activity.

Appropriate aliquots of the supernatants and pellets were assayed for MDH activity in 3 ml of 0.4 M sucrose, 20 mM HEPES, 0.25 mM NADH, 1 mM oxaloacetate, 1 mM KCN, pH 7.4. The oxidation of NADH was monitored by the change in absorbance at 340 nm.

ing as a K^+-specific ionophore (51). VAL induced similar responses in N and T mitochondria with respect to oxidation of exogenous NADH and mitochondrial swelling, although the latter was more marked in N mitochondria at low VAL concentrations (38,43).

Gramicidin D. Gramicidin (GRAM) causes swelling in KCl or NaCl media, uncouples oxidative phosphorylation, and stimulates NADH oxidation in KCl or sucrose media (52). GRAM caused more stimulation of NADH oxidation by T mitochondria than by N mitochondria, although GRAM-induced swelling of N mitochondria was always slightly greater than that in T mitochondria (43).

2,4-Dinitrophenol (DNP) uncouples oxidative phosphorylation. DNP caused dramatic stimulation of NADH respiration in both N and T mitochondria. However, at high levels of DNP, the stimulation was significantly higher for T mitochondria than for N mitochondria (43). Bednarski et al. (38) could not detect significant differences between T and N mitochondria in sensitivity to DNP.

Sodium azide is a terminal oxidase inhibitor. The degree of inhibition of NADH oxidation by sodium azide was the same for N and T mitochondria (43).

Nigericin (NIG) plus K^+ causes uncoupling in mitochondria. T mitochondria were more sensitive than were N mitochondria to uncoupling by NIG plus K^+ (38).

Decenylsuccinic Acid. Decenylsuccinic acid (DSA) causes mitochondrial swelling by mediating membrane disruption. DSA treatments also stimulate exogenous NADH oxidation and result in loss of oxidative phosphorylation (53). DSA stimulated NADH oxidation at low concentrations but caused inhibition at higher concentrations. The N and T mitochondria responded similarly to DSA with respect to NADH oxidation. However, DSA caused considerably more swelling in N mitochondria than in T mitochondria (43).

Calcium (in the presence and absence of phosphate). Calcium in the absence of inorganic phosphate stimulated NADH oxidation in N mitochondria somewhat more than in T mitochondria. Calcium plus phosphate induced similar stimulations of NADH oxidation in the N and T mitochondria (43).

Digitonin. Digitonin is a saponin which can disrupt biological membranes. At low concentrations digitonin is selective for steroid-rich membranes. Evidence for this selectivity comes from the finding of Schnaitman et al. (54) that the outer membrane of rat liver mitochondria is more digitonin-sensitive than the inner membrane. These authors suggested that as digitonin is known to combine with cholesterol (a major membrane steroid in animals) on a 1:1 basis, it was possible that the outer mitochondrial membranes were higher in cholesterol than the inner membrane. Subsequent studies with guinea pig liver mitochondria showed that the steroid content of the outer mitochondrial membrane was six times more concentrated, on a protein basis, than in the inner membrane (55). Gregory et al. (56) have recently studied the respiratory effects of low concentrations of digitonin on N, C, and T mitochondria (Figure 2). The T mitochondria were dramatically more digitonin-sensitive than the N or C mitochondria, as indicated by inhibition of malate and succinate respiration and uncoupling of oxidative phosphorylation (Figure 2). There were also differences between the N and C mitochondria with respect to digitonin sensitivity (Figure 2). These data were thought (56) to reflect differences in membrane steroid composition between the N, T and C mitochondria. It was noted (56) that the digitonin effects were somewhat similar to those of HmT toxin on T mitochondria.

The data outlined above showed that compounds other than HmT toxin can produce differential effects on various inner membrane functions in N and T mitochondria and thus strongly supports the concept that those mitochondria differ with respect to membrane structure.

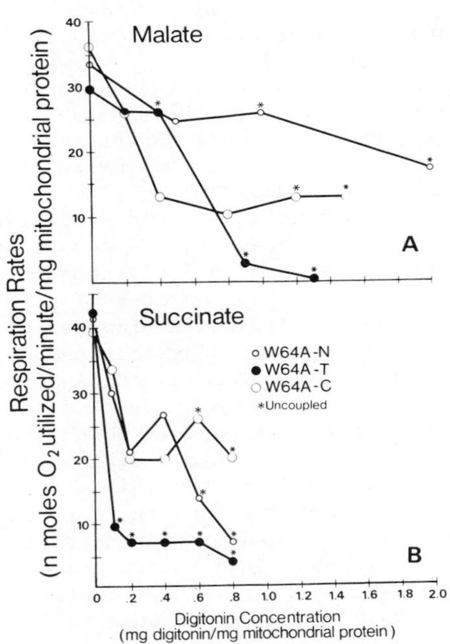

Figure 2. Effects of digitonin on respiration and oxidative phosphorylation in mitochondria isolated from W64A(N), W64A(T), and W64A(C) roots. Experimental conditions are found in Ref. 56.

Chemical Composition of N and T Mitochondria.

Nucleic Acids. Levings and Pring (57) have shown that N and T cytoplasms differ with respect to mitochondrial DNA (mt DNA). In these experiments, mt DNA was isolated from the N and T cytoplasms of several corn lines. The mt DNA's were subjected to restriction enzyme fragment analysis using each of the restriction endonucleases Eco RI (57,58), Hin D III (57,58), Bam I (57), and Sal I (57). Each of these enzymes cut DNA at different sequence specific sites and each yielded distinctions between the mt DNA from N and T cytoplasms (57). In addition, it was shown (57) that the male sterile cytoplasms other than T, namely S, C and EP, differed with respect to mt DNA. As these differences may have been due to contaminating DNA from nuclear, bacterial or viral sources, Levings and Pring (58) were careful to confirm the identity of their mt DNA preparations by using buoyant density determinations in neutral cesium chloride and showing that their mt DNA preparations appeared as a single band with a buoyant density of 1.706 g/cm^3 which is characteristic of higher plant mt DNA (59).

In addition, the upper and lower DNA bands obtained from cesium chloride-ethidium bromide gradients were in agreement with the expected value for mt DNA. There was no indication of nuclear DNA contamination because the upper and lower DNA bands on cesium chloride-ethidium bromide gradients were indistinguishable from each other and, in any case, restriction enzyme fragment analysis of the complex nuclear DNA does not yield distinct bands which could be confused with mt DNA bands (58). The possibility of contamination of the mt DNA preparations by viral DNA was not ruled out (58). Previous work involving buoyant density analysis of mt DNA's on neutral cesium chloride had failed to demonstrate differences between N and T cytoplasms (60). Differences between the mt DNA of susceptible and resistant cells are a prerequisite for theories involving T mitochondria as primary factors in cytoplasmically inherited susceptibility to southern corn leaf blight disease. Whether or not the observed mt DNA differences are associated with differences in disease susceptibility remains to be seen.

Proteins

1. Subunit fingerprints of total mitochondrial protein. The use of high resolution two-dimensional electrophoresis in our laboratory has failed to show qualitative differences between N, C and T mitochondria with respect to total protein subunits (61). The analytical technique, developed by O'Farrell (62), involves separation of mitochondrial protein according to isoelectric point by isoelectric focusing in the first dimension, followed by separation according to molecular weight by sodium dodecyl sulfate electrophoresis in the second dimension. The

result is a "fingerprint" of the mitochondrial protein subunits. The technique is highly reproducible so that each spot on one separation can be matched with a corresponding spot on a different separation. In addition the technique can resolve proteins differing in a single charge (62). In our experiments, visualization of the protein fingerprints was by means of staining rather than by autoradiography which is more sensitive. It is therefore very possible that one or more proteins were undetected in our experiments and that quantitative differences exist between the proteins of N, C and T mitochondria. Future experiments in our laboratory involving a combination of autoradiography and two-dimensional electrophoresis of mitochondrial proteins which are labelled with either ^{14}C or ^{35}S, will facilitate better sensitivity and quantification of the system.

2. Cytochromes. Pring (63) has reported that T mitochondria contain a full complement of cytochromes a + a_3, b and c as detected by difference spectra at 25°C. It has also been shown that T cytoplasm contains 7-12% more cytochrome b than does N (63) and that TRf mitochondria contain slightly more cytochrome a + a_3 than do N mitochondria (64) in each of five inbred lines.

Small differences in cytochrome b and c content of N and TRf mitochondria were noted, but these differences were not apparent for all of the inbred lines tested (64).

Steroids. Quantitative and qualitative determinations of mitochondrial steroids in our laboratory have yielded no reproducible differences between N and T cytoplasms (65). Lack of reproducibility may be a function of varying purity among the different mitochondrial preparations. The experiments involved extraction of a crude steroid preparation from the isolated mitochondria, purification of the steroids by silicic acid column chromatography, followed by GLC of the free steroids.

Three major steroids were found in the N and the T mitochondria: campesterol, stigmasterol and sitosterol. The identity of these compounds was confirmed by GLC-mass spectroscopy in collaboration with Dr. S. F. Osman, ARS-USDA, Eastern Regional Research Center, Wyndmoor, Pa. A minor steroid was also found, but the identity of this compound is not known at present. The fact that N and T mitochondria are differentially sensitive to digitonin (56) as explained previously, suggests that further experiments involving fractionation, purification, and steroid analyses of inner and outer mitochondrial membranes may yield differences between these mitochondria. It is of interest that sterol differences have been found between N and T tassels during development (66).

Studies on the Molecular Structure of *H. maydis* Race T Host Specific Toxin

The exact molecular structure of HmT toxin is not yet known

and there is disagreement as to which class of compounds HmT toxin belongs. Our current information on HmT toxin structure consists of published data by Karr et al. (24) and a recent personal communication from J. M. Daly of the University of Nebraska (67).

Karr et al. (24) isolated five toxic compounds from crude HmT culture filtrate. It was postulated that the toxic compounds were terpenoids, existing free or as glycosides, ranging in molecular weight from 350 to 600 and a chemical assay for HmT toxin was developed (68). This work involved fractionation and purification of the toxic components in combination with a leaf puncture bioassay for toxin activity in each fraction. The fractions which exhibited activity in the bioassay were analysed for chemical structure by chemical tests, chromatography, mass spectroscopy and nuclear magnetic resonance analysis. The chemical assay for HmT toxin was a generalized test for terpenoids using sulfuric acid - acetic anhydride reagent. The involvement of the five terpenoid compounds in HmT toxin specificity is open to question. A subsequent comparison (69) of the chemical assay described by Karr et al. (68) to measure the five terpenoid toxin components in partially purified preparations revealed that both HmT and HmO (which does not produce a host-specific toxin) produce compounds similar to the toxins identified by Karr et al. (24). Consequently it is possible that the five terpenoid compounds characterized by Karr et al. (24) are not associated with the host-specific activity of HmT toxin. It is interesting, however, that when Watrud et al. (70) treated T mitochondria with a mixture of toxins I and II (see reference 24 for terminology), the characteristic responses were elicited. Another problem in the work of Karr et al. (24) derives from the use of the insensitive leaf puncture bioassay in the assessment of toxic activity of the various fractions (15,69). It is possible that one or more of the discarded fractions contained non-terpenoid host-specific toxin activity.

A non-terpenoid structure for HmT toxin has been proposed recently by Kono and Daly (67). Some details of the structure have not been finalized, but a great deal of useful information has been gained. The molecule has an empirical formula of $C_{29}H_{50}O_9$ (542 daltons) and is extremely active against T corn, causing complete necrosis of the first true leaf at levels as low as 5 to 10 ng. The toxicity of the molecule was specific for T corn as no toxic effect was observed on N corn or on other plant species. The use of a very sensitive toxin bioassay, in which dark CO_2 fixation by corn leaf discs was measured (18) insured that no biologically active toxin fractions were discarded prior to structural analysis. In aqueous solvents the material chromatographed as a single diffuse spot, but in mixtures of methanol and non-polar organic compounds a major and two minor components with host specificity were observed. These additional compounds were probably isomeric or homologous because acetylation of toxin resulted in the formation of a tetraacetate of empirical formula $C_{37}H_{58}O_{13}$ (710 daltons) which also resolved into a major and two minor components. The IR, NMR and UV spectra of two of the acetylated derivatives could be distinguished only by the extinction coefficient in the UV. There were insufficient amounts of the third component to make comparisons. ^{13}C and proton NMR of the major acetylated derivative support the existence of four hydroxyl groups and five carbonyl groups in the original toxin. On the basis of IR, ^{13}C and proton decoupling NMR and mass spectra of the component isolated in the largest yield, two possible structures have been proposed for HmT toxin (Figure 3). Work to determine the exact structure of the molecule is still in progress. Whether the single purified toxin molecule of Kono and Daly (67) mimics the multiple mitochondrial effects induced by crude toxic filtrate and chloroform-toxin is a critical and exciting question.

Concluding Remarks

Mechanism of Southern Corn Leaf Blight Disease. In our opinion present data suggests that a major component (and perhaps the primary component) of southern corn leaf blight disease is the HmT toxin-induced damage of mitochondria in susceptible (T cytoplasm) corn. This opinion is based on two major pieces of experimental evidence. Firstly, several groups have shown that isolated susceptible T mitochondria (but not the resistant mitochondria) are severely damaged by HmT toxin. The physiological manifestation of this damage, should the damage also occur *in situ*, would lead to the death of corn cells which contain T cytoplasm. Secondly, ultrastructural studies by our group have shown that HmT toxin can damage T mitochondria *in situ* and that this damage is similar to that observed in toxin-treated, isolated T mitochondria. However, it is of extreme importance to elucidate whether or not HmT toxin affects T mitochondrial physiology *in vivo*, because at present there is no conclusive evidence in this area. Sensitive measurements of *in vivo* respiration and phosphorylation in toxin-treated corn protoplasts are in progress in our laboratory and may have great relevance to the nature of southern corn leaf blight disease. The cytoplasmic inheritance of susceptibility to the disease may also be consistent with the mitochondria (which contain DNA) being of primary importance. Should the transfer of resistant (N) mitochondria to susceptible (T) protoplasts result in a conferring of resistance to these protoplasts, there will be little doubt that mitochondria are a major factor in the resistance of corn to southern corn leaf blight disease. This experiment is in progress in our laboratory.

The Use of the Southern Corn Leaf Blight System in Studies on Mitochondrial Genetics in Higher Plants. Little is known about the mitochondrial genetics of higher plants. Although asexual genetic manipulation such as protoplast fusion and organelle transfer can bring together cytoplasmic components from diverse cell types, few mitochondrial markers are available in higher

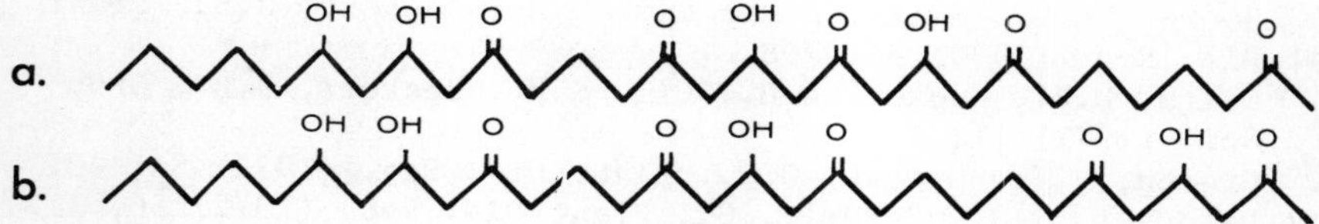

Figure 3. Two possible structures for HmT toxin recently proposed by Kono and Daly (67)

ecause HmT toxin rapidly and specifically destroys both n protoplasts and T mitochondria within protoplasts 33) the toxin will be a useful probe for following mitochondrial transfers. The response of protoplasts and mitochondria to HmT toxin after fusion of N and T cytoplasm protoplasts, or after transfer of isolated N mitochondria to T protoplasts, may answer several important questions: can "foreign" mitochondria be transferred to a different cytoplasm?; can such mitochondria survive, function, replicate, and alter the survival of host protoplasts?; do mitochondria carry all the genetic material that determines cytoplasmic response to toxin? Techniques developed using HmT toxin as a probe for successful genetic manipulations with mitochondria may facilitate a new experimental approach to the study of mitochondrial genetics in higher plants.

Literature Cited

1. Hooker, A.L., Smith, D.R., Lim, S.M., Beckett, J.B., Plant Dis. Rep. (1970) 54, 708.
2. Smith, D.R., Hooker, A.L., Lim, S.M., Beckett, J.B., Crop Sci. (1971) 11, 772.
3. Gracen, V.E., Grogan, C.O., Plant Dis. Rep. (1972) 56, 432.
4. Berquist, R.R., Peverly, G., Plant Dis. Rep. (1972) 56, 112.
5. Hallauer, A.R., Martinson, C.A., Agronomy J. (1975) 67, 497.
6. Lim, S.M., Plant Dis. Rep. (1974) 58, 811.
7. Caunter, I.G., Ph.D. thesis, 1977, Cornell University, "A Genetic Study of Nuclear Controlled Resistance to Southern and Yellow Leaf Blights in Maize" (*Zea mays* L.).
8. Payne, G.A., Yoder, O.C., Phytopathology (1977) 67, in press.
9. Lim, S.M., Hooker, A.L., Genetics (1971) 69, 115.
10. Yoder, O.C., Gracen, V.E., Phytopathology (1975) 65, 273.
11. Lim, S.M., Hooker, A.L., Phytopathology (1972) 62, 968.
12. Halloin, J.M., Comstock, J.C., Martinson, C.A., Tipton, C.L., Phytopathology (1973) 63, 640.
13. Gracen, V.E., Forster, M.J., Sayre, K.D., Grogan, C.O., Plant Dis. Rep. (1971) 55, 469.
14. Turner, M.T., Martinson, C.A., Plant Dis. Rep. (1972) 56, 29.
15. Yoder, O.C., Payne, G.A. Gregory, P., Gracen, V.E., Physiol. Plant Path. (1977) 10, 237.
16. Wheeler, H., Williams, A.S., Young, L.D., Plant Dis. Rep. (1971) 55, 667.
17. Arntzen, C.J., Koeppe, D.E., Miller, R.J., Peverly, J.H., Physiol. Plant Path. (1973) 3, 79.
18. Bhullar, B.S., Daly, J.M., Rehfeld, D.W., Plant Physiol. (1975) 56, 1.
19. Wheeler, H., Ammon, V.D., Phytopathology (1977) 67, 325.
20. Aldrich, H.C., Gracen, V.E., York, D., Earle, E.D., Yoder, O.C., Tissue and Cell (1977) 9, 167.
21. Koeppe, D.E., Malone, C.P., Miller, R.J., Plant Physiol. (supp.) (1973) 51, 10.

22. Mertz, S.M. Jr., Arntzen, C.J., Plant Physiol. (supp.) (1973) 51, 16.
23. Watrud, L.S., Hooker, A.L., Koeppe, D.E., Phytopathology (1975) 65, 178.
24. Karr, A.L. Jr., Karr, D.B., Strobel, G.A., Plant Physiol. (1974) 53, 250.
25. Earle, E.D., unpublished.
26. Arntzen, C.J., Haugh, M.F., Bobick, S., Plant Physiol. (1973) 52, 569.
27. Gracen, V.E., Grogan, C.O., Forster, M.J., Can. J. Bot. (1972) 50, 2167.
28. Frick, H., Bauman, L.F., Nicholson, R.L., Hodges, T.K., Plant Physiol. (1977) 59, 103.
29. Gengenbach, B.G., Green , C.E., Crop Sci. (1975) 15, 645.
30. Laughnan, J.R., Gabay, S.J., Crop Sci. (1973) 13, 681.
31. Pelcher, L.E., Kao, K.N., Gamborg, O.L., Yoder, O.C., Gracen, V.E., Can. J. Bot. (1975) 53, 427.
32. Earle, E.D., Gracen, V.E., Yoder, O.C., Gemmill, K.P., submitted to Plant Physiol.
33. York, D.W., unpublished.
34. Miller, R.J. , Koeppe, D.E., Science (1971) 173, 67.
35. Gengenbach, B.G., Miller, R.J., Koeppe, D.E., Arntzen, C.J., Can. J. Bot. (1973) 51, 2119.
36. Krueger, W.A., Josephson, L.M., Hilty, J.W., Phytopathology (1974) 64, 735.
37. Barratt, D.H.P., Flavell, R.B., Theoretical and Applied Genetics (1975) 45, 315.
38. Bednarski, M.A., Izawa, S., Scheffer, R.P., Plant Physiol. (1977) 59, 540.
39. Peterson, P.A., Flavell, R.B., Barratt, D.H.P., Theoretical and Applied Genetics (1975) 45, 309.
40. Flavell, R., Physiol. Plant Path. (1975) 6 , 107.
41. Peterson, P.A., Flavell, R.B., Barratt, D.H.P., Plant Dis. Rep. (1974) 58, 777.
42. Matthews, D.E., unpublished.
43. Gengenbach, B., Koeppe, D.E., Miller, R.J., Physiol. Plant. (1973) 29, 103.
44. Tipton, C.L., Mondal, M.H., Uhlig, J., Biochem. Biophys. Res. Comm. (1973) 51, 725.
45. Tipton, C.L., Mondal, M.H., Benson, M.J., Physiol. Plant Path. (1975) 7, 277.
46. Gracen, V.E., Forster, M.J., Grogan, C.O., Plant Dis. Rep. (1971) 55, 938.
47. Frick, H., Nicholson, R.L., Hodges, T.K., Bauman, L.F., Plant Physiol. (1976) 57, 171.
48. Hodges, T.K., Leonard, R.T., Bracker, C.E., Keenan, T.W., Proc. Nat. Acad. Sci. U.S.A. (1972) 69, 3307.
49. Scheffer, R.P. IN Biochemistry & Cytology of Plant-Parasite Interaction [K. Tomiyama, J.M. Daly, I. Uritani, H. Oku and S. Ouchi, eds], p. 118, Elsevier Scientific Publishing Company, 1976.

en, V.E., unpublished.
on, J.B., Bertognolli, B.L., Shepherd, W.D., Plant Phy-
. (1972) 50, 347.
52. Miller, R.J., Dumford, W.S., Koeppe, D.E., Plant Physiol. (1970) 46, 471.
53. Koeppe, D.E., Miller, R.J., Plant Physiol. (1971) 48, 659.
54. Schnaitman, C., Erwin, V.G., Greenawalt, J.W., J. Cell Biol. (1967) 32, 719.
55. Parsons, D.F., Yano, Y., Biochem. Biophys. Acta (1967) 135, 362.
56. Gregory, P., Gracen, V.E., Yoder, O.C., Steinkraus, N.A., Plant Sci. Lett. (1977) 9, 17.
57. Levings, C.S. III, Pring, D.R., Proc. 31st Annu. Corn and Sorghum Res. Conf. (1976) 110.
58. Levings, C.S. III, Pring, D.R., Science (1976) 193, 158.
59. Wells, R., Ingle, J., Plant Physiol. (1970) 46, 178.
60. Shah, D.M., Levings, C.S. III, Crop Sci. (1974) 14, 852.
61. Gemmill, R., Gregory, P., unpublished.
62. O'Farrell, P.H., J. Biol. Chem. (1975) 250, 4007.
63. Pring, D.R., Plant Physiol. (1975) 55, 203.
64. Watrud, L.S., Laughnan, J.R., Gabay, S.J., Koeppe, D.E., Can. J. Bot. (1976) 54, 2718.
65. Gregory, P., unpublished.
66. Comita, J.J., Klosterman, H.J., Phytochem. (1976) 15, 917.
67. Kono, Y., Daly, J.M., personal communication.
68. Karr, D.B., Karr, A.L., Strobel, G.A., Plant Physiol. (1975) 55, 727.
69. Yoder, O.C., Gracen, V.E., Plant Physiol. (1977) 59, 792.
70. Watrud, L.S., Baldwin, J.K., Miller, R.J., Koeppe, D.E., Plant Physiol. (1975) 56, 216.

Acknowledgements

The experiments performed in our laboratory were supported in part by grant 75002 from the Rockefeller Foundation. We gratefully acknowledge the help of Dr. D.E. Matthews, H. Pham, and D.W. York in the preparation of this paper.

8

Host Plant Resistance to Insects

A. C. WAISS, JR., B. G. CHAN, and C. A. ELLIGER

Western Regional Research Center, Agricultural Research Service, U.S. Department of Agriculture, Berkeley, CA 94710

In order to use chemical insecticides more advantageously, systems for integrated pest management were recommended in 1975 by the National Academy of Sciences (1, 2), U. S. Department of Agriculture (3) and the Entomological Soc. of America (4). Among the various alternatives to chemical insecticides, the use of insect resistant plants, in combination with good cultural practice, is perhaps the most effective, convenient, economical and environmentally acceptable method of insect control. In addition, it is a method that is completely compatible with both chemical and other biological control measures.

Host Plant Resistance (HPR) to Insects

Since many excellent reviews (5, 6, 7, 8, 9) have been written on the subject of plant-insect interactions, no attempt will be made in this paper to fully review past literature. Rather, we will use a few selected examples stressing the problems that may be encountered by chemists working in this relatively unexplored area of research, as well as future research needs of this very important field.

Host plant resistance (HPR) has been recognized since the early 1800's. Because plants, which are generally immobile, lack a circulatory system, the existence of the classical antigen-antibody immune system would not be possible. Through evolutionary processes plants have elaborated an array of "secondary metabolic products" of which many function in HPR. From the standpoint of natural selection in evolution HPR is a preadaptive characteristic of plants. Before being cultivated by man, plants, co-evolving with insects, either intrinsically possessed or have developed means of surviving attack by arthropods. Although the first significant economic contribution by HPR to agriculture occurred in 1890, this was the successful grafting of European grape vines onto resistant root stock from North America (10) to save the French wine industry from *Phylloxera vitifolia* (Fitch), the potential of

HPR as an insect control method, however, was not fully appreciated until the mid 1960's partly due to the availability of economical, effective and persistent insecticides which had been introduced some two decades earlier.

While over a million species of insects exist, only a few thousand may be classified as pests; and, of these, only about 500 species produce appreciable economic damage. The general asynchrony in space and time between the growth and development of plants and insects, caused by both geographical and other environmental factors is the major escape mechanism of the host. The "emphermerality" (9) or lack of "apparency" (8) of many deciduous and annual plants causes further selective pressure against specialized herbivores. In nature, HPR provides still another defense for plants to maintain ecological balance with their predators.

Three mechanisms to account for the plant's resistance to insect damage were proposed by Painter (11, 12):

1. Nonpreference - In co-evolution, insects have adapted to their hosts through source attractants and feeding and oviposition stimulants. Nonpreference is a result of the lack of these stimulants or, often, the presence of chemical and physical deterrents.

2. Tolerance - This is the ability of certain plants to withstand insect attack without appreciable loss in vigor or crop yield.

3. Antibiosis - This comprises the defensive mechanism of plants against their pests through adverse influence on growth, survival or reproduction of the insects by means of chemical or morphological factors.

HPR in Non-Agricultural Plants

For centuries plants have been known to possess insecticidal activities. Nicotine and other tobacco alkaloids were exported from America as early as 1690 for insect control (13). Other natural insecticides such as the pyrethoids and rotenoids have been used in the Middle East and far eastern tropics since ancient time (14, 15). A recent review (16) cited nearly 1500 species of plants containing toxicant and pest control agents.

Detailed studies of host plant resistance have been confined almost exclusively to non-agricultural plants. The intricate relationship between the oak tree and its primary predator, the winter moth, has been explained both in terms of escape in space and time as well as by chemical antibiosis (8, 17). Apparently the quantity of polyphenolic tannin and its precursors, changing from 0.5% of the dry leaf weight in April, to 5% in September, correspond to relative infestation and growth of the adapted pest. Similarily, the distribution of polyphenols helped to explain host plant resistance of creosote bush, Larrea tridentata (9), bracken fern, Pteridium aguilinum (18) and several other plants (37).

The subject of antifeedants, repellents and deterrents, and the role of alkaloids and terpenoids in insect control will be presented by other speakers in this symposium and so will not be treated here.

HPR in Agricultural Crops

The main reason for the recent surge of interest in HPR to insects for crops of economic importance is due to the fact that over 50% of the approximately 500 million lbs of pesticides applied to United States cropland are insecticides (2). While they are essential to agricultural production at the present time, these chemicals present serious questions in regard to environmental quality, public health, and the existence of many species of animals. The severe restriction on the use of DDT, aldrin, dieldrin, chlordane and heptachlor increases the need for alternatives.

Although, at present, over 100 insect resistant cultivars are grown in the USA (13, 14), little has been reported on the nature of resistance at the chemical or molecular level. HPR is an inheritable characteristic of plants; knowledge about its chemical basis would play an immense role in plant breeding and genetic engineering in the future. The following examples represent the challenge, and frustrations for chemists working in a biological field and illustrate the tremendous need for a multidisciplinary approach to solving this problem.

Resistance of Cotton to Tobacco Budworm, Heliothis virescens (F.) In the early 1900's it was suggested that the presence of dark internal glands in Gossypium plants was associated with resistance to insects (21, 22). Gossypol, I, isolated from these glands, was considered to be the main contributing factor against tobacco budworm, Heliothis virescens (F) (23, 24, 25, 26). Certain morphological characteristics as observed in glabrous and nectariless strains, also appeared to add to cotton's defense against this insect (27). A glandless variety of cotton was developed in 1959 (28) that contains no gossypol (29). Although glanded cottons, in general, show higher resistance to Heliothis (30), recent studies indicate that a similar level of resistance could be found in certain glandless varieties as well (31, 32). It is logical to assume therefore that if gossypol-free cotton were indeed resistant to Heliothis attack there may exist yet other resistance factors.

Successive solvent extraction of cotton flower buds (squares) in our laboratory (33) has indicated that while hexane extractables (containing gossypol) were toxic (Table 1), the subsequent methanol extract was even more active. An appreciable level of toxicity also appears in the exhaustively extracted residue.

Ether precipitation of the methanol extract, followed by chromatographic separation on Sephadex (G-25) gave a purified toxic fraction, identified as condensed tannin, II.

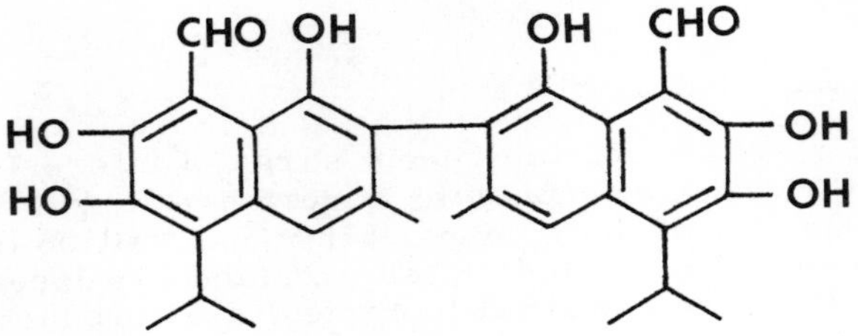

CHO OH OH CHO
HO OH
HO OH I

Table 1

Bioassay of successive solvent extracts from cotton flower buds (Texas -254) with <u>Heliothis virescens</u> (F).

Extract[a]	Larval weight (mg)[b]
Hexane	48
Acetone	280
Methanol	1
Water	260
Residue	1
Control	267

[a] Solvent extract from 6 g. of dried flower bud added to 30 g of synthetic diet.

[b] Average weight of 10 larvae after 14 days at 28°C .

When cotton tannin was added at 0.1 to 0.3% to an adequate synthetic diet, the larval weight of tobacco budworm was reduced correspondingly (Table II). The rate of pupation, percent of larvae pupating and percent of moth emergence from pupae were reduced even though the larvae were fed with tannin during only part of the growth period (7 days). In nature reduction of larval weight and vigor undoubtedly exposes the insects to further predation, diseases, and other mortality factors and, thereby, diminishes damage to the host plant.

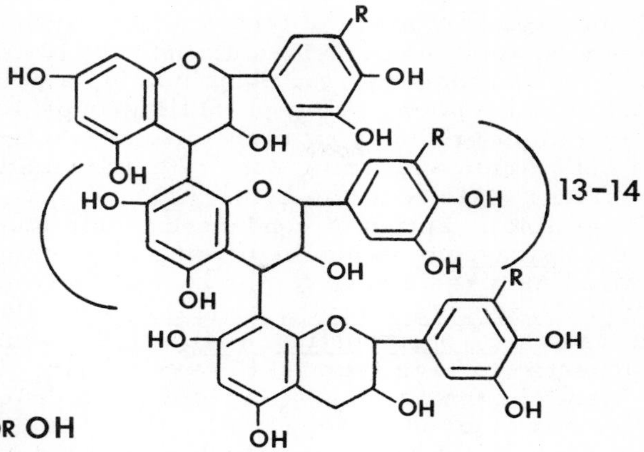

II

R = H OR OH

Table II

The residual effect of cotton tannin and tannic acid treatment on larval weight, days to pupation, percent pupation and percent emergence of Heliothis virescens (F)

Treatment (% in diet)	Mean larval weight in mg after 7 days	% Larval pupating[a]	Days to reach pupal stage[a]	% Emergence[a]
Cotton tannin				
0.0	299	81	14.9	100
0.1	160	69	15.2	91
0.2	84	50	17.4	75
0.3	37	63	18.1	80
Tannic acid				
0.0	290	75	14.9	100
0.1	15	69	19.1	82
0.2	14	38	19.8	83
0.3	15	31	19.8	40

[a] Larvae were transfered to adaquate synthetic diet after 7 days on tannin or tannic acid containing diet.

While the exact mode of tannin's toxic action to H. virescens has not been elucidated, there is little doubt that the phenolic function is involved. Tannic acid, a polygallic acid ester of glucose having ortho-phenolic groups similar to those of cotton tannin produces similar antibiotic activity. Methylation of both tannin and tannic acid with diazomethane drastically reduces their toxicity to the insect (23).

The antibiotic action of condensed tannin found in cotton against H. virescens is not surprising since tannin is well recognized as a resistance factor in the oak tree against the oak moth, Operophtera brumata, (17) and in alfalfa against the alfalfa weevil, Hypera postica, (34). In these instances tannin was suspected to have reduced the availability of nutrient to the insect by interacting with protein and deactivation of digestive and other enzymes (8, 9, 35).

Our work would have been much simplified at this point if the tannin contents in cotton flower buds (squares were in direct relation to the degree of susceptibility or "resistance". In fact, freeze dried square powders from both "resistant" and "susceptible" varieties showed appreciable toxicity at less than 10% equivalent weight of fresh plant material in the synthetic diet (29, 31, 36).

Preliminary experiments in our laboratory indicate that one or more of the following factors may help to explain the apparently high toxicity in cotton and differentiate the level of resistance among them:

1. The degree of polymerization in tannin and toxicity in synthetic diet may be increased by exposure to air oxidation, thereby overshadowing the difference between "resistant" and "susceptible" varieties.

2. The present method for tannin analysis (37) is innacurate and as it does not differentiate different molecular sizes of tannin.

3. Present knowledge of the feeding behavior of H. virescens is inadequate. It is suspected, at the present time, that first instar larvae feed on the young terminal leaves for several days before migrating to the flower bud (38). As young larvae are more susceptible to toxicants (39), a more careful observation of feeding behavior and accurate analyses of gossypol, tannin, or other toxicants and synergists as well as antagonists in the terminal leaves should be made. Preliminary experiments indicate that young cotton leaves contain little or no gossypol but substantial amounts of condensed tannin.

4. Estimations of tannin contents (40) in floral parts show as much as 4-fold difference between the anther where larvae of the second or older instar feed (38), and the total flower bud This observation may help to explain the excessive toxicity of the total square in the synthetic diet and the lack of correlation between gossypol (35) and tannin content with resistance of cotton varieties.

5. The possibility of induced resistance cannot be ignored. The increased concentration of phenolics and other phytoalexins in plants after fungal attack has been well documented (42, 43). The profile and biosynthesis of polyphenols in Pinus radiata are reported to be affected by the attack of the wood wasp, Sirex noctilio (44).

6. Significant variance in insect feeding behavior on different cultivars may occur due to differences in chemical attractants or morphological factors.

Resistance of Corn to European corn borer, Ostrinia nubilalis (H). - The most celebrated classical example of HPR in an economic crop is corn's chemical defense against the first generation of corn borer. It was established that 2, 4-dihydroxy-7-methoxy-1, 4-benzoxazin-3-one (DIMBOA), IV, was responsible for the inhibitory effect on corn borer (45, 46, 47). DIMBOA is the enzymatic hydrolysis product of its natural glycoside, III, and is liberated while the insect feeds on the whole tissue of corn. Direct correlation between the age of plant (48), and concentration of 6-methoxy-2-benzoxazolinone, (MBOA), V, a more stable degradation product of DIMBOA, and resistance to feeding, has been firmly established (49).

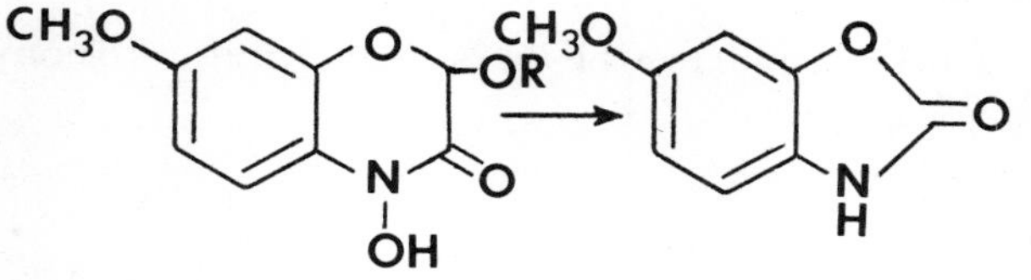

CH_3O ... OR ... N ... OH → CH_3O ... N H

III R = Gl

IV R = H

V

Much plant damage by corn borer, however, is inflicted by the second brood larvae feeding on the pollen, silk, collar, and sheath. Evidence indicates that some factors, other than DIMBOA, are responsible for resistance to 2nd-generation borers (50, 51). In addition, it has been established that resistance to 1st and 2nd generations of the European corn borer in maize is not conferred by the same genes.

Preliminary experiments in our laboratory indicate that the causative agent for 2nd-brood resistance may be much more polar than DIMBOA.

Resistance of corn to corn earworm, Heliothis zea (B). - It is generally accepted that the adult moths of corn earworm lays their eggs on the corn silk. As the larvae hatch and grow, they feed and travel toward the husk, and eventually inflicting damage to the kernels. It was observed in the early part of this century that there seemed to be a positive correlation between long, tight husks and earworm resistance in corn (52), perhaps because of increased probability of cannibalism (12), or prolonged exposure to lethal toxicants(53). Chemical antibiosis was demonstrated in lyophilized corn silk and plant extracts (54, 55).

Table III

Bioassay of successive solvent extracts from corn silk (var. Zopalote Chico) with Heliothis Zea (B.)[a]

Sequential extracts	Laval wt (mg)
Hexane	515
Acetone	468
Methanol	98
Hot water	65
Residue	45
Control	567

[a] Ave. of 10 larvae fed for 11 days with extracts from 6 g. of corn silk/30 g. diet.

Table IV

The effect of methylation on the toxicity of extract and purified fraction from corn silk (var. Z. Chico) to Heliothis Zea (B.)[a]

	Larval wt (mg)	
Extract	before methylation	after methylation
Methanol	88	243
Methanol, Chrom. on LH-20	50	562
Water	20	579
Control	581	581

[a] Average of 10 larvae fed for 12 days with extracts or purified extracts from 5 g of lypholized corn silk/30 g of synthetic diet.

Successive solvent extraction of silks from several corn varieties showed toxic materials in both methanol and water extracts, as well as in the residue after water extraction (Table III). The toxic factor in the methanol extract was purified by chromatographic separation. The toxicity of both the methanol and water fractions was drastically reduced by methylation with diazomethane (Table IV). Present evidence indicates that these toxic compounds are polyphenols.

HPR to insects in other agricultural crops. - Resistance of Sunflowers to sunflower moth, Homeosoma electellum, (H), has been reported to be associated with the inability of the insect to penetrate the seed coat of the resistant varieties (56). Recently two diterpene acids, trachyloban-19-oic acid (VI) and (--)-16-kauren-19-oic acid (VII), isolated from sunflower florets were shown to possess larvacidal activity to sunflower moth and several other Lepidoptera species (57, 58). While the concentration of these acids varied with age and variety of sunflower, their contribution to the host defense mechanism has not yet been fully elucidated.

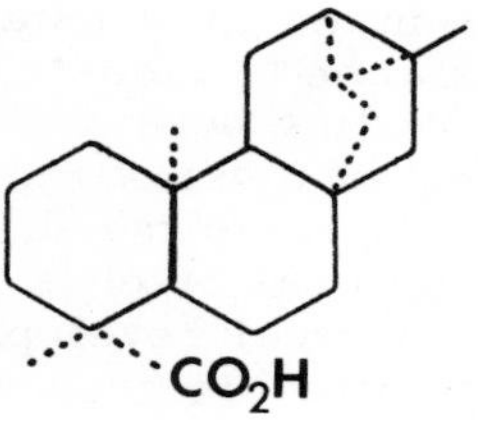

CO_2H CO_2H

VI 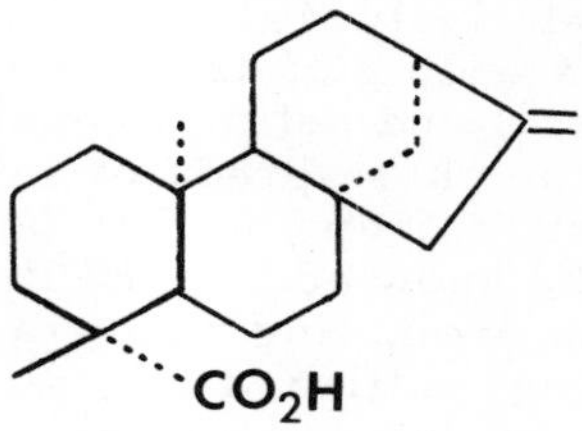 VII

In addition to the previous examples, germplasm of many other crops which are resistant to insects has been identified (19, 20, 59). Some of the more prominent examples are listed in Table V. While morphological or physical characteristics may be involved in the host-insect relationships in some of these cases, chemical antibiosis has been suggested to have the dominant role for the plants' resistance to their predators. It thus appears that there is fundamental need for the participation of the chemist in this important area of research.

Table V

Some commercial crops that exibit resistance to insect attack.

Crop	Insect
Alfalfa	Alfalfa aphids and weavil
Barley and other cereal grains	Greenbugs
Corn	Western rootworm
Cotton	Lygus bug
Potato	Potato beetle and leafhopper
Rice	Rice borer and hoppers
Soy	Mexican bean beetle and pot-worm
Wheat	Hessian fly and greenbug

Concluding Remarks

Relatively little effort has been invested in establishing the understanding of crop plants' resistance to insects at the chemical and cellular level. The feeding behavior of insects and other host-predator interactions in most of the economically important crops need to be thoroughly established.

The knowledge of HPR at the chemical level is imperative not only in promoting more rapid and effectual development of insect resistant cultivars but, also, to monitor the presence and distribution of these chemicals.

To most of us chemists, the isolation of an insect growth inhibitor, antifeedant, or other toxicant from a plant is not unsurmountable, providing a convenient and reliable bioassay is available to monitor the purification process. To elucidate the delicate balance between an insect and its host in the field is a much more formidable task. From our experience in studying the nature of HPR, we feel that the following suggestions may be worth considering by scientists interested in this area of research.

1. Since the chemical basis for HPR involves the chemical interactions between two biological systems, the host plant and its predator, a closely coordinated multidisciplinary approach to the investigation cannot be over-stressed. It is imperative that the chemist develops an appreciation for the almost infinite variability and constant change in the plant's chemical composition during its growth, development, and stress, as well as the insect's ability to develop, evolve, and coexist with its host, guided and stimulated by the plant's unique combination of chemicals for food, shelter, and reproduction. A biological system is a dynamic one, and in each insect-host plant interaction there exists a special biological relationship.

Thus, solely chemical approaches to these investigations are inadequate.

2. The establishment of an appropriate, convenient and reproducible bioassay procedure is a prerequisite to the success of HPR research. Although, by way of example, the black armyworm has been used extensively in bioassays for feeding deterrents and other allelochemics, the appropriate insect should be used in the study of plant-insects interaction. While gossypol in cotton, for example, is toxic to several lepidopterous insects, it is a feeding stimulant for the monophagous cotton boll weevil *Anthonomus grandis* B. (60) A convenient procedure to incorporate plant extracts and purified fractions into small quantities of diet is yet to be found, although several synthetic diets for *Lepidoptera* have been reported. In addition, there is still need for improvement in bioassay procedures for other arthropods such as *Coleoptera* (beetle), *Diptera* (flies and mosquitos), *Heteroptera* (true bugs) and *Arachnids* (spiders).

3. The development of gc - mass spectrometry has greatly facilitated the purification and identification of volatile organic compounds in the last decade. At present, improved techniques are needed for separation and purification of higher molecular weight polar compounds. Techniques of protein, and other high polymer chemistry may be modified and employed here.

4. From our experience with cotton's resistance to bollworms (*Heliothis* spp.) we feel that unless the antibiotic factor (gossypol in this case) is located in the plant part in which insect feeding actually occurs, and is present in an effective concentration at the time of insect damage, the presence of a toxicant in "resistant" varieties (glanded lines) and absence in "susceptible" cultivars (glandless lines) may not be sufficient proof for the toxicants role in HPR.

5. It may be useful to consider, in HPR research, that on a scale of 0 to 10 (Figure 1), where 10 represents a plant totally immune to certain specific insect attack and "0" represents a plant which supports maximum insect growth, development, and reproduction, resistance and susceptibility are relative phenomena ranging between the two extremes.

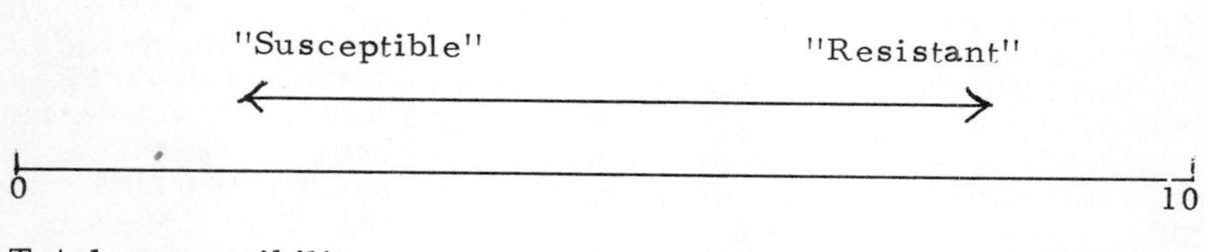

Figure 1. Relative scale of insect growth and plant's response

By definition, a plant is no longer a host to an insect when it is totally immune to its enemy's attack. One should not expect, therefore, in HPR research to find a toxicant only in the "resistant" varieties (a qualitative phenomenon) but, rather, more often a varying concentration (quantitative difference) of antibiotic factor in "resistant" and "susceptible" lines, complicated further by both insect behavior and plant development.

Literature Cited

1. Smith, E. H. "Pest Control Strategies for the Future". National Academy of Sciences, Washington, D.C. 1972.
2. Kennedy, D., chairman, "Pest Control: An Assessment of Present and Alternative Technology" 5 vol., National Academy of Sciences, Washington, D.C. 1975.
3. Klassen, W. in "Insect, Science and Society" (D. Pimentel, ed.) Academic Press, New York, 1975.
4. Glass, E. H. "Integrated Pest Management: Rational Potential, Needs and Implementation" Entomological Soc. Amer. Baltimore, MD, 1975.
5. Beck, S. D., Ann. Entomol. Soc. Amer. (1965) 53 207
6. Van Emden, H. F. and May, M. J. in "Insect-Plant Relationship" H. F. Van Emden, ed.) Wiley, 1973.
7. Kogan, M. In "Introduction to Insect Pest Management (W. H. Luckmann and R. H. Metcaff, ed.) Wiley, New York 1975.
8. Feeny, P. in "Recent Adv. in Phytochem. - Biochemical Interaction Between Plants and Insects" (J. W. Wallace and R. L. Mansell, ed.), Plenum, New York, 1976.
9. Rhodes, D. F. and Cates, R. G., ibid.
10. Balachowsky, A. S. "La Lutte Contre les Insects Payot", Paris, 1951.
11. Painter, R. H., J. Econ, Entomol. (1941) 34:358-67.
12. Painter, R. H. "Insect Resistance in Crop Plants, MacMillan, New York, 1951.
13. Schweltz, I. in "Naturally Ocurring Insecticides" (M. Jacobson and D. G. Crosby, ed.) M. Dekker, Inc. 1971.
14. Matsui, M. and Yamamoto, I. In ibid.
15. Fukami, H. and Nakajima, M. In ibid.
16. Jacobson, M. "Insecticides from Plants - A Review of the Literature 1954-1971". Agric. Handbook No. 461.
17. Feeny, P. O. J. Insect. Physiol. (1968) 14, 805-17.
18. Lawton, J. H. in "The Biology of Bracken" (F. H. Perring ed.) Academic Press, New YOrk 1976.
19. Gallun, R. L., J. Environ. Quality (1972) 1 259-65
20. Schalk, J. M. and Ratcliffe, R. H. Bull. Ent. Soc. Am. (1976) 3. 7-10.
21. Quaintance, L. L. and Brues, C. T. USDA Bur. Entom. Bull. (1905) 50:150.

22. Cook, O. F. USDA Tech Bull. (1906) 88.
23. Adams, R., Geissman, T. A. and Edwards, J. D. Chem. Rev. (1960) 60, 555-74.
24. Bottger, G. T., Sheehan, E. T. and Lukefahr, M. J., J. Econ. Entomol. (1964) 57, 283-5.
25. Lukefahr, M. J. and Mattin, D. F. Ibid (1969) 59, 176-9.
26. Lukefahr, M. J. and Houghtaling, J. E. ibid. (1969).
27. Lukefahr, M. J., Houghtaling, J. E. and Cruhm, D. G. ibid. (1975) 68, 743-6.
28. McMichael, S. C. Agron. J. (1959) 51 630.
29. Waiss, A. C., Jr., Chan, B. G., Benson, M. and Lukefahr, M. J. J. Ass. Off. Anal. Chem., in press.
30. Lukefahr, M. J., Noble, L. W. and Houghtaling, J. E. J. Econ. Entomol. (1966) 59, 817-20.
31. Shaver, T. N., Garcaa, J. A., and Dilday, R. H., Environ. Entomol. (1977) 6, 82-4.
32. Malm, N. R., et al., Am. Rep., Agric. Exp. Station, Las Cruces, N. Mexico (1976).
33. Chan, B. G., Waiss, A. C., Jr., and Lukefahr, M. J. Insect. Physol., in press.
34. Bennett, S. E. J. Econ. Entomol. (1966) 58, 372-3.
35. Pridham, J. B., "Enzyme Chemistry of Phenolic Compounds". MacMillan. New York. 1963.
36. Shaver, T. N. and Lukfahr, M. J. J. Econ. Entomol. (1971) 64, 1274-7.
37. Swain, T. and Hillis, W. E. J. Sci. Food Agri. (1959) 10, 63-68.
38. Private communication from Drs. M. J. Lukefahr and W. L. Parrott.
39. Shaver, T. H. and Parrott, W. L. J. Econ. Entomol. (1970) 63, 1802-4.
40. Data to be published.
41. Lukefahr, M. J., Shaver, T. N., Cruhm, D. E. and Houghtaling, J. E. Beltwide Cotton Prod. Res. Conf. Proced. 1974.
42. Levin, D. A. Amer. Naturalist (1971) 105, 157-81.
43. Ingham, J. L. Bot. Rev. (1972) 38 343-417.
44. Hillis, W. E. and Inone, T. Phytochem. (1968) 7, 13-22.
45. Beck, S. D. J. Insect Physiol. (1957) 1, 158-77.
46. Beck, S. D. and Stauffer, J. F. Ann. Entomol. Soc. Amer. (1957) 50, 166-70.
47. Klun, J. A., Tipton, C. L. and Brindley, T. A. J. Econ. Entomol. (1967) 60, 1529-33.
48. Beck, S. D., Ann. Rev. Entom. (1965) 10, 207-32.
49. Klun, J. A., and Brindley, T. A. J. Econ. Entomol. (1966) 59, 711-8.
50. Sulivan, S. L., Gracen, V. E. and Ortega, A. Environ. Entomol. (1974) 3, 718-20.
51. Russell, W. A., Guthrie, W. D., Klun, J. A. and Grindeland, R. J. Econ. Entomol. (1975) 68, 31-34.

Colling, G. N. and Kempton, H. H. J. Agr. Res. (1917) 11, 549-72.

53. Guthrie, W. D. and Walter, E. V. J. Econ. Entomol. (1961) 54, 1248-50.

54. McMillan, W. W., Wiseman, B. R. and Sekul, A. A. Am. Entomol. Soc. Amer, (1970) 63, 371-78

55. Straub, R. W. and Fairchield, M. L. J. Econ. Entomol. (1970) 63 1901-3.

9

The Effects of Plant Biochemicals on Insect Growth and Nutritional Physiology

JOHN C. REESE

Department of Entomology, 237 Russell Labs, University of Wisconsin, Madison, WI 53706

In terms of their effects on insects, plant biochemicals may be divided into nutrients and non-nutrients. Nutrients have received a great deal of attention over the years. They were even thought to be the basis for host plant specificity at one time. Research in recent decades has shown, however, that most species of insects do not differ greatly in their qualitative requirements for nutrients. Thus, although the host plant obviously has to satisfy the nutritional requirements of the insect, it does not seem likely that the insects' nutritional requirements play more than a minor role in host plant specificity (1).

Non-nutrients, or allelochemics (non-nutritional chemicals produced by an organism of one species, and which affect the growth, health, behavior, or population biology of another species (2)), can be extremely important factors in host plant resistance. I will concentrate my discussion of the effects of these allelochemics on insect growth and nutritional physiology, how these non-nutrients may be interacting with nutrients, and on recent work I have done in Dr. Beck's laboratory at the University of Wisconsin on the effects of various allelochemics and dietary moisture levels on insect nutritional physiology.

Current Areas of Active Research.

Plant Apparency. Recent investigations have indicated that evolutionary strategies in plant defense mechnisms may be based on host plant specificity as well as on the population densities and successional status of the plant species (3-5). Within a particular ecosystem, some species of plants will be predictable or apparent in both time and space (i.g. an oak tree). These plants can be easily found by herbivores. Other species are less predictable (and thus less apparent) and so are less likely to be found by herbivores. The predictable species are probably subjected to greater feeding pressure by herbivores and tend to contain high concentration of dosage-dependent (quantitative) inhibitors of digestion and assimilation. These substances are usually not highly toxic, but slow the growth of insects. The less apparent species tend to contain lower concentrations of more highly toxic (qualitative) allelochemics. Insects tend to be more likely to evolve detoxifying mechanisms for these substances, and so to become adapted or specialized for feeding on certain species. Such plant species rely mainly on escape as a defense against the adapted species of insects. The toxins and deterrents in these plants are most effective against the non-adapted generalized feeders. Unfortunately, our agricultural practices have taken many plants that probably evolved under the unapparent category (and so contain the chemical defenses characteristic of this group) and have made these plants highly predictable and therefore apparent by planting them in huge fields and by planting the same species year after year on the same ground. Perhaps a productive direction for host plant resistance research would be to attempt to increase concentrations of dosage-dependent factors characteristic of apparent plants in agronomically acceptable varieties of our crop plants.

Metabolic Effects. Such hypotheses as the plant apparency hypothesis are examples of a general shift in emphasis from behavioral effects toward metabolic effects. For a number of years the token stimuli (6) theory of host selection formed the basis of a great deal of research. A rough idea of the research emphasis in past years can be gained by comparing the approximately 400 references cited by Hedin et al (7) in their compilation of behavioral chemicals, to the approximately 100 references cited by Beck and Reese (1) in their compilation of chemicals having metabolic effects. Despite this approximately four-to-one ratio of work on behavioral aspects to work on metabolic effects, there are really very few good examples of a resistance variety utilizing a behavioral chemical for its resistance mechanism (8).

Chronic Effects Hypothesis. For several years I have been testing the hypothesis that some allelochemics may have, in addition to immediate effects on survival and feeding behavior, subtle chronic effects, even at low concentrations, on rate of growth, utilization of food and pupation (1, 9-12). In other words, I have been interested in the things Feeny would call "dosage-dependent" factors. The inhibition of growth may be due to an inhibition of ingestion, assimilation, or efficiency of conversion of assimilated or ingested food. The more biologically active compounds relied on inhibition of various combinations of the above processes for their effects. The results of these experiments will be discussed in more detail in a later section.

Insect Dietetics. Over the years, each of the major ideas concerning host plant specificity (nutrition, token stimuli, and metabolic effects) have played roles in broadening our understanding of the biochemical basis (or better, bases) of plant-insect interactions. In general, each idea was considered more or less separate from each of the others, and certainly most research projects have been largely based on only one of these ideas. We are now starting to recognize that any given plant-insect interaction must surely depend upon all of these factors (1, 13, 14, 17). Under the concept of insect dietetics (15, 16) it is recognized that the feeding insect must ingest food "that not only meets its nutritional requirements, but that is also capable of being assimilated and converted into the energy and structural substances required for normal activity and development" (1). Therefore, insect dietetics encompasses insect nutrition in the classical sense, any allelochemic effects on feeding behavior, the effects of any allelochemics on survival, and the effects of any compounds which have chronic affects on growth, development, or reproduction. All of these aspects are crucial to a given plant-insect interaction. A deleterious effect in any one of these areas could cause a plant to become more resistant to the attack of insects. These different areas of insect dietetics undoubtedly interact with each other in ways unknown to us at present. Some of these interactions may prove to be very important in host plant resistance research of future years.

In addition to physiological interactions between different aspects of insect dietetics, there may be some important biological interactions in relation to the permanence of a resistant variety. There are some outstanding examples of resistant varieties retaining their resistance over long periods of time. Such exceptionally stable resistant varieties must surely have a number of resistance mechanisms such that the likelihood of a resistant insect biotype arising is lessened. A resistant cultivar probably has a greater probability of relative permanence, if it contains allelochemics which affect both behavior and metabolism. Thus behavioral errancy is punished by metabolic effects (17, 18). Such effects may come from the same or different compounds. For an insect to counteradapt to such a situation would require at least two mutations, one behavioral and one metabolic.

Interactions between Allelochemics and Nutrients. Allelochemics may interact with essential nutrients of insect food. Indeed, many of the deleterious metabolic or chronic effects of plant allelochemis may be due primarily to various interactions between allelochemics and nutrients.

Except for a few examples that will be discussed later, little work has been done on the interactions between allelochemics and nutrients in insects and the possible utilization of such knowledge in host plant resistance. More investigations have been conducted on vertebrates. A brief review of some of this work may prove to be useful in giving us clues as to how certain allelochemics inhibit insect growth.

Hatfield (21) has reviewed several examples from the vertebrate literature in which there seemed to be interactions between allelochemics and nutrients. Chestnut tannins have been found to reduce the availability of lysine in various diets. Interactions between gossypol, cyclopropene fatty acids, and protein have been reported. Allelochemics may serve as antioxidants and thus protect certain nutrients from oxidation (22), so interactions are not always negative.

Various toxic factors, especially enzyme inhibitors, from soybeans have been known for many years (23). Some of these factors render certain elements such as nitrogen or sulfur unavailable. Since several of these factors are heat labile, and so in many cases are really not much of a problem in human nutrition (24), such factors might prove especially useful in resistance programs. Unlike many potential resistance factors, their deleterious effects might be felt primarily by the insect pests, rather than by the human ccnsumer. This is an area that certainly warrants further research.

Phytate is a compound which can be found in relatively high amounts in cottonseed meal (23). It may form complexes with such minerals as zinc, thus rendering them unavailable to monogastric animals. Certain proteins may also be complexed by phytate. Experiments in which cottonseed meal was treated with the enzyme phytase suggest that there are interactions between phytate and such things as phosphorous and gossypol.

The interactions between dicoumarol in sweet clover hay and vitamin K have been known for a number of years. Supplementing the diet with vitamin K can lessen the hemorrhagic effects of the dicoumarol (23).

From the preceding discussion of interactions between allelochemics and nutrients in vertebrates, it is apparent that entomologists may profit from testing some of the same types of hypotheses that have been tested with vertebrates. Certainly, some of the deleterious effects of plant allelochemics may be due to interactions between the allelochemic and some nutrient. Several such interactions have already been demonstrated. For example, certain allelochemics resemble certain nutrients so closely that they may compete metabolically. L-Canavanie is quite similar to L-arginine. Its toxicity (25-29) may be due to the formation of defective canavanyl proteins. L-Canavanine and various other amino acid analogues may also act as inhibitors of insect reproduction (30), possibly through this same mechanism.

Certain allelochemis may interfere with nutrients by blocking their availability. This seems to be the case for oak leaf tannins. The tannins apparently form a complex with proteins such that the proteins are less available to winter moth (*Operophthera brumata*) larvae (31, 32). Similarly, the "digestibility-reducing" factors in creosote resins seem to somehow block digestibility (5). Recently, interactions between certain diterpene acids and cholesterol have been demonstrated (20). The partial reversal of growth inhibition in the presence of relatively large amounts of cholesterol suggested to these workers that these diterpene acids (e.g. levopimaric acid) affect the insects' hormonal system.

As in the vertebrate literature discussed above, protease inhibitors have been shown to have deleterious effects on various insects. Most have dealt with trypsin inhibitors. Ryan (33) and Ryan and Green (34), have reviewed this subject in detail.

Birk and Applebaum (35) studied the effects of soybean trypsin inhibitors on development and proteiolytic activity of *Tribolium castaneum*. Su et al (36) found that relatively high doses of soybean trypsin inhibitor caused increased adult mortality in *Sitophilus oryzae*. A particularly exciting aspect of this area of research is the possibility that the plant may be able to produce much higher levels of enzyme inhibitors after being attacked. This feature gives the plant an adaptive advantage (and an agronomic advantage) in that it does not expend energy for the synthesis of such materials until they are actually needed for defense. Green and Ryan (37) made the exciting discovery that the wounding of the leaves of potato or tomato plants by adult Colorado potato beetles induces a rapid accumulation of protease inhibitor. Further, this response was not confined to immediate area of the attack, but spread to other parts of the plant.

Recently, some work has been done on the effects of specific allelochemics on assimilation of food (passage across the gut wall), efficiency of conversion of assimilated food into insect tissue, and efficiency of conversion of ingested food. A reduction in nutritional indices, such as those mentioned above, resulting from the ingestion of deleterious allelochemics may be due to a number of factors, most of which relate in one way or another to interactions between allelochemics and nutrients. For example, some deleterious allelochemics may bind to a specific nutrient. They may also bind to and inactivate digestive enzymes or membrane carrier proteins. Allelochemics with hydroxyl groups on adjacent carbon atoms may chelate certain essential minerals.

The literature dealing with the effects of specific allelochemics on these nutritional indices reamins sparse, but is increasing. Shaver et al (19) found that gossypol decreased assimilation by bollworm larvae, *Heliothis zea*, but had no measurable effect on utilization by tobacco budworm larvae, *H. virescens*. Erickson and Feeny (18) tested the hypothesis that the larval feeding niche of *Papilio polyxenes asterius* is partially bounded by allelochemics which are not required for perception of host or non-host plants. They demonstrated that sinigrin, or one of its breakdown products, reduced assimilation but did not reduce the efficiency with which assimilated food was converted into tissue. Dr. Beck and I have examined the effects of a number of plant allelochemics on growth and development of black cutworm (*Agrotis ipsilon*) larvae (1, 9-12). We have shown that growth inhibition may be due to inhibition of either assimilation, or efficiency of conversion of assimilated food, or a combination of both. Inhibition of either of these processes will, of course, inhibit the efficiency of conversion of ingested food. The specific results of these experiments will be summarized in a later section.

Potential of Nutritional Index Techniques

As discussed in the previous section, nutritional index techniques have been used infrequently to demonstrate possible nutrient-allelochemic interactions or mechanisms by which deleterious plant compounds may inhibit herbivore growth. Plant-feeding insects certainly do not grow equally well on all plants or plant tissues, even when there are no apparent behavioral barriers to their feeding. The differences are much greater than can be explained in terms of possible differences in nutrient content as such of the different plant tissues. It seems likely, therefore, that allelochemics characteristics of various plant species exert strong influences on the growth and development of insects.

The nutritional indices I found most useful are assimilation (AD), efficiency of conversion of assimilated food (ECD), and efficiency of conversion of ingested food (ECI).

$$AD = \frac{\text{amount ingested (mg)} - \text{feces (mg)}}{\text{amount ingested (mg)}} \times 100$$

$$ECD = \frac{\text{weight gain (mg)}}{\text{amount ingested (mg)} - \text{feces (mg)}} \times 100$$

$$ECI = \frac{\text{weight gain (mg)}}{\text{amount ingested (mg)}} \times 100$$

AD (approximate digestibility) measures the assimilation of food and is termed "approximate" because it does not subtract the weight of waste products in the feces or such metabolic products as the peritrophic membrane (38) and exuviae from the total weight of the feces. ECD measures the efficiency with which assimilated food is converted into insect tissue. This index will decrease as the proportion of assimilated food metabolized for energy increases (38). ECI measures the overall ability of the insect to convert ingested food into tissue (38).

Nutritional indices for insects on host and non-host plants vary a great deal. Using the southern armyworm, *Prodenia eridania*, Soo Hoo and Fraenkel (39, 40) compared the feeding behavior, growth, survival, and nutritional indices of the insect on plant tissues representing 32 families. Most of the plants tested were fed upon, but host suitability was quite varied, ranging from normal rapid larvae growth to complete mortality. Nutritional indices were determined for larve feeding on fresh plant tissues representing 12 different plant families. AD ranged from 36% on a "poor" host to 73% on a "good" host. ECD values ranged from 16% to 57%; ECI values ranged from 10% to 38%. Waldbauer (41) determined nutritional indices of the tobacco hornworm, *Manduca sexta*, larvae on both host plants and on non-host plants. In the latter case, the larvae were induced to feed on non-host plant tissues by removing the sense organs that enable the insect to distinguish host plants from non-host plants.

The use of nutritional indices as a basis for comparing host plant utilization by a phytophagous insect being reared on different host plants has some limitations (38). Variations in water and fiber contents may result in variations in the indices (39-42) (see later section of this paper). A relatively low AD might be the result of high fiber content of the diet, but might also be due to wound-induced proteolytic enzyme inhibitors (34, 37) which would reduce the digestibility of the food. A low ECD could result from antimetabolites present in the dietary material, but could also be produced by unfavorable amino acid ratios that would prevent the synthesis of normal structural proteins. Plant material containing acute toxins preclude the determination of meaningful nutritional indices because the larvae would sicken and/or die during the experimental period. Accordingly, nutritional indices are useful in a preliminary assessment of host plant utilization, but are not sufficient to identify the specific factors influencing the efficiency of that utilization. The effects of chemical plant factors, including both allelochemics and nutritional factors, on the efficiency of dietary utilization can best be investigated by incorporating them into a standardized artificial diet and then determining nutritional indices. In this way chemical factors can be studied with much less equivocation. Also, by using the technique of perfusion of known amounts of a compound into plant material, most of the limitations discussed above can be avoided. Erickson and Feeny (18) were quite successful in their use of this technique. They demonstrated that *Papilio polyxenes asterius* larvae ate just as much wen fed on plant tissue perfused with sinigrin as when fed control tissue. The sinigrin inhibited assimilation, though.

In addition to the nutritional index experiments discussed above, I also studied the effects of allelochemics on some other aspects of dietetics of the black cutworm. The insects were reared on an artificial diet (43). Allelochemics were added at known concentrations. A number of plant allelochemics with possible allomone functions were tested over a 10,000-fold range of concentrations for effects on larval survival, growth, pupation rate, and pupal weight. If incorporation of a compound into the diet resulted in a statistically significant correlation between its dietary concentration and one of the above parameters, then the nutritional indices were determined for that compound at 3.75 x 10^{-2} M. A more detailed description of the methods used for these experiments are described elsewhere (9).

The experimental results (Table I) (summarized from Reese and Beck (9-12) plus some material being prepared for publication) show a number of things. p-Benzoquinone (Fig. 1) reduced the amount of ingestion and assimilation. The reduced assimilation appeared to be compensated for by an increase in ECD. The interaction of these two parameters resulted in an overall ECI

Table I. Effects of some allelochemics on black cutworm survival, weight at 10 days, pupation at 28 days, pupal weight, pupation at 35 days, ingestion, dry weight gain, AD, ECD, and ECI. + indicates significant correlation between compound concentration and parameter measured for survival, weight, 28-day pupation, and 35-day pupation. + indicates significant difference between experimental and control insects for ingestion, dry weight gain, AD, ECD, and ECI.

	Survival	10-Day Wt.	28-Day Wt.	Pupal Wt.	35-Day Pup.	Ingestion	Dry Wt. Gain	AD	ECD	ECI
p-Benzoquinone			+	+		+	+	+	+	
Duroquinone	+	+	+		+	+	+		+	+
Hydroquinone										
Catechol		+	+				+		+	+
L-Dopa	+	+	+	+			+	+	+	+
Dopamine										
Chlorogenic Acid										
Resorcinol		+				+	+			
Phloroglucinol	+	+			+		+		+	+
Gallic Acid	+		+					+		
p-Coumaric Acid				+					+	
Cinnamic Acid			+							
Ferulic Acid										
Benzyl Alcohol				+						+
Pyrogallol	+	+	+		+		+			
Orcinol										

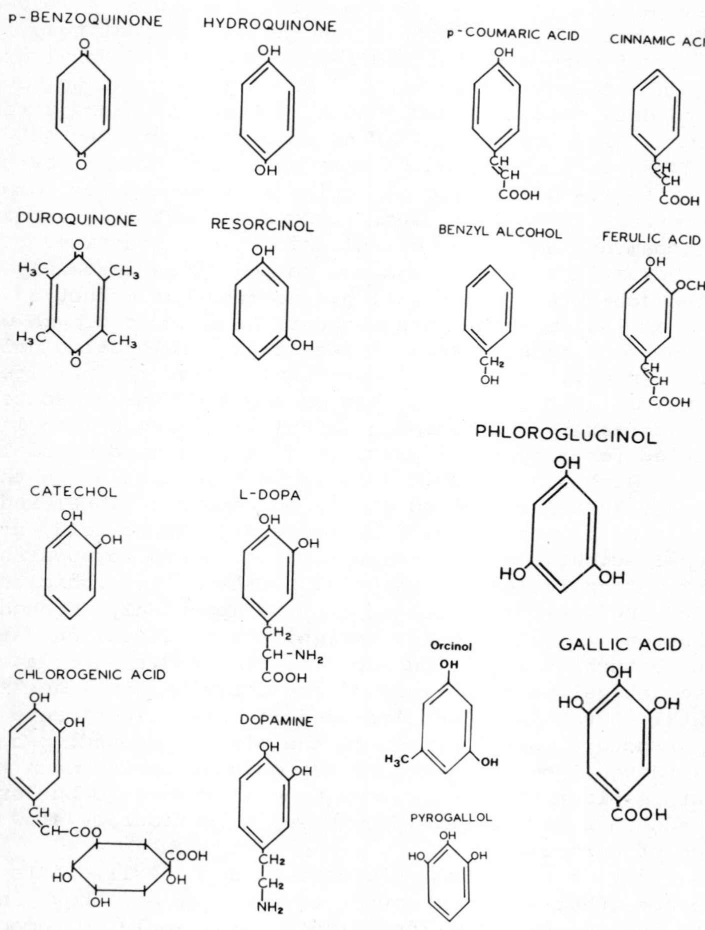

Figure 1. Plant allelochemics bioassayed with black cutworm larvae (duroquinone is probably not found in plants)

that was similar to the control larvae. The reduced form of p-benzoquinone, hydroquinone, showed little biological activity. Catechol inhibited ECD, resulting in a low ECI, while the amount ingested was not affected. L-Dopa also exerted no effect on ingestion, but inhibited both AD and ECD. Interestingly, the structures of both catechol and L-dopa include hydroxyl groups located on adjacent carbon atoms. No compound that we have tested to data that included this structure has had any effect on ingestion. Also, compounds having this structure may act as chelating agents, thus binding some essential mineral or trace element. L-Dopa might also be acting as some type of competitor with tyrosine or phenylalanine. L-Dopa without the carboxyl group (dopamine) had no detectable biological activity. Thus, having hydroxyl groups on adjacent carbon atoms certainly did not guarantee activity. Resorcinol had no apparent effect on the nutritional indices; the data suggest the reduced growth was mainly due to a reduced rate of ingestion. Phloroglucinol (with one more hydroxyl group than resorcinol) (Fig. 1) inhibited growth mainly through a reduction of ECD and thus a reduced ECI. Orcinol, differing from phloroglucinol in having a methyl group substituted for a hydroxyl group (Fig. 1), showed no biological activity. Although cinnamic acid inhibited pupation in the preliminary tests, it had no effect on the nutritional indices. p-Coumaric acid, structurally similar to cinnamic acid, affected only pupal weight and ECD. Ferulic acid showed no apparent activity. Benzyl alcohol inhibited pupation, and inhibited ECD slightly. Pyrogallol inhibited growth, apparently through an accumulation of statistically insignificant effects on various, parameters such as ingestion and ECI. The addition of a carboxyl group to get gallic acid lessened the activity, at least in terms of growth. Duroquinone was included in these experiments because of its structural similarities to the plant compound, p-benzoquinone, although I am not aware of duroquinone having been reported to occur in plant tissues. Duroquinone increased AD but strongly inhibited other parameters. Growth was exceptionally slow in the presence of duroquinone.

In relation to the chronic effects hypothesis, it is interesting to note that of 37 compounds tested (not all shown in Table I due to incomplete nutritional index data), only 10 compounds reduce survival (toxicity), whereas 25 compounds reduced growth, pupation, or pupal weight. Thus, the chronic effects hypothesis is supported.

I think that nutritional index data are extremely useful in starting to identify physiological processes influenced by plant allelochemics, particularly if the chemicals are tested singly under defined experimental conditions. Unfortunately, this type of data does not elucidate specific biochemical modes of action. Also, it does not eliminate the possibility of additional or non-nutritional effects. However, nutritional index techniques will surely help us make progress in these directions. In addition, these techniques have already shown that growth can be inhibited by plant allelochemics in different ways. Also, they clearly demonstrate that the nutritional physiology ("dietetics) of insects can be influenced by non-nutritional substances.

Interrelationships of Nutritional Indices and Dietary Moisture Levels

Recently, I examined the interrelationships between various growth-related parameters and the effects of dietary moisture levels on these parameters. Although water is of obvious importance to animal nutrition, little quantitative work has been done on how water intake affects the dynamics of growth. Stored-products insects have remarkable abilities to conserve water (44), while a number of phytophagous insects suffer deleterious effects if dietary moisture is not relatively high (38, 41, 45). House (46) found that as the nutrients of food were diluted by water, *Celerio euphorbiae* larvae tended to eat more. House did not find statistically significant differences in the efficiencies of conversion of ingested food between the control and experimental insects over the range of moisure levels he tested. However, I calculated the correlation coefficient between percentage of nutrients (ingredients other than agar, cellulose flour, water, and foliage extract) and the mean efficiency of conversion of ingested food in House's experiment, and I found it to be negative and highly significant. Soo Hoo and Fraenkel (39, 40) suggested that water content of the diet was important to efficiency of conversion in *Prodenia eridania*; dilution of the diet caused an apparent increase in efficiency. Feeny (3) found that efficiency of conversion of assimilated food decreased with decreasing moisture of the respective food plants of various lepiodopterous larvae. Hoekstra and Beenakkers (47) suggested that part of the difference in efficiencies of conversion for *Locusta migratoria* were due to differences in moisture content of the different species of plants they were fed. Scriber (48) recently investigated the effects of varied moisture levels in host leaves on *Hyalophora cecropia* larvae. He, too, found that lower moisture levels decrease the efficiency of conversion of food into biomass, as well as the efficiency of conversion of the caloric content and the efficiency of conversion of the nitrogenous content of the food.

The following abbreviations will be used in this section:

IFrWtL -- initial fresh weight of larvae (weight of larvae at beginning of experiment)
DWG -- dry weight gain
DWE -- dry weight eaten
FWE -- fresh weight eaten
DML -- percent dry matter of larvae at end of experiment
DWF -- dry weight of feces

DMF -- percent dry matter of feces at end of experiment
DMD -- percent dry matter of diet at beginning of experiment

A stock culture of the black cutworm was maintained using the methods of Reese et al (43) as modified by Reese and Beck (9). Correlation coefficients were calculated between all possible combinations of pairs of growth-related parameters from the controls of published and unpublished allelochemic experiments (9-12). The mean values from each of 18 experiments were used in this portion of the study. In each of these 18 experiments, there were 20 larvae in each group and the experiments were performed from the 10th through the 20th day of larval life. In a second series of experiments, the dietary moisture levels were varied. There were 20 larvae in each control group and 20 larvae in each experimental group. The experimental period was again from the 10th through the 20th day of larval life. Remaining details of the nutritional index techniques employed have been discussed earlier in this paper.

Interrelationships of Indices. IFrWtL varied from experiment to experiment, apparently due to slight variations in rearing conditions and variations within the population; the importance of running precise controls along with experiments is apparent. In such a situation, it would be reasonable to expect that larvae that started the experiment at a heavier weight would have a higher DWE, DWG, and DWF during the experiment. This is clearly shown in Table II. Likewise, it would be reasonable to assume that DWG (other factors being equal), should be directly proportional to DWE. This too proved to be the case (Table II and Fig. 2) and confirmed the work of Kogan and Cope (42) who used *Pseudoplusia includens* feeding on soybeans. Kogan and Cope presented the linear regression characteristics for the relationship they found between DWE and DWG. I used these characteristics to calculate DWG values in our cutworm experiments (Fig. 2). The DWG values calculated from the Kogan and Cope regression were higher than those for the black cutworm. This indicates that *P. includens* was more efficient at converting soybeans into insect tissue than the black cutworm was at converting the artificial diet used in my experiments. Since IFrWtL and DWG were positively correlated, and since IFrWtL and DWF were positively correlated, it is reasonable to suspect that a relationship might exist between DWG and DWF. This was the case, as shown in Table II.

Dr. Beck and I have stated (9) that it may not be necessary to use specific instarts for nutritional index experiments, especially when controls are always run. It was therefore interesting that there was no significant correlation between IFrWtL and any of the three nutritional indices (AD, ECD, and ECI); in fact, the correlations were exceptionally low (Table II). In other words, as long as the experiments were started with 10 day old larvae,

Table II. Coefficients of correlation (17 degrees of freedom in each case) between pairs of nutritional index parameters.

* indicates statistical significance at $P < 0.05$ level.
** significance at $P < 0.01$ level.
*** significance at $P < 0.0005$ level.

	DWE	DML	DWG	DWF	DMF	AD	ECD	ECI
IFrWtL	+0.866 ***	+0.541 *	+0.795 ***	+0.822 ***	-0.409	-0.001	-0.076	-0.093
DWE		+0.611 **	+0.912 ***	+0.958 ***	-0.453	-0.037	-0.065	-0.072
DML			+0.703 **	+0.652 **	+0.179	-0.151	+0.206	+0.236
DWG				+0.892 ***	-0.397	-0.106	+0.209	+0.331
DWF					-0.344	-0.311	+0.166	+0.001
DMF						-0.142	+0.126	+0.004
AD							-0.828 ***	-0.291
ECD								+0.749 ***

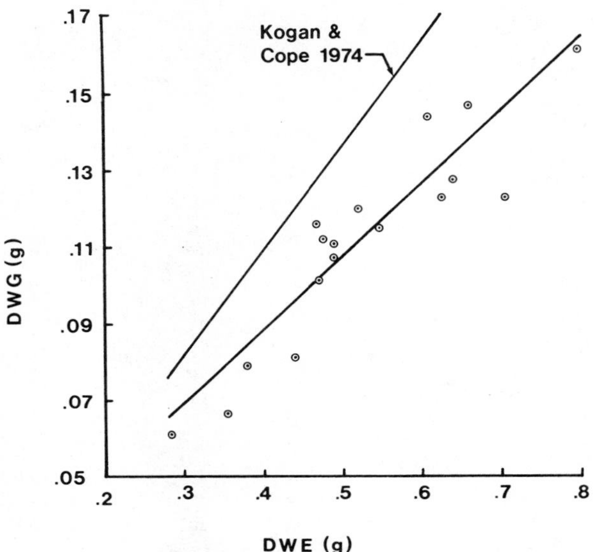

Figure 2. Relationship between DWG and DWE in black cutworm larvae. Kogan and Cope 1974 line refers to line calculated from data in Ref. 42.

differences in weight (and in instars) within the range studied, had no apparent relationship to AD, ECD, or ECI. A similar lack of relationships is suggested by Kogan and Cope (42). They found ECI to be roughly similar from day 6 through day 12, when calculated in terms of fresh weights. It was more eratic when calculated in terms of dry weights, but there was still little apparent trend up or down, as the insect grew. All of these lines of evidence suggest that efficiency changes little during the log phase of growth. This further strengthens our proposal (9) that it is during this log phase that it is appropriate to conduct experiments on growth-reducing dietary factors.

AD increased as ECD decreased (Table II). In other words, as more food was assimilated, a smaller proportion of the assimilated food was converted into insect biomass. Since ECI depends directly upon assimilation and the efficiency with which assimilated food is converted, this compensating mechanism had the effect of decreasing the variability of ECI. ECI had a range of only 7.6 percentage points, while ECD and AD had ranges of 39.5 and 19.0 percentage points, respectively, over the 18 experiments.

DML and DMF Interrelationships. Larvae that started the experimental period at a heavier weight tended to end the experiment at a higher DML (Table II). It would appear from this that black cutworm larvae may have some ability to conserve moisture. This is substantiated by the fact that DML for controls was usually around 16%, whereas DMD at the beginning of experiments was usually close to 21%. Also, it fits well with the fact that many insects increase in percent dry matter as they increase in age (49, 50). Given that black cutworm larvae can apparently conserve moisture when feeding on a diet of 21% DMD, how much ability do they have to maintain a constant moisture level when fed diets of varying moisture levels? This and other questions prompted us to vary the dietary moisture level. The results of these experiments will be discussed in a subsequent section.

Considering the relationship between IFrWtL and DML, and considering the relationships between IFrWtL and DWE, DWG, and DWF, it was not surprising to find positive correlations between DML and DWE, DWG, and DWF (Table II). DMF, however, was apparently independent of any other parameter, including DML. This was somewhat unexpected, since if there were a regulatory mechanism for DML, one would assume that it would operate by absorbing various amounts of water from the hindgut.

Interestingly, DML and DMF were both very precise parameters in any given experiment. Standard error of the mean values were almost always less than 1% and were usually less than 0.5%. The range of DMF values over the 18 experiments was 15.1 percentage points, while the range of DML values was only 3.4 percentage points. Compared to the wide individual variation of such things as larval weight, this variation in DML seemed exceptionally small. We have also noted that in certain of our allelochemic experiments, DMF and DML seemed to be among the most sensitive parameters. In some experiments in which parameters (i.g. DWG) were not significantly different from the controls, DMF or DML showed signficant differences. Also, such small variation over many experiments implies that black cutworm larvae have the ability to regulate DML. To investigate this possibility and to try to answer questions discussed earlier, a series of experiments were performed in which the experimental insects were fed on diets of varying moisture content.

Effects of Various Dietary Moisture Levels. FWE was not affected by dilution with water (Fig. 3). Unlike *Celerio euphorbiae* larvae (46), black cutworm larvae seem to have little ability to compensate for a dilution of the diet. Diets drier than the controls were consumed much more poorly (Fig. 3).

With FWE remaining essentially constant for the diluted diets, the actual DWE necessarily decreased with increasing moisture from the control level (Fig. 4). With diets drier than the controls, DWE also decreased. Thus, the optimal moisture content in terms of DWE was a diet slightly drier than the controls. DWF followed a similar pattern.

As DMD increased, ECD decreased over the entire range of concentrations tested (Fig. 5). Note that the optimal DMD in terms of ECD is not at all close to the control (100%) level. These data confirmed what Feeny (3), Scriber (48), and House (46) found with other lepidopterous larvae. I feel this is a significant finding since Feeny (3) and Scriber (48) used different plants (among which water content happened to vary), or the water content of a plant was changed artificially. In either case, many other biochemical factors may also have been different. In my experiments, a standardized diet was used, therefore, I was confident when I changed the dietary moisture level, that this was the only thing that was varied. In addition, I was able to study moisture levels above the optimal growth level, as well as those below this level.

As might be expected from the relationship between ECD and ECI, ECI was also inversely proprtional to DMD (Fig. 6). As discussed above, ECD and AD were inversely proportional to each other; this apparently had a stabilizing effect on ECI. In the experiments in which dietary moisture was varied, ECD was inversely correlated with DMD, so if AD and ECD were related to each other, one would expect DMD and AD to be correlated. This was not the case. Thus, over a winde range of dietary moisture levels, the compensating mechanism between ECD and AD apparently does not operate at a detectable level. Also, the fact that AD was not affected by dietary moisture was in itself somewhat surprising, since one would think that the moisture content of food would have some effect on how readily it passes across the gut wall.

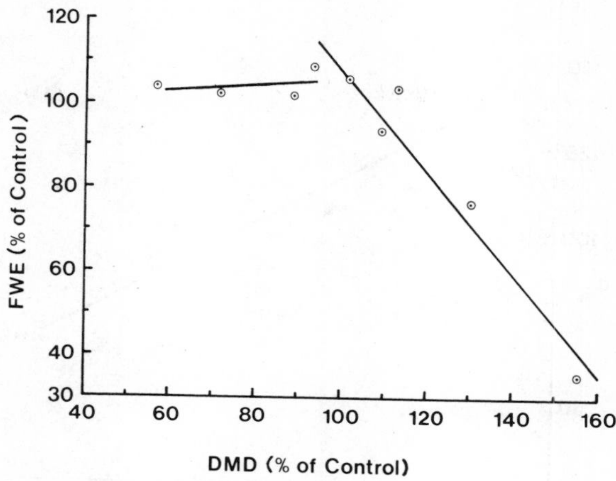

Figure 3. Effect of DMD on FWE in black cutworm larvae

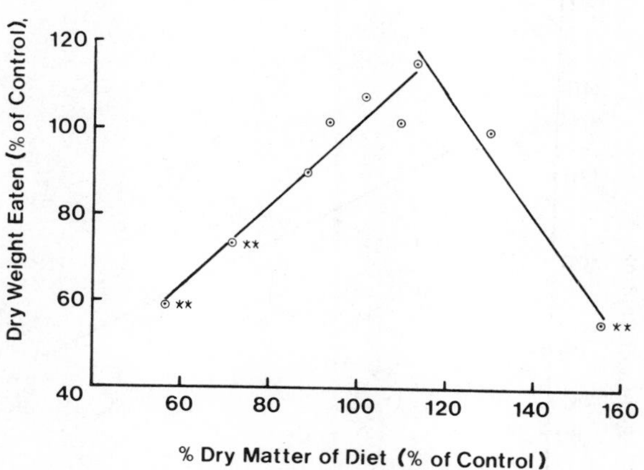

Figure 4. Effect of DMD on DWE in black cutworm larvae

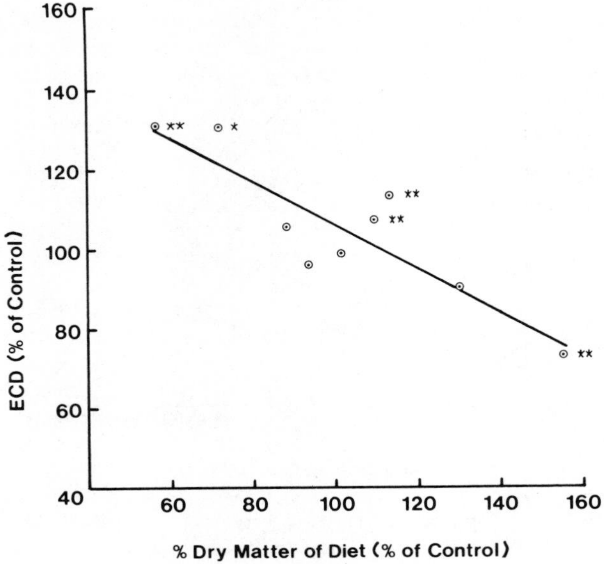

Figure 5. Effect of DMD on ECD in black cutwork larvae

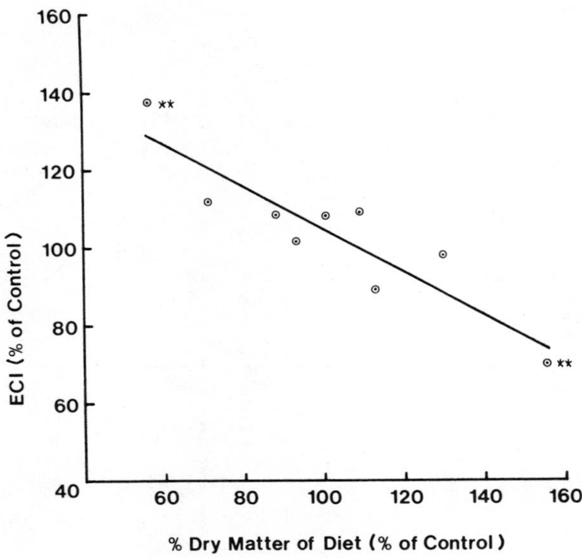

Figure 6. Effect of DMD on ECI in black cutworm larvae

Growth depends upon a combination of how much an organism eats and how efficiently that material is converted into tissue. This is clearly demonstrated and quantified by our data. Figs. 4 and 6 show the relationship between DMD and DWE, and between DMD and ECI, respectively. The relationship of DMD and DWG (Fig. 7) can be viewed as a resultant of Figs. 4 and 6. On low DMD diets, the larvae ate less, but converted it more efficiently, giving net DWG values that were somewhat lower than the controls. Had the influence of a higher ECI and ability of the larvae to regulate moisure levels to some extent (discussed in a later section) not been factors, the curve for the moist diets in Fig. 7 might have followed the dashed line more closely. This dashed line was calculated on the basis of the three points between 80% and 100% DMD. For diets drier than the controls, both ECI and DWE dropped sharply, and combined to give greatly reduced growth. Note that although the optimal DWE was a point above the controls (higher DMD than controls), that the optimal DWG was quite close to the controls, due to the influences of ECI. The diet developed empirically by Reese et al (43) was thus shown to be very close to the optimum in terms of moisture level.

These experiments were originally conducted for the purpose of investigating the ability of black cutworm larvae to deal physiologically with various dietary moisture levels. As has been shown, dietary moisture has striking effects on growth and efficiency of conversion, as well as related parameters. It also had rather precise effects on DMF, as shown in Fig. 8. Fecal percent dry matter in terms of percent of the controls, was essentially the same as dietary dry matter (Fig. 8). The dashed line in Fig. 8 illustrates the situation if DMF were exactly equal to DMD, in terms of percent of the controls. The larvae seemed to have little ability to regulate dry matter compared to many stored-products insects. Nevertheless, evidence for a certain amount of water conserving ability is presented in Fig. 9. Note the position of the curve in relation to the dashed line, which shows the situation if DML were exactly equal to DMD, in terms of percent of the controls. The fact that the observed data falls above the dashed line for low dry matter diets, but lags behind the dashed line for drier diets, indicated a degree of regulatory ability. Also, when viewed in terms of actual DML and DMD values (rather than in terms of percent of controls), DML was always slightly lower than DMD, demonstrating some ability to conserve water. The ability is not very great, though, as indicated by the fact that as DMD increased, DML also increased over the range of dietary moisture levels tested. Apparently the mechanism was not putting excess moisture back into the gut tract. If this were happening, the feces from larvae on low dry matter diets should have been much lower in dry matter. From these data the mechanism seems to be a water conservation mechanism in the sense of maintaining DML lower than DMD, but not a regulatory mechanism in the sense of maintaining a constant DML over a range of DMD values.

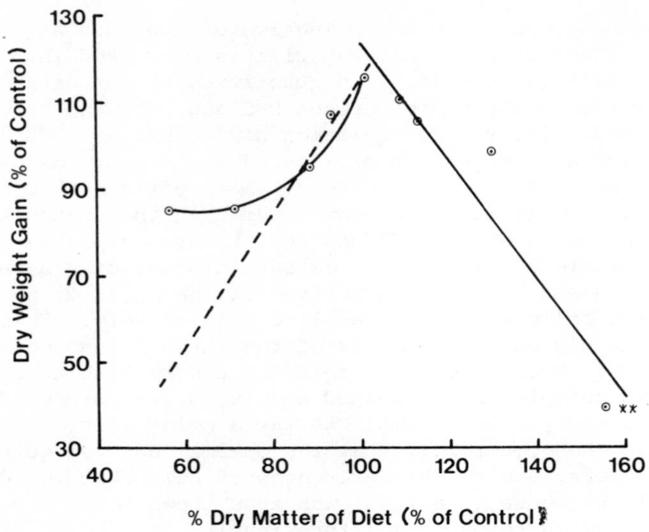

Figure 7. Effect of DMD on DWG in black cutworm larvae. Dashed line based on three points closest to it.

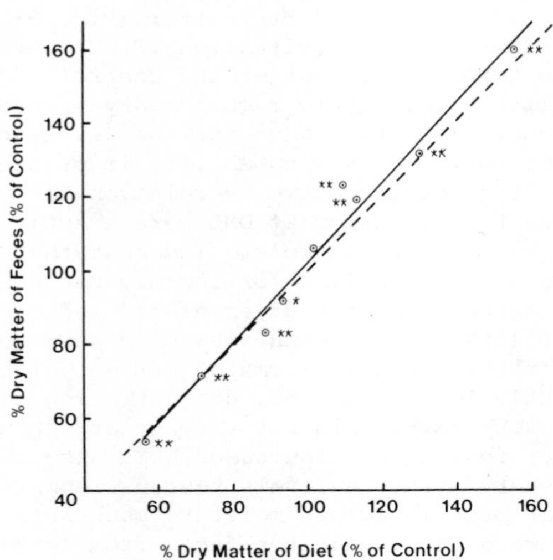

Figure 8. Effect of DMD on DMF in black cutworm larvae. Dashed line represents situation if DMF equaled DMD, in terms of percent of control.

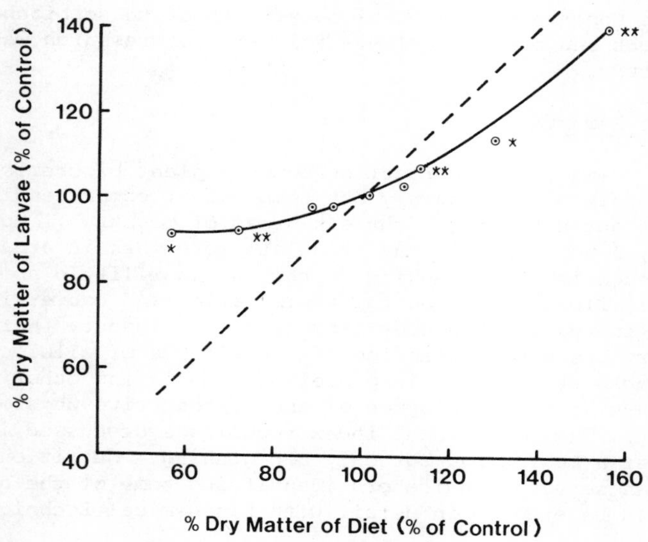

Figure 9. Effect of DMD on DML in black cutworm larvae. Dashed line represents situation if DML equaled DMD, in terms of percent of control.

Moisture level is certainly an important factor in plant-insect interactions. Black cutworm larvae, although highly polyphagous, are undoubtedly capable of successfully living only on plants with fairly high moisture levels. Outbreaks of this pest are associated with floods and unusually wet years. Their ability to regulate their moisture level is limited. Moisture level may be a useful factor in host plant resistance. If this proves to be the case, it would certainly simplify some of the problems encountered in biochemically-based resistance, in which we must be concerned with possible effects of these biochemicals on livestock and man. I believe this area of research warrants further study.

Concluding Remarks

This summary of interactions between plant biochemicals and insect dietetics is certainly not complete or comprehensive. It does point out some of the current areas of research. More important, I hope that it may stimulate more work in areas where our knowledge is still lacking depth. We know little about the effects of allelochemics on the assimilation and conversion processes in the gut and body tissues of the insect. Nor do we have a very broad understanding of the effects of allelochemics on amino acid utilization in protein synthesis and other synthetic pathways in the developmental and reproductive physiology of the insect. The nutritional index techniques discussed in this paper offer a beginning, but only a beginning. Nutritional indices can serve as indicators identifying some of the aspects that should be studied in detail with biochemical techniques.

Acknowledgement

This research was supported by the College of Agricultural and Life Sciences, University of Wisconsin and by a research grant (PCM 74-24001) to Dr. Stanley D. Beck from the National Science Foundation. I would like to thank Dr. Dale M. Norris for helpful suggestions for improving this manuscript. I also thank Holly Beermann for technical assistance.

Literature Cited

1. Beck, S. D., Reese, J. C., Recent Adv. Phytochem, (1976), 10, 4192.
2. Whittaker, R. H., pp. 43-70, "Chemical Ecology", Sondheimer, E., Simeone, J. B., Eds., Academic Press, New York, 1970.
3. Feeny, P. P., pp. 3-19, "Coevolution of Animals and Plants". Gilbert, L. E., Raven, P. H., Eds., University of Texas Press, Austin, 1975.
4. Feeny, P. P., Recent Adv. Phytochem., (1976), 10, 1-40.
5. Rhoades, D. F., Cates, R. G., Recent Adv. Phytochem, (1976), 10, 168-213.

6. Fraenkel, G. S., Science, (1959), 129, 1466-1470.
7. Hedin, P. A., Maxwell, F. G., Jenkins, J. N., pp. 494-527, "Proceedings of the Summer Institute on Biological Control of Plant Insects and Diseases", Maxwell, F. G., Harris, F. A., Eds., University Press of Mississippi, Jackson, 1974.
8. Kogan, M., pp. 103-146, "Introduction to Insect Pest Management", Metcalf, R. L., Luckmann, W., Eds., John Wiley and Sons, New York, 1975.
9. Reese, J. C., Beck, S. D., Ann. Entomol. Soc. Am., (1976a), 69, 59-67.
10. Reese, J. C., Beck, S. D., Ann. Entomol. Soc. Am., (1976b), 69, 68-72.
11. Reese, J. C., Beck, S. D., Ann. Entomol. Soc. Am., (1976c), 69, 999-1003.
12. Reese, J. C., Beck, S. D., pp. 217-221, "The Host-Plant in Relation to Insect Behaviour and Reproduction", Jermy, T., Ed., Plenum Press, New York, 1976.
13. Kogan, M., pp. 211-227, "Proc. XV Internat. Congress Ent.", Packer, J. S., D. White, Eds., 1977.
14. Beck, S. D., Ann. Rev. Entomol., (1965), 10, 207-232.
15. Beck, S. D., pp. 1-6, "Insect and Mite Nutrition", Rodriguez, J. G., Ed., North Holland Publishing Co., Amsterdam, 1972.
16. Beck, S. D., pp. 290-311, "Proceedings of the Summer Institute on Biological Control of Plant Insects and Diseases", Maxwell, F. G., Harris, F. A., Eds., University Press of Mississippi, Jackson, 1974.
17. Beck, S. D., Schoonhoven, L. M., "Breeding Plants Resistant to Insects", Maxwell, F. G., Jennings, P. R., Eds., John Wiley and Sons, New York, (In Press).
18. Erickson, J. M., Feeny, P. P., Ecology, (1974), 55: 103-111.
19. Shaver, T. N., Lukefahr, M. J., Garcia, J. A., J. Econ. Entomol., (1970), 63: 1544-1546.
20. Elliger, C. A., Zinkel, D. F., Chan, B. G., Waiss, A. C., Jr., Experientia, (1976), 32: 1364-1365.
21. Hatfield, E. E., pp. 171-179, "Effect of Processing on the Nutritional Value of Feeds", National Acad. Sci., Washington, D. C., 1973.
22. Cheeke, P. R., Nutrition Reports Internat., (1972), 5: 159-170.
23. Stephenson, E. L., pp. 67-71, "Effect of Processing on the Nutritional Value of Feeds", National Acad. Sci., Washington, D. C., 1973.
24. Liener, I. E., J. Food. Sci., (1976), 41: 1076-1081.
25. Rosenthal, G. A., Janzen, D. H., Dahlman, D. L., Science (1976), 192: 256-258.
26. Vandersant, E. S., Chremos, J. H., Ann. Entomol. Soc. Am., (1971), 64: 480-485.
27. Isogai, A., Chang, C., Murakoshi, S., Suzuki, A., Tamura, S., J. Agr. Chem. Soc. Japan, (1973a), 47: 443-447.
28. Isogai, A., Murakoshi, S., Suzuki, A., Tamura, S., J. Agr. Chem. Soc. Japan, (1973b), 47: 449-453.

29. Dahlman, D. L., Rosenthal, G. A., Comp. Biochem. Physiol., (1975), 51A: 33-36.
30. Hegdekar, D. M., J. Econ. Entomol., (1970), 63: 1950-1956.
31. Feeny, P. P., J. Insect Physiol., (1968), 14: 805-817.
32. Feeny, P. P., Ecology, (1970), 51: 565-581.
33. Ryan, C. A., Ann. Rev. Plant Physiol., (1973), 24: 173-196.
34. Ryan, C. A., Green, T. R., Recent, Adv. Phytochem., (1974), 8: 123-140.
35. Birk, Y., Applebaum, S. W., Enzymologia, (1960), 22: 318-326.
36. Su, H. C. F., Speirs, R. D., Mahany, P. G., J. Georgia Entomol. Soc., (1974), 9: 86-87.
37. Green, T. R., Ryan, C. A., Science, (1972), 175: 776-777.
38. Waldbauer, G. P., Adv. Insect Physiol., (1968), 5: 229-288.
39. Soo Hoo, C. F., Fraenkel, G., J. Insect Physiol., (1966a), 12: 693-709.
40. Soo Hoo, C. F., Fraenkel, G., J. Insect Physiol., (1966b), 12: 711-730.
41. Waldbauer, G. P., Entomol. Exp. Appl., (1964), 7: 253-269.
42. Kogan, M., Cope, D., Ann. Entomol. Soc. Am., (1974), 67: 66-72.
43. Reese, J. C., English, L. M., Yonke, T. R., Fairchild, M. L., J. Econ. Entomol., (1972), 65: 1047-1050.
44. Fraenkel, G., Blewett, M., Bull. Entomol. Res., (1944), 35: 127-139.
45. Waldbauer, G. P., Entomol. Exp. Appl., (1962), 5: 147-158.
46. House, H. L., Canad. Entomol., (1965), 97: 62-68.
47. Hoekstra, A., Beenakkers, A. M. T., Entomol. Exp. Appl., (1976), 19: 130-138.
48. Scriber, J. M., Oecologia (Berl.), (1977), 28: 269-287.
49. Wigglesworth, V. B., "The Principles of Insect Physiology", 741 pp., Methuen and Co. Ltd., London, 1965.
50. Beck, S. D., Hanec, W., J. Insect Physiol., (1960), 4: 304-418.

10

Isolation and Identification of Toxic Agents from Plants

MARTIN JACOBSON

Biologically Active Natural Products Laboratory, Agricultural Research Service, U.S. Department of Agriculture, Beltsville, MD 20705

Methods, techniques, and instrumentation for the improved, rapid isolation and identification of physiologically active plant constituents are available today that were lacking 25, 10, or even 5 years ago. The time period between the isolation of a pure compound and its complete identification has often been frustratingly protracted, but that period has recently been narrowed. Perhaps one of the classical examples of the lengthy time span between isolation and complete structural determination is the case of the well known insect toxicant pyrethrum, whose active components—the pyrethrins, cinerins, and jasmolins—were isolated in 1910-1916 by Staudinger and Ruzicka (1), in 1944 by LaForge and Barthel (2), and in 1965 by Godin et al. (3), respectively; however, their complete structures were not obtained until years after they were isolated.

No attempt has been made here to present a comprehensive review of all of the methods useful for isolating and identifying pesticidal toxicants from plants. Rather those methods that are most generally useful and those pesticidal compounds that are perhaps of major interest are treated. These include pyrethrum, rotenoids, nicotine and nicotinoids, other well-known alkaloids such as those from *Veratrum*, unsaturated isobutylamides, bitter substances (especially insect feeding deterrents), fungal peptides, gossypol, mycotoxins, and insect growth regulators. The first section treats the isolation and general methods of structure elucidation for the toxicants in each of these categories; the second section cites additional information for each of the most significant methods.

For further general reference your attention is directed to the two recent volumes by Nakanishi et al. (4,5) entitled "Natural Products Chemistry" and to "Naturally Occurring Insecticides" (6) edited by Jacobson and Crosby.

Pyrethrum

The term "pyrethrum" usually refers to the dried flower heads or to an extract of the dried heads of Chrysanthemum cinerariifolium Vis. (family Compositae). The active components are toxic to house flies, mosquitoes, cockroaches, lice, and a number of stored product species.

Isolation. Partition of a petroleum ether solution of petroleum ether extract from dried flower heads between the hydrocarbon and nitromethane, followed by passage of the nitromethane solution through a short column of activated charcoal gives an active concentrate of 90% purity or above (7). Molecular distillation can then be used to purify and decolorize the concentrate (8). The insecticidal activity of pyrethrum is attributed to the action of 6 constituents, namely pyrethrins I and II, cinerins I and II, and jasmolins I and II. Pyrethrin I and cinerin I are fairly easily separated from pyrethrin II and cinerin II by column chromatography on alumina or silica gel, and good separation may also be obtained by paper and thin-layer chromatography (TLC) (9).

Identification. The pyrethrins and cinerins form crystalline 2,4-dinitrophenylhydrazones and are further identified by degradation, hydrogenolysis, ozonolysis, ultraviolet (UV), infrared (IR), and nuclear magnetic resonance (NMR) spectroscopy, and optical rotation (9). The jasmolins are identified by UV, IR, and NMR spectroscopy (10).

Rotenone and Rotenoids

This group of compounds comprises mainly rotenone, though it includes minor amounts of deguelin, tephrosin, toxicarol, sumatrol, malaccol, and elliptone. The materials are obtained from various parts (usually the roots) of species of Derris, Lonchocarpus, Tephrosia, and Mundulea (family Leguminosae). They act as contact or stomach poisons for aphids, house flies, and various species of chewing insects.

Isolation. The active compounds are obtained by extracting the crushed, dried root with ether or methylene chloride, followed by overnight cooling of a carbon tetrachloride solution of the concentrate at 0°. The carbon tetrachloride solvate of rotenone crystallizes out and is filtered off and taken up in trichloroethylene. When this solution is diluted with methanol, nearly pure rotenone separates out and can be further purified by crystallization from trichloroethylene. Isolation of rotenone from seeds is simplified if the seeds are defatted with petroleum ether prior to extraction with ether (11).

The resolution of rotenoid mixtures can be effected by column chromatography, but the use of alkaline alumina must be avoided since it causes racemization; neutral or acid grades of adsorbents must be used. Countercurrent distribution (CCD) is useful for dealing with intractable gums, and paper, TLC, and high-performance liquid chromatography (HPLC) have also been used (11,12).

Identification. The structures are identified by chemical degradation and by the use of IR and NMR spectroscopy, and optical rotatory dispersion (ORD). The optically active forms of rotenone are much more toxic to insects than the racemic form (13).

Nicotine and Nicotinoids

Nicotine and nornicotine occur in the leaves of the tobacco plant, *Nicotiana tabacum* L. Although these compounds were used to a considerable extent as insect stomach poisons before the advent of DDT, they are seldom used as insecticides today in the United States. Anabasine, obtained from the roots of *Anabasis aphylla* L. and *N. glauca* Graham (tree tobacco), is extensively used in the Soviet Union to combat thrips, mites, aphids, leafhoppers, sawflies, lice, and mosquitoes (14). *N. tabacum* and *N. glauca* are members of the family Solanaceae, and *A. aphylla* is in the family Chenopodiaceae.

Isolation. The tertiary amine, nicotine, was first obtained in pure form in 1828 by steam distilling the basified ether extract of tobacco leaves (15). Although nicotine sometimes occurs in the plant in the free state, it is usually present as the monoacidic base of citric or malic acid (16).

Anabasine is obtained by mixing ground tree tobacco with ether and distilling the alkaline extract. Since this compound and nornicotine have a secondary amino N group, they may be separated from nicotine by converting them to N-derivatives.

Identification. The structures of nicotine, nornicotine, and anabasine were determined by degrading the molecules with acids, as well as by TLC and ORD. Natural nicotine is levorotatory; dextrorotatory nicotine is much less toxic to insects (16).

Other Alkaloids

The major insecticidal alkaloids other than those from tobacco are jervine, veratramine, cevine, germine, zygadenine, and protoverine; they are present in many species of *Veratrum* and *Zygadenus*, and in sabadilla (*Schoenocaulon officinale* A. Gray), members of the family Liliaceae.

Isolation. The alkaloids are extracted from the powdered plant with acidic alcohol or an organic solvent with or without ammonia. The individual alkaloids are separated by fractional crystallization, chromatography on alumina, silica gel, kieselguhr, or an ion-exchange resin, or by CCD (17).

Identification. The veratrum alkaloids are identified by infrared and NMR spectroscopy, and by mass spectrometry (MS).

Unsaturated Isobutylamides

A number of insecticidal isobutylamides of unsaturated, aliphatic, straight-chain C-10 to C-18 acids have been isolated from plants of the families Compositae and Rutaceae. Although most of these compounds, which are highly toxic to flies and mosquitoes, have been identified, and in some cases synthesized, others have been only partially characterized. They possess two properties in common with one another and with the pyrethrins— pungency, and rapid knockdown and kill of flying insects. Examples of these compounds are anacyclin from *Anacyclus pyrethrum* DC, spilanthol from *Spilanthes* spp., affinin from *Heliopsis longipes* (A. Gray) Blake, scabrin and heliopsin from *H. scabra* Dunal., echinacein from *Echinacea angustifolia* DC, and herculin from *Zanthoxylum clava-herculis* L. (18).

Isolation. These compounds are isolated by the same procedures applicable to the pyrethrins; namely, by partition of a petroleum ether extract of the plant between this solvent and nitromethane followed by passage of the nitromethane solution through activated charcoal and either high-vacuum distillation of the filtrate or column chromatography on alumina or silica gel (18).

Identification. The structures are determined by hydrolysis to the component amine and acid moieties and by UV, IR, and NMR spectroscopy, and MS (18).

Bitter Substances

The limonoid bitter principles are a class of C_{26} triterpenes believed to arise in plants of the Meliaceae and Rutaceae families as oxidation products of tetracyclic triterpenes. They include such insecticides and feeding deterrents as limonin, nomilin, melianone, nimbin, azidirone, nimbalide, and salannin; melianone occurs in *Melia azedarach* L. and the others occur in neem, *Azadirachta indica* (L.) Juss. Other limonoid bitter principles are rather plentiful among the Simaroubaceae. Limonoids occurring among the Meliaceae show, in general, much greater structural variation and complexity than those in the Rutaceae. The chemistry of the bitter substances was well reviewed by Korte (19) in 1959 and by Dreyer (20) in 1968.

Isolation. Column chromatography of the flavonoids and peptides with various types of packings has been compared (21). The most recently described method for separating flavonoids and pigments from citrus oils involves gel permeation chromatography followed by thin-layer chromatography (22).

One of the most potent insect feeding deterrents, azadirachtin, occurs in various parts (mainly the seed kernels) of A. indica. Although first isolated from neem in 1968 (23,24) by tedious column chromatography of the ethanol extract, it was obtained in much higher yield by Nakanishi's group at Columbia University (25) in 1974 by means of a single column chromatography of the extract on silica gel followed by preparative thin-layer chromatography.

Identification. Limonin, the major limonoid in citrus seeds, has been known for over 100 years and, as a reasonably accessible material, has been the most extensively studied member of the limonoids. Its structure determination was reported in 1960 and confirmed soon afterward by X-ray crystallographic studies on the iodoacetate of epilimonol. The structure determination of many limonoids subsequently isolated has followed the same pattern, that is, degradation, IR, and NMR spectroscopy, and MS, as well as ORD (20).

The determination of the exceedingly complex structure of azadirachtin would not have been possible without the simultaneous use of partially relaxed Fourier transform and continuous wave decoupling carbon-13 NMR techniques (25).

Toxic Fungal Peptides

Isolation. Insecticidal peptides such as phalloidin, phalloin, amanitin, and amanin were isolated from Amanita pantherina (Fr.) Quelet (deadly agaric) and A. muscaria (Fr.) S. F. Gray (fly agaric) as described by Wieland (26) and Eugster (27). The mushrooms were steeped in methanol, the steepate was evaporated to a small volume, and the precipitated inorganic substances were filtered off. The filtrate was freed of solvent, dissolved in water, treated with lead acetate and filtered; the filtrate, freed of lead with sulfuric acid, was saturated with ammonium sulfate at pH 4 to precipitate the peptides. These toxic peptides are separable by partition chromatography following their separation into lipophilic and hydrophilic fractions by treatment with methyl ethyl ketone-acetone-water (20:6:5). Paper and TLC are also satisfactory. Amanitin is separable into its α-, β-, and γ- forms (26).

The peptides of A. muscaria, which are highly toxic to house flies and mosquitoes, may also be isolated by chromatography of an alcoholic extract on a column of Dowex 50 or Amberlite IR-120 (H^+), washing first with water and then with 2N formic acid. Purification on Amberlite IRC-50(H^+) yields ibotenic acid and muscazone; decarboxylation of ibotenic acid gives muscinol. Electrophoresis may also be used. An improved method giving larger yields of the toxic compounds involves chromatography of a butanol extract of the fungus on alumina, in which water-saturated butanol is used as solvent (27).

Identification. The peptides are identified by degradation, by UV, IR, and NMR spectroscopy, and by MS (27).

Gossypol

Gossypol is a yellow coloring matter occurring in species of the genus *Gossypium* that is responsible for the resistance of certain varieties of cotton, *G. hirsutum* L. (family Malvaceae) to cotton insects such as the boll weevil. It is also toxic to rats, guinea pigs, and rabbits, but not to cattle feeding on cottonseed meal.

Isolation. Cottonseed kernels are extracted with petroleum ether to remove the oil, and the gossypol is then extracted with ether. Addition of acetic acid to the ether solution gives a crystalline gossypol-acetic acid complex (28).

Identification. Gossypol was identified by degradation and combination of various forms of spectroscopy, including MS (28).

Mycotoxins

The most important of the mycotoxins today is aflatoxin, which is found in peanuts, cottonseed, and stored products contaminated with the common mold *Aspergillus flavus* Link. Soon after attempts were made to isolate aflatoxin it became clear that at least 4 major aflatoxins, as determined by thin-layer chromatography of the blue- and green-fluorescing compounds, were involved; these were designated aflatoxins B_1, B_2, G_1, and G_2. At least 8 minor aflatoxins are now known. The aflatoxins are highly toxic to warm-blooded animals feeding on contaminated foodstuffs; they are also carcinogenic.

Isolation. A rapid procedure for separating the aflatoxins in roasted peanuts utilizes thin-layer plates coated with Adsorbosil No. 1 and the solvent system benzene-ethanol-water (40:6:3) (29).

Especially recommended for the solvent extraction of aflatoxins from mold-damaged commodities is the use of the azeotrope of 2-propanol-water (87.7:12.3) (30).

Identification. The aflatoxins are identified by chemical degradation, UV, IR, and NMR spectroscopy, and fluorometry. An excellent, entertaining discussion of the entire aflatoxin problem, including methods for isolation and identification, appeared this year in a report by Goldblatt (31).

Insect Growth Regulators (Juvenile Hormones)

Many species of plants have been extracted and tested for the presence of juvenile hormone (JH)-mimicking activity (see references 32 and 33 for recent reviews). These compounds have profound effects on many species of insects when applied to the larval or pupal stage; if adults do develop from such species they are almost always deformed in one or several ways and may be incapable of feeding and(or) reproducing.

The classical example of a plant-derived insect juvenoid is the so-called "paper factor" (juvabione) reported by Slama and Williams (34) in 1965 and subsequently isolated by Bowers et al. (35) from the wood of balsam fir, *Abies balsamea* (L.) Mill. (family Pinaceae). Juvabione was identified as the methyl ester of todomatuic acid following chromatographic isolation.

Isolation. A more recent example of a juvenoid obtained from a plant is echinolone [(+)-(*E*)-10-hydroxy-4,10-dimethyl-4,11-dodecadien-2-one], isolated from the roots of the common American coneflower, *Echinacea angustifolia* DC (family Compositae) (36). Echinolone was obtained in pure form by partitioning a pentane extract of the roots between pentane and nitromethane, separating the pentane-soluble portion into neutral, acidic, and basic fractions, and chromatographing the neutral fraction on successive columns of Florisil, silica gel, and silver nitrate-coated silica gel prior to preparative gas chromatography.

Identification. Echinolone was identified by IR and NMR spectroscopy, MS, microozonolysis, and ORD measurements (36).

Insect Growth Regulators (Molting Hormones)

The best-known examples of these compounds are the polyhydroxy sterol ecdysone and its analogs isolated from insects and plants (37). Treatment of insect pupae with these compounds causes greatly accelerated growth and development that result in premature breaking of diapause and, in some cases, giant adults, conditions that are detrimental to the normal life of the insect.

Isolation. Since the polyhydroxy sterols are highly soluble in polar solvents and sparingly soluble in nonpolar solvents, the plant materials are extracted with alcohols. Purification of the extracts is obtained by a combination of partition, liquid chromatography, preparative TLC, and crystallization. CCD is also used. A mixture of ecdysterols that is difficult to separate may be acetylated to give a mixture of acetates whose separation may be much easier, affording pure acetates from which the free ecdysterols can be regenerated by hydrolysis (38).

The isolation of ecdysterols is now most commonly carried out by TLC on silica gel-coated plates or by column chromatography on Amberlite XAD-2 with gradient elution by using 30 to 70% ethanol while monitoring the eluates by absorption at 254 nm (38).

A simplified procedure for fractionating plant materials by sequential extraction with methanol-chloroform-water and phenol-acetic acid-water mixtures gives water-soluble, low molecular weight compounds with little chemical damage (39).

Identification. Structural elucidation of ecdysone with UV, IR, and NMR spectroscopy resulted in a partial structure, which was completed only with the help of its X-ray diffraction pattern and MS. A comprehensive review of ecdysone chemistry is that by Horn (37).

Additional Information on Methods of Isolation and Identification

Thin-Layer Chromatography. New procedures for separating compounds by programmed multiple development TLC equivalent to 5,000-8,000 theoretical plates have recently been described (40). The usual solvent system is ethyl acetate-ethylene dichloride (1:10).

The most recent book on TLC is that entitled "HPTLC. High Performance Thin Layer Chromatography," edited by Zlatkis and Kaiser, which just appeared (41).

High-Performance Liquid Chromatography (HPLC). Hostettmann et al. (42) have very recently reported a method for obtaining pure compounds directly from crude plant extracts by filtration of a hexane extract through a column of silica gel with ether-hexane (1:1) followed by preparative HPLC on silica gel with ether-hexane (1:9) and two runs with ethyl acetate-hexane (1:4). This procedure required only 2-3 hours compared with conventional column chromatography, which required 2 weeks. No degradation of unstable compounds was observed.

HPLC on silver nitrate-coated silica gel (43) and a reverse phase method on μ Bondapak C-18 (44) is of value in separating cis and trans isomers of unsaturated compounds.

Infrared Spectroscopy. The application of IR spectroscopy to the identification of natural products was reported by Cole (45) in 1956. The use of Fourier transform IR in research has brought new and extended capabilities in spectral sensitivity and resolution.

Vibrational frequency assignments were published in book form by Silverstein et al. (46) in 1974, by Bellamy (47) in 1975, and by Pearse and Gaydon (48) in 1976.

NMR Spectroscopy. In 1939, Linus Pauling (49) published a review on the configuration and electronic structure of molecules with application to natural products. This may be considered the forerunner of the use of NMR for the structural determination of natural products. In 1965, Jackman (50) reviewed the use of NMR spectroscopy for determining empirical formulae, the classes of protons in a molecule, the sequence of groups in a molecule, relative stereochemistry, and conformation as applied to natural products.

The use of ^{13}C NMR spectroscopy has helped immensely in determining natural products structure, since the ^{13}C nuclei are usually completely decoupled from all of the ^{1}H nuclei by the use of double resonance. The spectrum is thus simply a series of singlets corresponding to each variety of carbon atom present. Textbooks on ^{13}C NMR spectroscopy oriented toward organic chemists are those by Stothers (51) and by Levy and Nelson (52), both published in 1972, by Levy (53) in 1974 and 1975, and by Muellen and Pregosin (54), published in 1977.

Mass Spectrometry. In 1966, Biemann (55) published a review of MS as applied to natural products, especially various types of alkaloids. Steroids have been very thoroughly investigated by this technique by Djerassi's group, and the information gained has helped considerably with the complicated electron impact induced fragmentation reactions of organic molecules. The polycyclic nature of many physiologically active natural products practically excludes simple fragmentation processes and requires cleavage of a number of bonds for the production of a fragment. However, consultation of available tables of frequently encountered fragment ions and collections of mass spectra for the class of compound under consideration makes structural determination much easier. Field desorption and electron impact MS are recent innovations of considerable importance. An excellent book for aid in interpreting mass spectra is that published by McLafferty (56) in 1973.

A novel representation of data obtainable from double-focussing mass spectrometers has been developed. It can display on a single three-dimensional surface the normal mass spectrum together with peaks due to all metastable transitions occurring in the instrument. Developed by Drs. Macdonald and Lacey, of the C.S.I.R.O. Division of Entomology in Canberra, Australia, this three-dimensional representation is more sensitive to molecular structure than any of the two-dimensional representations commonly used by mass spectrometrists (57).

An excellent review of microanalytical methodology useful in identifying unsaturated compounds is that published in 1975 by Beroza (58). It includes brief discussions of spectral analysis, chemical reagents useful in determining functional groups, ozonolysis to determine double bond position, and carbon-skeleton chromatography.

Disclaimer

The use of trade or proprietary names does not necessarily imply the endorsement of these products by the U.S. Department of Agriculture.

Literature Cited

1. Staudinger, H., and Ruzicka, L. Helv. Chim. Acta (1924), 7, 177-83.
2. LaForge, F. B., and Barthel, W. F. J. Org. Chem. (1944), 9, 242-9.
3. Godin, P. J., Sleeman, R. J., Snarey, M., and Thain, E. M. J. Chem. Soc. (C) (1966), 332-4.
4. Nakanishi, K., Goto, T., Ito, S., Natori, S., and Nozoe, S., eds., "Natural Products Chemistry," vol. 1, 562 pp., Academic, New York (1974).
5. Nakanishi, K., Goto, T., Ito, S., Natori, S., and Nozoe, S., eds., "Natural Products Chemistry," vol. 2, 586 pp., Academic, New York (1975).
6. Jacobson, M., and Crosby, D. G., eds., "Naturally Occurring Insecticides," 585 pp., Marcel Dekker, New York (1971).
7. Barthel, W. F., and Haller, H. L. U.S. Patent 2,372,183 (1945).
8. Elliott, M., Olejniczak, J. S., and Garner, J. J. Pyrethrum Post (1959), 5(2), 8-11.
9. Crombie, L., and Elliott, M. Fortschr. Chem. Org. Naturstoffe (1961), 19, 120-64.
10. Matsui, M., and Yamamoto, I. In reference 6, pp. 3-70.
11. Crombie, L. Fortschr. Chem. Org. Naturstoffe (1963), 21, 275-325.
12. Freudenthal, R. I., Emmerling, D. C., and Baron, R. L. J. Chromatog. (1977), 134, 207-10.
13. Fukami, H., and Nakajima, M. In reference 6, pp. 81-97.
14. Feinstein, L., and Jacobson, M. Fortschr. Chem. Org. Naturstoffe (1953), 10, 423-76.
15. Spaeth, E., and Kuffner, F. Fortschr. Chem. Org. Naturstoffe (1939), 2, 248-300.
16. Schmeltz, I. In reference 6, pp. 99-116.
17. Narayanan, C. R. Fortschr. Chem. Org. Naturstoffe (1962), 20, 298-371.
18. Jacobson, M. In reference 6, pp. 137-76.
19. Korte, F., Barkemeyer, H., and Korte, I. Fortschr. Chem. Org. Naturstoffe (1959), 17, 124-82.
20. Dreyer, D. L. Fortschr. Chem. Org. Naturstoffe (1968), 26, 190-244.
21. Ward, R. S., and Pelter, A. J. Chromatog. Sci. (1974), 12, 570-4.
22. Wilson, C. W., III, and Shaw, P. E. J. Agr. Food Chem. (1977), 25, 211-4.
23. Butterworth, J. H., and Morgan, E. D. Chem. Commun. (1968), 23-4.
24. Butterworth, J. H., Morgan, E. D., and Percy, G. R. J. Chem. Soc. Perkin Trans. (1972), 1, 2445-50.
25. Zanno, P. R., Miura, I., Nakanishi, K., and Elder, D. L. J. Am. Chem. Soc. (1975), 97, 1975-7.

26. Wieland, T. Fortschr. Chem. Org. Naturstoffe (1967), 25, 214-50.
27. Eugster, C. H. Fortschr. Chem. Org. Naturstoffe (1969), 27, 261-321.
28. Adams, R., Geissman, T. A., and Edwards, J. D. Chem. Rev. (1960), 555-74.
29. Waltking, A. E., Bleffert, G. W., Chick, M., and Fogerty, M. Oils Oilseeds J. (Bombay) (1975), 28(2), 32-3.
30. Rayner, E. T., Koltun, S. P., and Dollear, F. G. J. Am. Oil Chem. Soc. (1977), 54, 242A-4A.
31. Goldblatt, L. A. J. Am. Oil Chem. Soc. (1977), 54, 302A-9A.
32. Jacobson, M. Mitt. Schweiz. Entomol. Ges. (1971), 44, 73-7.
33. Jacobson, M., Redfern, R. E., and Mills, G. D., Jr. Lloydia (J. Nat. Prod.) (1975), 38, 455-72.
34. Slama, K., and Williams, C. M. Proc. Natl. Acad. Sci. U.S. (1965), 154, 411-14.
35. Bowers, W. S., Fales, H. M., Thompson, M. J., and Uebel, E. C. Science (1966), 154, 1020-2.
36. Jacobson, M., Redfern, R. E., and Mills, G. D., Jr. Lloydia (J. Nat. Prod.) (1975), 38, 473-6.
37. Horn, D. H. S. In referency 6, pp. 333-459.
38. Hikino, H., and Hikino, Y. Fortschr. Chem. Org. Naturstoffe (1970), 28, 256-312.
39. Laird, W. M., Mbadiwe, E. I., and Synge, R. L. M. J. Sci. Food Agr. (1976), 27, 127-30.
40. Jupille, T. H. J. Am. Oil Chem. Soc. (1977), 54, 179-82.
41. Zlatkis, A., and Kaiser, R. E., eds., "HPTLC. High Pressure Thin Layer Chromatography," 240 pp., Elsevier, New York (1977).
42. Hostettmann, K., Pettei, M. J., Kubo, I., and Nakanishi, K. Helv. Chim. Acta (1977), 60, 670-2.
43. Heath, R. R., Tumlinson, J. H., Doolittle, R. E., and Proveaux, A. T. J. Chromatog. Sci. (1975), 13, 380-2.
44. Warthen, J. D., Jr. J. Am. Oil Chem. Soc. (1975), 52, 151-3.
45. Cole, A. R. H. Fortschr. Chem. Org. Naturstoffe (1956), 13, 1-69.
46. Silverstein, R. M., Bassler, G. C., and Morrill, T. C. "Spectrometric Identification of Organic Compounds," 3rd ed., pp. 73-119, Wiley, New York (1974).
47. Bellamy, L. J. "The Infra-red Spectra of Complex Molecules," 3rd ed., 433 pp., Wiley, New York (1975).
48. Pearse, R. W. B., and Gaydon, A. G. "The Identification of Molecular Spectra," 4th ed., 408 pp., Halsted (Wiley), New York (1976).
49. Pauling, L. Fortschr. Chem. Org. Naturstoffe (1939), 3, 203-35.
50. Jackman, L. M. Fortschr. Chem. Org. Naturstoffe (1965), 21, 275-325.
51. Stothers, J. B. "Carbon-13 NMR Spectroscopy," Academic, New York (1972).

52. Levy, G. C., and Nelson, G. L. "Organic Carbon-13 Nuclear Magnetic Resonance for Organic Chemists," Wiley, New York (1972).
53. Levy, G. C. (ed.), "Topics in Carbon-13 NMR Spectrometry," Wiley, New York, vol. 1 (1974); vol. II (1975).
54. Muellen, K., and Pregosin, P. S. "Fourier Transform NMR Techniques," Academic, London (1977).
55. Biemann, K. Fortschr. Chem. Org. Naturstoffe (1966), 24, 1-98.
56. McLafferty, F. W. "Interpretation of Mass Spectra," 2nd ed., W. A. Benjamin, New York (1973).
57. Lacey, M. J. and Macdonald, C. G. "Organic Mass Spectrometry," 1977, in press.
58. Beroza, M. J. Chromatog. Sci. (1975), 13, 314-21.

11

Insect Antifeedants and Repellents from African Plants

I. KUBO and K. NAKANISHI

Department of Chemistry, Columbia University, New York, NY 10027

There is little doubt that the tropical flora which are constantly exposed to attack by various parasites such as viruses, bacteria, protozoans, fungi, and insects are confronted with much harsher conditions for survival than their temperate counterparts. This necessarily leads to efficient built-in defense mechanisms and it is presumably for this reason that tropical flora offer a rich and intriguing source for isolating natural products possessing attractive pesticidal or medicinal properties. The results outlined in the following stem from the studies initiated in 1974 and more systematically in mid-1975 at the International Centre of Insect Physiology and Ecology (ICIPE), Nairobi, Kenya.

The genesis of this unique center goes back to 1968 when Carl Djerassi (1) proposed a plan for establishing an international institute in a developing country and Thomas R. Odhiambo, Professor of Entomology at the University of Nairobi, responded enthusiastically to initiate an insect oriented institute in Nairobi. ICIPE was inaugurated in early 1970 with generous support from the United Nations Development Program, various other foundations, and federal governments. Of a total of about 130 personnel currently at the Centre, 25 are Research Scientists (post Ph.D.'s), mostly biologists, from all over the world who come for a period of several years to be engaged in basic and interdisciplinary studies of arthropods of agricultural and medical importance. An unusual system adopted at ICIPE is that of Visiting Directors of Research. These 10-15 VDR's from various institutions throughout the world pay one to several visits a year to the Centre to consult and direct (?) the research. The Chemistry unit has had two Research Scientists (ex-members, Drs. D.L. Elder, T. Gebreysus, W.F. Wood, I. Kubo, G.D. Prestwich; and current members, Drs. A. Maradufu and K. Wilson) and one Senior Technician, A. Chapya; the two VDR's have been J. Meinwald and K. Nakanishi. The studies on plant constituents comprise one half of the work carried out by the Chemistry unit, the other half being studies on insect pheromones and defense secretions (2).

Bioassay of Phagostimulants and Antifeedants

Two species of the African armyworm, *Spodoptera exempta* and *S. littoralis*, were mainly employed for the bioassays, either by electrophysiology or by the leaf disk method. The monophagous *S. exempta* has long been known as a major graminaceous crop pest in East and South Africa, whereas the polyphagous *S. littoralis* is a major cotton pest. Other species of the genus *Spodoptera* are distributed throughout the world and constitute a major pest. However, most of the antifeedants isolated so far appear to be specific to certain insects. For instance, warburganal (see below) (3), which is one of the most potent antifeedants against *S. exempta* was hardly active against *Manduca sexta* (tobacco hornworm) and *Schistocerca vaga* (vagrant grasshopper) (4). Hence two insects which seem to have complementary taste senses are currently used in our laboratory, namely, *Epilachnia varivestis* (Mexican bean beetle) and *S. eridania* (southern armyworm).

The most revealing bioassay for following phagostimulants and deterrents is that of electrophysiology with the eight sensilla at the tip of the maxillary palp (5). These correspond to the taste buds and are shown in Figure 1 (scanning electromicrograph). As depicted schematically in Figure 2, a microelectrode is inserted into the maxillary palp and another into one of the sensillum. The tip of the sensillum is brought into contact with a filter paper impregnated with the test solution, and the scoring, which is recorded as impulses per second, is done with an oscilloscope. The maximum number of impulses that can be evoked is of the order of 200. The functions of the eight sensilla appear to differ but the details have not yet been clarified (5).

Most of the compounds described in this article were isolated by following the fractionation with more conventional bioassays. The leaf disk method is shown in Figure 3 (see 6). Leaf disks made with a 20 mm cork borer are immersed for 2 seconds in the test acetone solution; immersion for longer than 2 seconds should be avoided because the phagostimulants (see below) are extracted out. These are placed in a Petri dish with con-

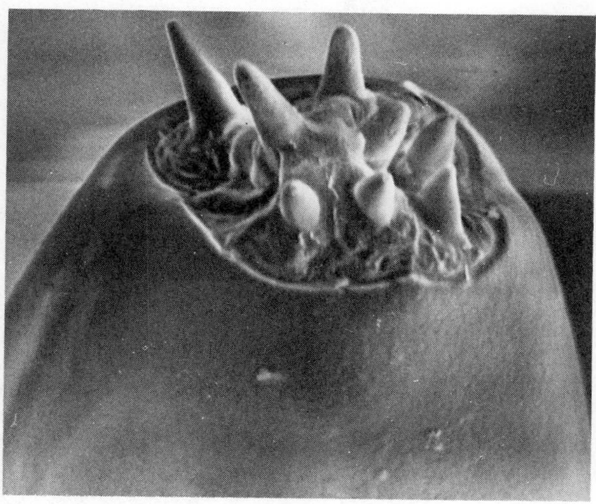

*Figure 1. Scanning electromicrograph (×1650) of tip region of African armyworm (*Spodoptera exempta*) larva. The maxillary palp of eight sensilla are seen.*

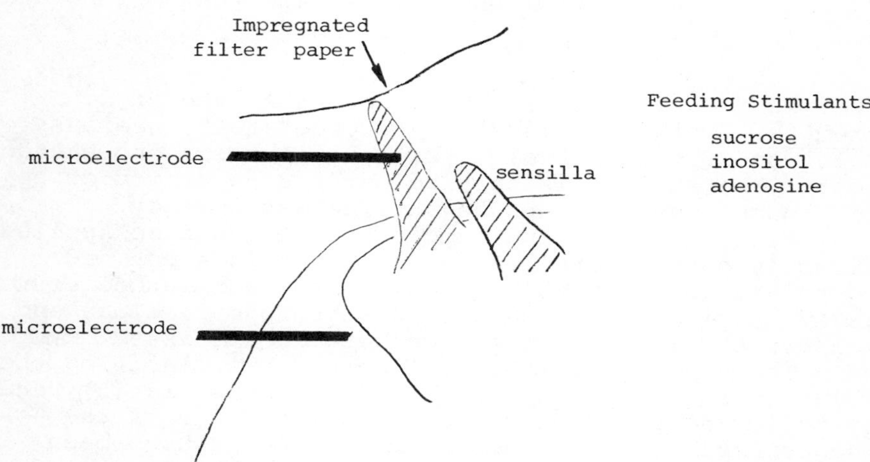

Figure 2. Schematic of electrophycological experiment with sensilla

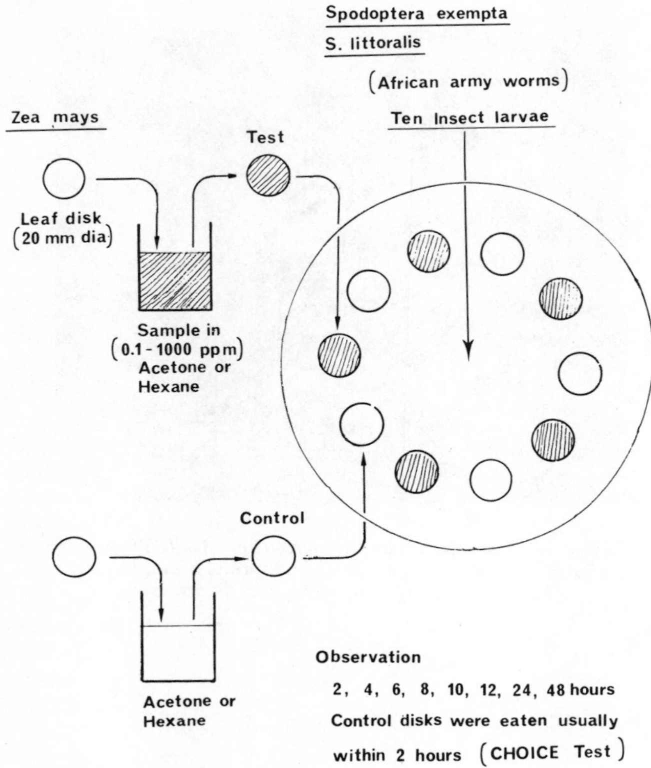

Figure 3. The leaf disk bioassay—a "choice" method

trol leaf disks and ten third instar larvae of S. exempta. This constitutes a "choice" test, whereas deletion of the control disks constitutes a "no choice" test. The scoring is carried out over a period of up to 2 days depending on the antifeedant potency.

Alternatively, the sample solution can be applied directly onto several leaves of plant in a pot and scored against control leaves. A further method of bioassay is to coat the sample on styrofoam lamellae and measure the weight difference between untreated lamellae.

The electrophysiological method has clarified the sugar receptors on S. exempta larval sensilla (5), and by following systematic fractionations, two of the phagostimulants contained in maize leaves have been isolated and chemically characterized as sucrose and adenosine (7). An important finding from the sugar receptor experiments is the fact that a few contacts of the sensilla with filter paper impregnated with potent antifeedants such as warburganal lead to irreversible loss of taste response; in addition, the antifeedant activity of warburganal is inhibited by addition of an equimolar amount of the SH-containing L-cysteine (7).

Antifeedants

For the sake of self-defense, it is not at all surprising that plants would contain insect antifeedants, as well as pesticides and repellents (7a). Since the life cycle of an insect is deranged if the up-take of exogenous moulting hormones (ecdysones) far exceeds the biosynthetic level, the wide occurrence in plants of compounds with moulting hormone activity (phytoecdysones) (8,9,10) presumably is also a self-defending evolutionary development. With the understanding that the term feeding is not clarified with insects, i.e., it may involve biting and/or swallowing factors, we would like to broadly define insect antifeedants as substances which when tasted can result in cessation of feeding either temporarily or permanently depending upon potency (3). Some naturally occurring antifeedants which have been published are: glycosides of steroidal alkaloids, demissine (11), solacauline (12), tomatine (13), leptines I and II (14); ring A enone and/or ring D aromatic steroids, the nicandrenones and others (15, 16,17,19); juglone (5-hydroxynaphthoquinone) (19); the isoquinoline alkaloid isoboldine (6); phenylpropanoids (20); germacrane sesquiterpenes shiromodiol and shiromol (21,22); ent-clerodane and clerodane diterpenes, clerodendrin (23), caryoptin and others (24), and the hydroxylated steroid meliantriol (25).

The collections in East Africa were carried out on the basis of information gathered from books (26,27) and especially from local people on insect-resistant plants. In fact, some of the plants are widely used to control pest insects or for medicinal purposes. The extractions were followed by bioassays, and structures were determined by a combination of various spectral methods combined occasionally with chemical reactions.

a) Azadirachtin from Leaves and Berries of *Azadirachta indica* (Indian neem tree) and *Melia azedarach* (28). The trees occur commonly in India and East and West Africa and are widely used for chewing sticks for cleaning the teeth and as a remedy against malaria. The *Schistocerca gregaria* (desert locust) antifeedant meliantriol (25) was isolated from the leaves. Azadirachtin was a noncrystalline compound reknown for its

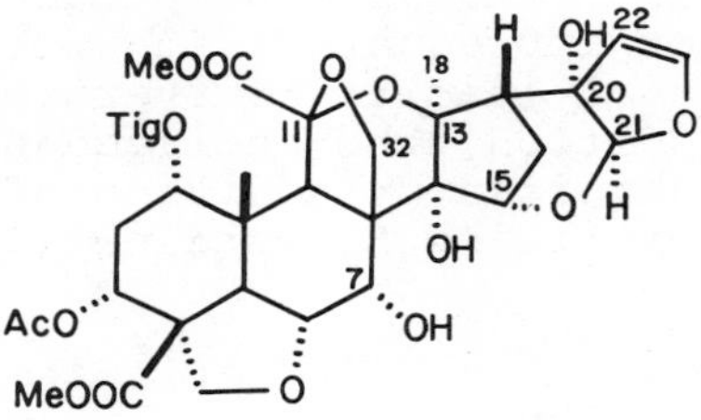

Azadirachtin

very strong antifeedant properties against S. gregaria, i.e., 100% inhibition of feeding is caused by 40 μg/l, or when impregnated on filter paper by 1 ng/cm^2 (29,30). Although partial structures and the correct molecular formula $C_{35}H_{44}O_{16}$ were forwarded (29), complexity of the structure eluded elucidation without the usage of CMR. The structure is based on extensive NMR studies on azadirachtin and its 14-monoacetate (probably formed by acyl migration from 7-OH), in particular the technique of simultaneous partially relaxed Fourier transform (PRFT) and off-resonance decoupling, which was used for the first time. It is a limonoid containing a cleaved C ring and an unstable dihydrofuran moiety. It is one of the most potent antifeedants against S. exempta (roughly one-half of the warburganal potency). The structure is too complex to be synthesized on a practical level but since the yield is quite high (we were able to obtain 800 mg from 300 g of seeds) and the tree is easy to cultivate, practical usage of the neem tree may not be out of the question.

b) Xylomolin from Unripe Fruits of Xylocarpus moluccensis Roem. (Meliaceae) (31,32). The unripe fruits (weighing about 200 g) are bitter but the ripe fruits are treasured by East Africans and are reputed to have aphrodisiac properties. Since all of the skeletal protons are contiguous, PMR elucidated the entire proton system. The configurations at chiral centers are based on the facts that no J values were larger than 3 Hz, and neither NOE nor W-type couplings could be observed. Presence of the acetalic methoxyl group was deduced from the presence of a doublet of quartets signal in the undecoupled CMR spectrum; this is due to additional coupling of the methoxyl carbon to 3-H. It is a secoiridoid with a structure closely related to secologanin, the origin of the terpenoid derived moieties in indole alkaloids. It is possible that xylomolin is a progenitor of alkaloids which may be contained in the ripe fruits but this has not been investigated. The antifeedant level is moder-

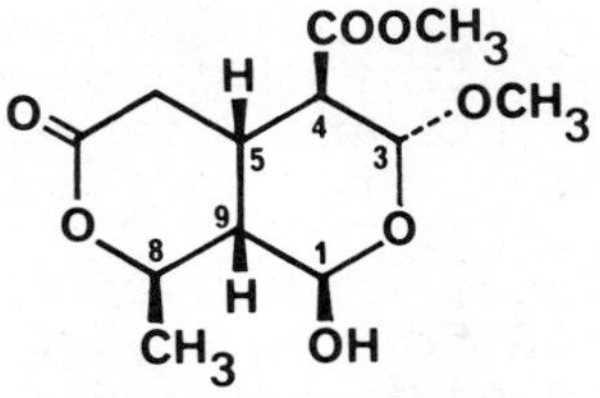

Xylomolin

Ar-I Ar-II Ar-III

Ar-I

CMR	
Me	4
CH_2	9
CH	3
4°C	3
C=	2
C=O	3
	24

ate, i.e., 100 ppm against S. exempta; it strongly inhibits the respiration of rat liver mitochondria.

c) Ajugarins from Ajuga remota (Labiatae) Leaves (33). As quoted above, other clerodane antifeedants have been isolated previously (23,24). However, it is interesting to note that both antipodal skeletons are active, as exemplified by clerodendrin (23), an ent-clerodane diterpene, and caryoptin (24), a clerodane diterpene. The absolute configuration of the ajugarins (ent-clerodane) was determined (33) by conversion to a 6-keto compound and comparing its CD with a reference compound derived from clerodin (24). The antifeedant activity is 100 ppm against S. exempta.

d) Harrisonin from Leaves of Harrisonia abyssinica Oiv (Simarubaceae) ("Msabubini" or "Mpapura-doko" in Swahili) (34). This shrub is used widely in East African folk remedies including treatment for bubonic plague, hemorrhoid, snake bite, etc. The chopped root (650 g) gave 70 mg of harrisonin and 25 mg of obacunone, a related limonoid bitter principle (35, 36).. The structure was determined by extensive NMR studies, especially CMR (PND, PRFT and undecoupled) and comparisons of peaks with those of obacu-

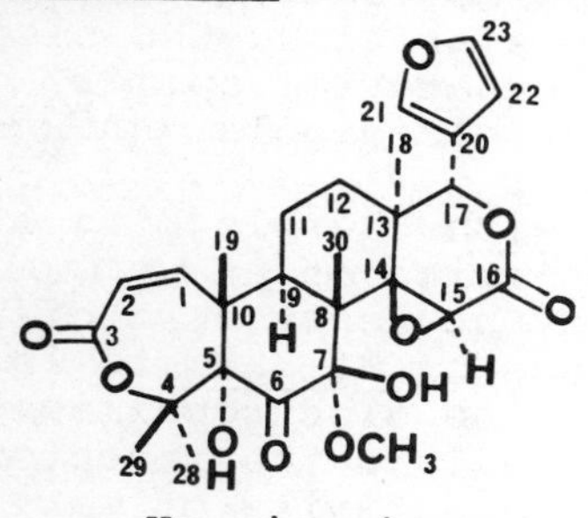

Harrisonin

none. Although it was not clear whether the hemiketal function at C-7 was an artefact or not, subsequent studies have shown that the natural product is a 6,7-alpha-diketone. The S. exempta antifeedant activity is 20 ppm; it also exhibits antibiotic activity (5 μg/ml against Bacillus subtilis) (37) and cytotoxicity (2.2 μg/ml, KB test) (38).

e) Polygodial, Ugandensidial and Warburganal from the Bark of Warburgia stuhlmannii and W. ugandensis (39).

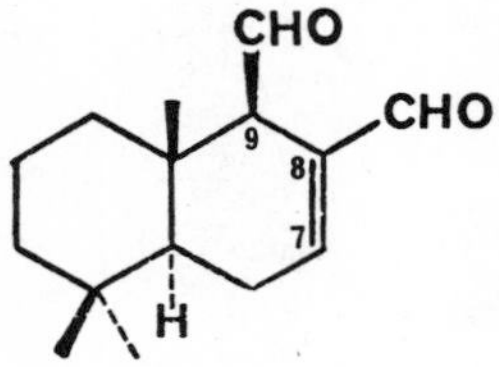

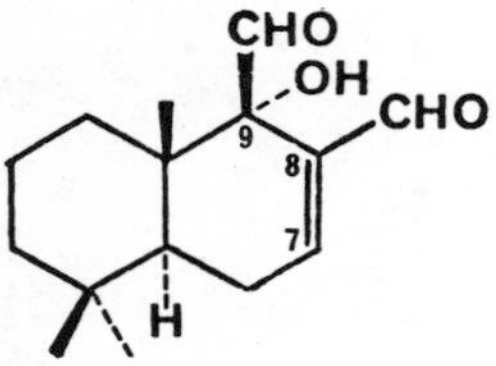

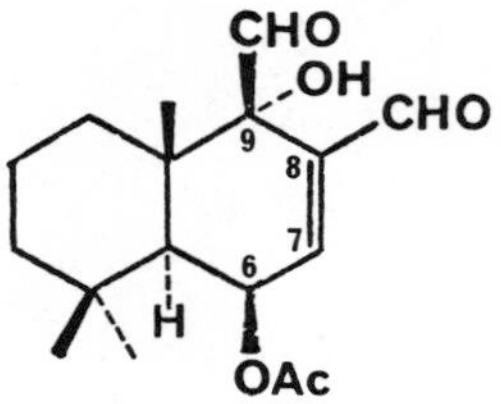

Polygodial Warburganal Ugandensidial

The barks of both trees are treasured in folk medicine and also as food spice. W. stuhlmanni gave the known polygodial (40,41,42) in moderate yield while W. ugandensis gave the known ugandensidial (43) and new warburganal. The yield of the latter two are not only low but they are difficult to separate, even by hplc, from another constituent and this led to confusion in structural studies. Of the three, warburganal is by far the most active antifeedant (0.1 ppm against S. exempta), but more extensive studies will have to depend on synthetic material because of the limited quantity. Some structure/activity studies carried out with the simple polygodial indicated that the activity was lost when one or both aldehydes were reduced to hydroxyl groups or oxidized to carboxyl groups. Epimerization of the equatorial 9β-CHO to the more stable axial 9α-CHO also resulted in loss of activity (the equatorial epimer probably owes its instability to dipole-dipole repulsion between the two aldehyde functions).

These results in conjunction with the blocking of antifeedant activity upon administration of an equimolar amount of cysteine (7) suggest that at least part of the antifeedant activity in the case of S. exempta is caused by Michael type addition of the electrophilic antifeedant to a nucleophilic terminal in the receptor site, and that, as shown by the high activity of warburganal, the structural complexity of which is in between those of the other two, the steric requirement of the receptor site is fairly selective. Interestingly, the antifeedants against the armyworm taste spicy to the human tongue.

f) Inflexin from Leaves of *Isodon inflexus* (Labiateae) (44) and Isodomedin from Leaves of *I. shikokianus* var *intermedius* (45). These two compounds were isolated

inflexin

isodomedin

from Japanese sources but are quoted here because they both exhibit antifeedant activity as tested against the African armyworm. They are also cytotoxic, the levels of inflexin and isodomedin being LD_{50} 5.4 µg/ml and 4.0 µg/ml, respectively. They are *ent*-kaurenoid diterpenes with closely related structures.

g) Crotepoxide from Leaves of *Croton macrostachys* (46). The antifeedant principle was isolated according to the bioassay with *S. exempta* and the structure was determined by detailed NMR and other methods. To our surprise the structure was that of crotepoxide (47,48,49), a compound exhibiting tumor-inhibitory, antileukemic and antibiotic activity. The cytotoxicity and antifeedant activity are both dependent, at least partially, on the reaction between the nucleo-

Crotepoxide

philic receptor and electrophilic moiety of the active molecule; isolation of the same molecule by two totally different bioassays proves the similarity in certain aspects of these two seemingly different activities.

h) Unedoside from the bark of Canthium euroides (50). Structure determination of the antifeedant contained in this plant turned out to be the known udeside which had been isolated from Arbutus unedo (51).

Unedoside

i) Prep LC of Crude Plant Extracts (52).

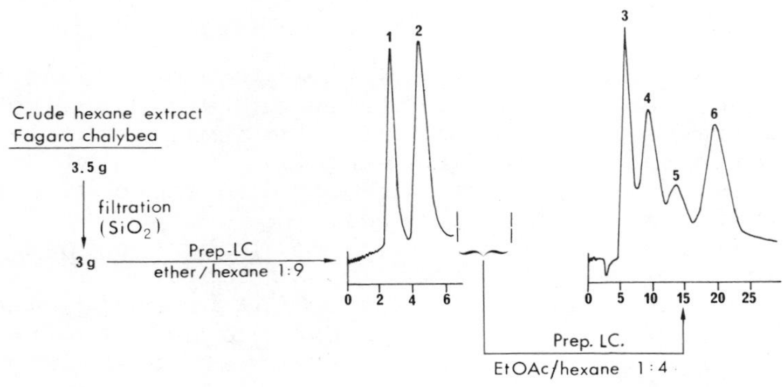

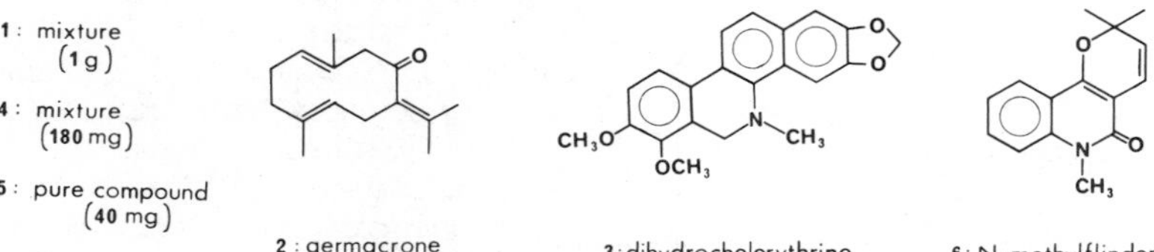

Helvetica Chimica Acta

Figure 4. Direct obtention of pure compounds from crude plant extracts by preparative liquid chromatography (52)

The results shown in Figure 4 were obtained with a Waters Prep LC-500 system. The dried bark (150 g) of the Indian and African medicinal plant Fagara chalybea ENGL. (Rutaceae), which exhibited antifeedant, antibacterial and antifungal activity was extracted with hexane. The 3.5 g of sticky oil remaining after evaporation of the solvent was prefiltered through 100 g of silica gel and the filtrate was injected directly into the prep LC. Peaks 1 (1 g of a mixture of non-polar materials) and 2 (850 mg of pure germacrone) were eluted in 6 min. The remainder was collected as one fraction, and was rechromatographed by prep LC employing a more polar solvent system, upon which peaks 3 (120 mg of dihydrochelerythrine), 4 (180 mg of a mixture of several components), 5 (40 mg of pure compound, structure to be published, ref. 53) and 6 (125 mg of N-methylflindersine) were obtained in another 25 min. The entire operation was achieved in 2 hours as compared to 2 weeks required by conventional open column chromatography employing a variety of chromatographic conditions including gradient elution.

The more polar fractions are also being submitted to prep LC. This direct application of prep LC to crude plant explants clearly results in great shortening of the tedious extraction procedure. Furthermore, the reduced time on the column minimizes deterioration of sample on the column. For example, the relative amount of dihydrochelerythrine (peak 3) obtained from a regular open column operation was much less.

Acknowledgment

This research has been supported by NIH Grant AI 10187 (to K. N.) and U.N. Development Program (to ICIPE). We are greatly indebted to our colleagues cited in the references, namely, Dr. V. Balogh-Nair, Dr. D.L. Elder, Dr. K. Hostettmann, Y.-W. Lee, I. Miura, M. Pettei, Dr. F. Pilkiewicz, S. Tanis, F. Yeh, Dr. P. Zanno (Columbia University) and A. Chapya, and Dr. W.-C. Ma (ICIPE). We also acknowledge the collaboration with Prof. T. Kubota and his group (Kinki University and Osaka City University).

Literature Cited

1. Djerassi, C., Bull. Atomic Sci. (1968) 24, 22.
2. Meinwald, J., Prestwich, G., Kubo, I., and Nakanishi, K., Science, in press.
3. Kubo, I., Lee, Y.-W., Pettei, M., Pilkiewicz, F., Nakanishi, K., J.C.S. Chem. Comm. (1976), 1013.

4. These bioassays were carried out by Dr. G.B. Staal and K.J. Judy, Zoecon Corporation.
5. Ma, W.-C., unpublished results. However, see ICIPE Annual Reports (1974), p. 7-11, and (1975), p. 22-30 for preliminary accounts. These Annual Reports are available on request to Librarian, ICIPE, P.O. Box 30772, Nairobi, Kenya.
6. cf. Wada, K. and Munakata, K., Agr. Food Chem. (1968), 17, 471.
7. Ma, W.-C. and Kubo, I., unpublished results. cf. ICIPE Annual Reports (1975), p. 22-30 for preliminary account.
7a. Jacobson, M. and Crosby, D.G. ed., "Naturally Occurring Insecticides", Marcel Dekker Inc., New York (1971).
8. Reviews: Hikino, H. and Hikino, Y. in Zechmeister, L. ed., "Fortschritte der Chemie Organischer Naturstoffe" Springer Verlag (1970), 28, 256.
9. Horn, D.H.S. in Ref. 7a, p. 330.
10. Nakanishi, K., Pure and Applied Chem. (1971), 25, 167, and XXIII rd Intern. Congress Pure and Applied Chem. (1971), Vol. III, 29.
11. Kuhn, R. and Guhe, Z., Z. Nat. Forsch. (1947), 26, 467.
12. Schreiber, K., Zuchter (1957), 27, 289.
13. Buhr, R., Toball, R., and Schreiber, K., Ent. Exp. Appl. (1958), 1, 209.
14. Kuhn, R. and Low, I., Chem. Ber. (1961), 94, 1096.
15. Bates, R.B. and Eckert, D.J., J. Am. Chem. Soc. (1972), 94, 8258.
16. Begley, M.J., Crombie, L., Ham, P.J., and Whiting, D.A., J.C.S. Chem. Comm. (1972), 1108.
17. Begley, M.J., Crombie, L., Ham, P.J., and Whiting, D.A., J.C.S. Chem. Comm. (1972), 1250.
18. Bates, R.B. and Morehead, S.R., J.C.S. Chem. Comm. (1974), 125.
19. Gilbert, B.L. and Norris, D.M., J. Insect Physiol. (1968), 14, 1063.
20. Isogai, A., Murakoshi, S., Suzuki, A., and Tamura, S., Agri. Biol. Chem(1973), 37, 889.
21. Wada, K., Enomoto, Y., and Munakata, K., Agr. Biol. Chem. (1970), 34, 941.
22. Wada, K., Enomoto, Y., and Munakata, K., Agr. Biol. Chem. (1970), 34, 946.
23. Kato, N., Shibayama, S., Munakata, K., and Katayama, C., J.C.S. Chem. (1971), 1632.
24. Hosozawa, S., Kato, N., and Munakata, K., Tetrahedron Letters (1974), 3753.
25. Lavie, D., Jain, M.K., and Shpan-Gabrielith, S.R., Chem. Comm. (1967), 910.

26. Watt, J.M. and Breyer-Brandwijk, M.G., "Medicinal and Poisonous Plants of Southern and Eastern Africa", E.S. Livingstone Ltd., Edinburgh and London (1962).
27. Kokwaro, J.O. "Medicinal Plants of East Africa", East African Literature Bureau, Nairobi (1976).
28. Zanno, P.R., Miura, I., Nakanishi, K., and Elder, D.L., J. Am. Chem. Soc. (1975), 2, 1975.
29. Butterworth, J.H., Morgan, E.D., and Percy, G.R., J. Chem., Soc. Perkin Trans. (1972), 1, 2445.
30. Morgan, E.D. and Thornton, M.D., Phytochem. (1973), 12, 391.
31. Kubo, I., Miura, I., and Nakanishi, K., J. Am. Chem. Soc. (1976), 98, 6704.
32. According to Prof. D.A.H. Taylor, it should be X. granatum. However, the literature is confused and the identity is being checked.
33. Kubo, I., Lee, Y.-W., Balogh-Nair, V., Nakanishi, K., and C hapya, A., J.C.S. Chem. Comm (1976), 949.
34. Kubo, I., Tanis, S.P., Lee, Y.-W., Miura, I., Nakanishi, K., and Chapya, A., Heterocycles (1976), 5, 485.
35. Arigoni, D., Barton, D.H.R., Corey, E.J., Jeger, O., Caglioti, L., Dev, S., Ferrini, P.G., Glazier, E.R., Melera, A., Pradham, S.K., Schaffner, K., Sternhell, S., Templeton, J.E., and Tobinaga, S., Experientia (1960), 16, 41.
36. Kubota, T., Matsuura, T., Tokoroyama, T., Kamikawa, T., and Matsumoto, T., Tetrahedron Letters (1961), 325.
37. Kindly measured by Dr. M. Taniguchi, Osaka City University.
38. Kindly measured by Dr. F.J. Schmitz, University of Oklahoma.
39. Kubo, I., Lee, Y.-W., Pettei, M., Pilkiewicz, F., and Nakanishi, K., J.C.S. Chem. Comm. (1976), 1013.
40. Barnes, C.S. and Loder, J.W., Austral. J. Chem. (1962), 15, 222.
41. cf. also Ohsuka, A., Nippon Kagaku Zasshi (1962), 83, 757.
42. Synthesis: Kato, T., Suzuki, T., Tanemura, M., Kumanireng, A.S., Ototani, N., and Kitahara, Y., Tetrahedron Letters (1971), 1961.
43. Brook, C.J.W. and Draffan, G.H., Tetrahedron (1969), 25, 2887.
44. Kubo, I., Nakanishi, K., Kamikawa, T., Isobe, T., and Kubota, T., Chem. Letters (1977), 99.
45. Kubo, I., Miura, I., Nakanishi, K., Kamikawa, T., Isobe, T., and Kubota, T., J.C.S. Chem. Comm. (1977), in press.

46. Kubo, I., Miura, I., and Nakanishi, K., to be published.
47. Kupchan, S.M., Hemingway, R.J., and Smith, R.M., J. Org. Chem. (1969), 34, 3898.
48. Synthesis: Oda, K., Ichihara, A., and Sakamura, S., Tetrahedron Letters (1975), 3187.
49. Synthesis: Demuth, M.R., Garrett, P.E., and White, J.D., J. Am. Chem. Soc. (1976), 98, 634.
50. Kubo, I., Miura, I., and Nakanishi, K., unpublished results.
51. Geisman, T.A., Knaack, W.F.Jr., and Knight, J.O., Tetrahedron Letters (1966), 1245.
52. Hostettmann, K., Pettei, M.J., Kubo, I., and Nakanishi, K., Helv. Chim. Acta (1977), 60, 670.
53. Yeh, F., Miura, I., Hostettmann, K., Kubo, I., and Nakanishi, K., to be published.

12

Antifeedant Sesquiterpene Lactones in the Compositae

TOM J. MABRY and JAMES E. GILL
Department of Botany, University of Texas, Austin, TX 78712
WILLIAM C. BURNETT, JR. and SAMUEL B. JONES, JR.
Department of Botany, University of Georgia, Athens, GA 30602

Most of the known 700 or so sesquiterpene lactones are characteristic constituents of many members of the Compositae, occurring only infrequently in a few other higher and lower plants (1,2). Whether or not the α,β-unsaturated γ-lactone function present in most of these terpenoids is responsible for their antifeedant properties is not known; however, this functionality has been implicated in most of the other biological activities known for these substances: e.g. antitumor, cytotoxicity, contact dermatitis, and antifungal properties. Since we recently described in detail the role of sesquiterpene lactones in plant-animal interactions (2), the present account will only summarize our specific findings concerning the way the germacranolide glaucolide-A (I) controls the level of feeding of insects and mammals on members of the genus *Vernonia* (3) and our general conclusions concerning the function of sesquiterpene lactones as antifeedants.

In connection with establishing the antifeedant and growth inhibition properties of glaucolide-A (I), a germacranolide-type sesquiterpene lactone from *Vernonia* (Compositae), six species of Lepidoptera as well as two species of mammals were utilized in a series of laboratory and field tests with *V. gigantea* and *V. glauca*, both of which contain glaucolide-A, and with *V. flaccidifolia*, a species which does not produce any sesquiterpene lactones. Some of the major findings of these tests are summarized in Table I and Figure 1.

Insect Antifeedant Tests

All of the six lepidopterous larvae used in testing the feeding deterrent properties of glaucolide-A in the laboratory tests are sympatric with the species of *Vernonia* tested and three of them, the yellowstriped armyworm, (*Spodoptera ornithogalli*), cabbage looper (*Trichoplusia ni*), and the yellow woollybear (*Diacrisia virginica*), feed in nature on Vernonias containing glaucolide-A; a fourth species, the saddleback caterpillar

(*Sibine stimulea*), feeds on *V. flaccidifolia*, a species lacking glaucolide-A. Two generalist feeders, the fall and southern armyworms (*Spodoptera eridania* and *S. frugiperda*), were not detected on Vernonias in the field.

The laboratory tests established two trends (Table I): firstly, when the insects had a choice of *Vernonia flaccidifolia* leaf material prepared in the form of a pellet with and without added glaucolide-A, all the larvae preferred the diet without the sesquiterpene lactone; secondly, when the larvae were offered a choice of fresh leaves of glaucolide-A species versus fresh leaves of *V. flaccidifolia*, they tended to exhibit the same feeding patterns observed in nature (*4*). That is, except for the yellowstriped armyworm, those species observed in the field on Vernonias containing glaucolide-A preferred these same plants over *V. flaccidifolia* in the laboratory; similarly, the one species observed on *V. flaccidifolia* under natural conditions still preferred this species in the laboratory. Two generalist feeders, the southern and fall armyworms, always avoided the glaucolide-A plants. In tests with the fall armyworm larvae, it was further determined that levels of about 1% glaucolide-A (in a pellet) gave maximum deterrence; this is roughly the quantity of lactone material synthesized in the leaves of many sesquiterpene lactone-containing Compositaes.

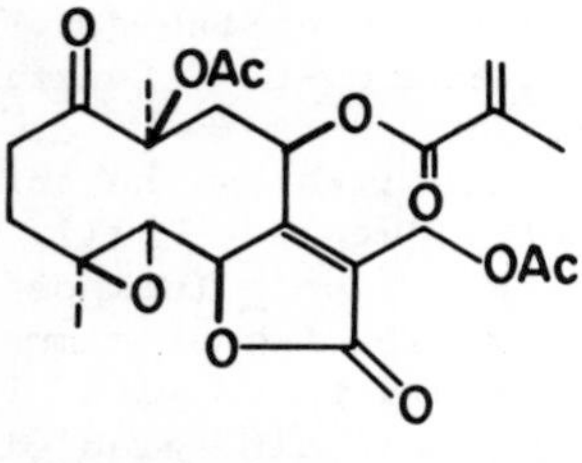

Glaucolide-A (*Vernonia*)

(I)

Larval Growth Inhibition Tests

Newly hatched larvae of the southern, fall, and yellowstriped armyworms, cabbage looper, and yellow woollybear were reared on diets containing glaucolide-A in amounts ranging up to 0.5%. The growth of the three armyworms was greatly reduced (Figure 1) and the reduction in growth for the fall and southern armyworms was considerably greater than the decrease in feeding. Moreover, except for the yellow woollybear, the

Table I. Insect Antifeedant Tests

Insects:	Diet Preference in Laboratory: Leaf Choice: Glaucolide-A Species	Leaf Choice: V. flaccidifolia	Pellet Choice: V. flac. With Added Glaucolide-A	Pellet Choice: V. flac. Without Glaucolide-A
Observed on Glaucolide-A Vernonias				
1. Yellow Woollybear	x (usually)			x
2. Cabbage Looper	x			x
3. Yellowstriped Armyworm		x		x
Not Observed On Glaucolide-A Vernonias				
1. Saddleback Caterpiller		x		*
2. Fall Armyworm		x		x
3. Southern Armyworm		x		x

*Not tested; culture died.

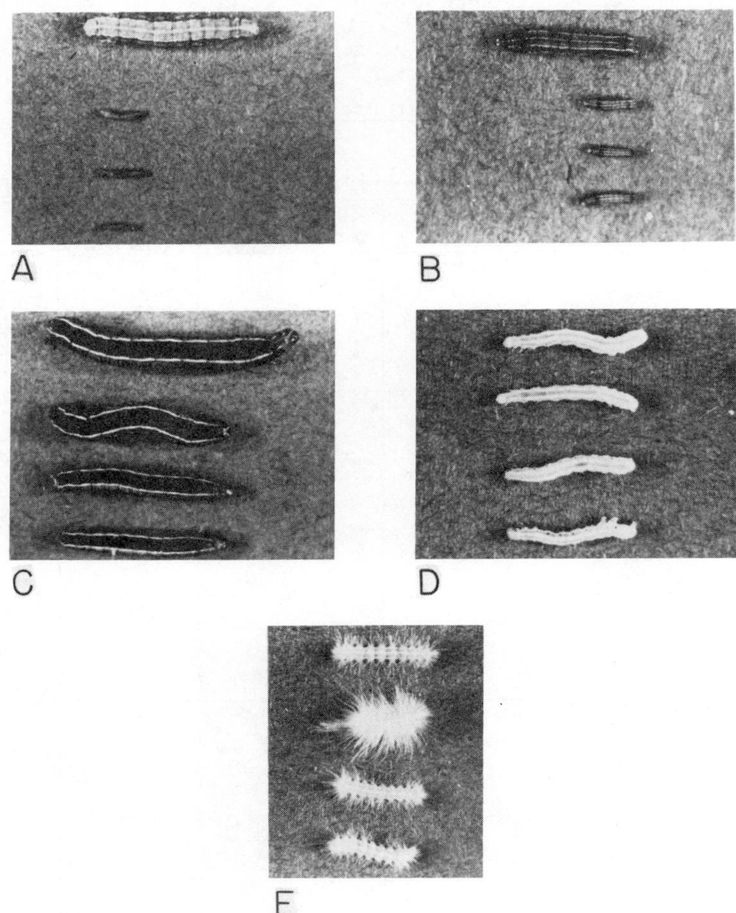

Figure 1. Relative size of larvae reared on diets containing glaucolide-A. Top insect fed on control diet without added lactone; others were fed on diets containing 1/8%, 1/4%, and 1/2% glaucolide-A by dry weight. (A) Fall armyworm; (B) southern armyworm; (C) yellow-striped armyworm; (D) cabbage looper; (E) yellow woolybear.

presence of glaucolide-A in the diets increased the number of days to pupation for all species. Reduced survival was especially evident for the southern and fall armyworms during the early larval instars (6).

Altogether, the growth inhibition tests demonstrated that glaucolide-A is detrimental to certain stages in the life cycle of several lepidopterous larvae.

In a separate series of experiments, it was also established that except for the fall armyworm, the presence of glaucolide-A on the leaves of Vernonias did not deter oviposition by the species of Lepidoptera tested (7).

Having established that glaucolide-A could deter the feeding of some but not all insect larvae in laboratory tests, both transplant garden and natural field plots of Vernonias were observed for the relative amounts of herbivory with respect to those plants which do and those plants which do not produce sesquiterpene lactones. Surprisingly, in all cases, *Vernonia flaccidifolia* was fed upon less than glaucolide-A containing species, a result for which we do not yet have a satisfactory explanation; however, we believe that this puzzling observation must be viewed in light of the mammalian antifeedant experiments described below.

Mammalian Antifeedant Tests

At the outset, it may be noted that cows avoided all Vernonias presumably on the basis of having previously determined by trial and error that most of these morphologically similar plants were bitter (due to the presence of sesquiterpene lactones).

Eastern cottontail rabbits and the whitetail deer were both found to feed primarily on *Vernonia flaccidifolia*, not the glaucolide-A species in controlled tests. Moreover, both avoided the *V. flaccidifolia* after its leaves had been coated with glaucolide-A; in this latter test, the deer would taste the coated *V. flaccidifolia*, detect the bitter compound, and then refuse to eat the plant. It was difficult to monitor mammalian feeding in the field since the glaucolide-A species are apparently not eaten at all by mammals; in contrast, we assume on the basis of our controlled feeding experiments with deer and rabbits that *V. flaccidifolia* in the field would be totally consumed if encountered.

From the laboratory insect feeding tests, it is evident that glaucolide-A can deter the feeding and induce growth reduction of some but not all insect larvae. Nevertheless, in the field its absence from *Vernonia flaccidifolia* did not result in heavier feeding by insects relative to *V. gigantea* and *V. glauca* (8). It is of interest to note that the two species of *Vernonia* in North America lacking sesquiterpene lactones, *V. flaccidifolia* and *V. pulchella*, are also species with restricted ranges; we suggest that mammalian feeding may account for these restricted ranges and small population sizes.

Summary

In summary, glaucolide-A apparently protects such species as *Vernonia gigantea* and *V. glauca* against all sympatric animals and some insects. In contrast, *V. flaccidifolia* has no chemical protection against mammals but nevertheless has evolved a still unknown but satisfactory defense against insects. Most sesquiterpene lactones are bitter suggesting that many members of this class of natural products serve as mammalian antifeedants. Nevertheless, insect antifeedant properties must have been important in the widespread selection of sesquiterpene lactones in the largest of flowering plant families, the Compositae.

Acknowledgements

Part of the research described here was supported by the National Science Foundation (Grant DEB 76-09320 to T.J.M.; GB-20687A and GB-39712 to S.B.J.), National Institutes of Health (Grant HDO 04488 to T.J.M.), and the Robert A. Welch Foundation (Grant F-130 to T.J.M.).

Literature Cited

1. Yoshioka, H., Mabry, T. J., and Timmermann, B. N., "Sesquiterpene Lactones", Univ. of Tokyo Press, Tokyo, 1973.
2. Burnett, W. C., Jr., Mabry, T. J. and Jones, S. B., Jr., "The Role of Sesquiterpene Lactones in Plant-Animal Coevolution", in "Biochemistry of Plant-Animal Coevolution", J. B. Harborne, ed., Academic Press, London, (in press).
3. Burnett, W. C., Jr., "Sesquiterpene Lactones -- Herbivore Feeding Deterrents in *Vernonia* (Compositae)". Ph. D. Dissertation, University of Georgia, Athens, 1974.
4. Burnett, W. C., Jr., Jones, S. B., Jr., Mabry, T. J., and Padolina, W. G., Biochem. Syst. Ecol., (1974), 2:25-29 .
5. Burnett, W. C., Jr., Jones, S. B., Jr., and Mabry, T. J., Taxon (in press), (1977a).
6. Burnett, W. C., Jr., Jones, S. B., Jr., and Mabry, T. J., Amer. Midl. Natur. (in press), (1977b).
7. Burnett, W. C., Jr., Jones, S. B., Jr., Mabry, T. J., and Betkouski, M. F., Oecologia (submitted), (1977c).
8. Burnett, W. C., Jr., Jones, S. B., Jr., and Mabry, T. J., Plant Syst. Evol. (submitted), (1977d).

13

Insect Antifeedants of *Spodoptera Litura* in Plants

KATSURA MUNAKATA

Laboratory of Pesticides Chemistry, Nagoya University, Nagoya, Japan

I. Introduction

The feeding behaviour of insects can be divided into four steps: (1) host plant recognition and orientation, (2) initiation of feeding, (3) maintenance of feeding and (4) cessation of feeding. Antifeedant is concerned with steps (2) and (3). The term antifeedant is difined as a chemical that inhibits feeding but does not kill the insect directly, the insect remaining near the treated leaves and dying through starvation. Gustatory repellent, feeding deterrent and rejectant are synonymous with antifeedant. Studies of the natural phenomenon of antifeeding (no consumption of plant material by insect) could reveal the presence of new antifeeding substances in plants, and provide correlations between chemical structure and antifeeding activities. Antifeeding could be of great value in protecting crops from noxious insects as an alternative insect control means to insecticides.

There are many investigations on synthetic chemicals which have insect antifeeding activity, but in this lecture the author would like to discuss the natural antifeeding substances contained in plants.

The host selection in phytophagous insects is governed by the presence or absence of attractants and repellents in plants [1]. For example, Buhr, Toball and Schreiber [2] reported that some of the alkaloid glycosides in plants of Solanaceae acted as repellents to the larvae of the Colorado potato beetle, Leptinotarse decemlineata Say. 2-Phenylethy-isothiocyanate from edible parts of turnip (Brassica rapa L.) and 5-allyl-1-methoxy-2, 3-methylene-dioxybenzene or myristicin from edible parts of parsnip (Pastinaca sativa L.) have been shown to act as naturally occuring antifeedants [3,4]. Moreover, one of the resistant factors in corn plants to the European corn borer, Pyrausta nubilalis (Hübner), was identified as 6-methoxybenzoxazolinone [5]. A waxy fraction from the extracts of wood of West Indian mahogany showed a high termite repellency[6]. Rudman and Gay [7] noted that 2-methyl-, 2-hydroxymethyl- and 2-formy-anthraquinones present in the extracts of teak heart-wood were all effective in inhibiting termite activity. The alkaloidal glycosides, such as leptine II and III, demmissine and tomatine, inhibited feeding of tomato beetles [8].

Jermy [9] studied extensively the botanical distribution of antifeeding substances in plants of 43 families in relation to eight insect species, and suggested that there was a fundamental difference in the function of chemoreceptors reacting upon deterrents.

The seeds of neem, the Indian lilac, _Melia azadirachta_ L. or _Melia indica_, have been shown to contain an antifeeding substance against locust [10]. The active principle was isolated and the chemical structure of the triterpenoid, named meliantriol, was established. However, the other antifeedant, named azadirachtin, from the same tree was isolated and its partial structure apart from meliantriol was reported by Butterworth and his group[11], and the whole structure was spectroscopically elucidated by Nakanishi and his group,[12] who also reported extensively army worm antifeedants from the east african _Warburgia_ plants to be the known poligodial, ugandensidiol, and the new warburganal which exhibit very strong antifeeding activityes against African army worms [13]

An extensibe survey for chemicals which as feeding deterrents against the tobacco cutworm, _Spodoptera litura_, has been conducted at the Laboratory of Pesticide Chemistry, Nagoya University, Nagoya Japan. The Tobacco cutworm is a noxious pest of sweet potatoes, sugar cane, crucifers, taro, and legumes, and control of this worm is very important to Japanese agriculture. However, certain plants have feeding-deterrent activity and thus resistance against this pest. Extracts of these plants were investigated in research directed toward the isolation and identification of the active principles.

II. Screening Method and Activities of Plant Extracts

Leaf discs were punched out with a cork borer from leaves of susceptible food plants, usually sweet potato, _Ipomoea batatas_. The discs were immersed in acetone solutions of test samples or in pure acetone as a control. After air-drying, the discs were placed in polyethylene dishes with test larvae, usually third instar, of the tobacco cutworm. About half the area of the control discs was usually eaten within 2 hr, at which time the consumed areas of all discs were measured by Dethier's method [14]. The consumed area of treated discs expressed as a percentage of the consumed area of control discs was used as an index of the antifeeding activity of the samples. Also, in some cases the minimum concentration of test samples needed to cause 100% antifeeding activity was determined[15]. Many kinds of plant leaves were collected, and their benzene extrats were assayed for their insect antifeeding activities by the " leaf-disk method" as above. The screening results are shown in Table I.

Table 1. Antifeeding activities of plant leaf extracts (conc. 5% acetone soln.)

Family	Species	Activity
Araceae	Symplocarpus foetidus Nutt. forma latissimus Makino	-
Araliaceae	Hedera rhombea Sieb. et Zucc.	+
Aristolochiaceae	Aristolochia Kaempferi Willd.	
	Aristolochia debilis Sieb. et Zucc	-
Caryophyllaceae	Dianthus japonicus Thunb	-
Caprifoliaceae	Sambucus Sieboldiana Blume.	-
Celastraceae	Euonymus japonica Thunb.	+
	Euonymus oxyphylla Mig.	-
Cephalotaxaceae	Cephalotaxus drupacea Sieb. et Zucc	-
Chloranthaceae	Chloranthus serratus Roem. et Schult.	-
Clethraceae	Clethra Barbinervis Sieb. et Zucc	-
Compositae	Achillea sibirica Ledeb.	+
	Crepidiastrum Keiskeahum Nakai	-
	Siegesbeckia pubescens Makino	-
	Erigeron linifolius Willd.	+
	Petasites japonicus Miq	+
Cornaceae	Helwingia japonica Dietr.	-
Elaegnaceae	Elaeagnus glabra Thunb.	-
Ericaceae	Enkianthus perulatus C.K. Schn	+
Euphorbiaceae	Euphorbia Sieboldiana Morr. Rt Decne.	-
	Sapium japonicum Paz et K. Hoffm	-
Labiatae	keiskea japonica Miq	+
	Teucrium viscisum Blume var. Miquelianum Makino	+
	Lamium album L. var. Barbatum Franch. et Sav.	-
Lauraceae	Cinnamomum Camphora Sieb.	-
	Parabenzoin praecox Nakai.	+
	Litsea glauca Sieb	-
	Benzoin umbellatum Rehd	+
	Benzoin obtusilobum O. Kuntze	+
Leguminosae	Pisum sativum L. var. arvense Poir.	
	Pueraria Thunbergiana Benth.	-
Liliaceae	Ophiopogon japonicus Ker. Gawl.	-
	Smilax China L.	-
Magnoliaceae	Magnolia obovata Thunb.	-
Menispermaceae	Stephania japonica Miers.	-
Moraceae	Ficus Carica L.	+
Osmundaceae	Osmunda Japonica Thunb.	-
Phytolaccaceae	Phytolacca esculenta Van Houtte	-

Table 1. Continued

Piperaceae	Piper Futokadzura Sieb. et Zucc	+
Polypodiaceae	Coniogramme intermedia hieron	-
	Cyrtomium falcatum L. Presl	-
	Cytomium Fortunei J. sm.	-
	Polypodium Thunbergianum Makino	-
	Pteridium aquilinum Kuhn.	-
Rosaceae	Agrimonia Eupatoria L. var. pilosa Makino	+
Rutaceae	Boenninghausenia albiflora. Reichb.	
Saxifragaceae	Hydranbea macrophylla Seringe subsp. serrata Makino var. acuminata Makino.	-
	Hydrangea macrophylla Seringe subsp. serrata Makino var. angustata Makino	+
Styracaceae	Styrax japonica Sieb. et Zucc.	-
Theaceae	Eurya japonica Thunb.	-
Verbenaceae	Caryopteris divardcata Maxim.	+
	Callicarpa japonica Thunb	+
	Lantana camara L.	-
	Vitextrifolia L. var. rotundirolia	-
Vitaceae	Ampelopsis Brevipedunculata Trautv.	+

III. Antifeeding substances in Plants

A. Antifeedants from Cocculus trilobus DC

Cocculus trilobus DC, which is well known as the host plant of Japanese fruit-piercing moths, Oraesia excavata Butler and O. emarginata Fabricius, is scarcely attacked by other insects in nature. Therefore, it was assumed that C. trilobus contained toxins or feeding inhibitors against other insects.

Two alkaloids were isolated as crystalline forms from the fresh leaves of this plant. An insecticidal alkaliod was named cocculolidine (II), and an antifeedant alkaloid was identified as isoboldine (I) by the comparison of its physical properties with those of an authentic sample. The authors proposed structure II for cocculolidine.

To determine the threshold concentration of isoboldine for feeding inhibition, serial acetone dilutions of pure isoboldine were applied to leaves of Euonymus japonicus Thunberg, and the treated leaves were submitted to the leaf-disc test with Abraxas miranda Butler. The feeding ratios were near zero at concentrations of 200 p.p.m. or greater; at concentrations of 100 and 10 p.p.m. the feeding ratios were 40-59% and 48-126%, respectively.

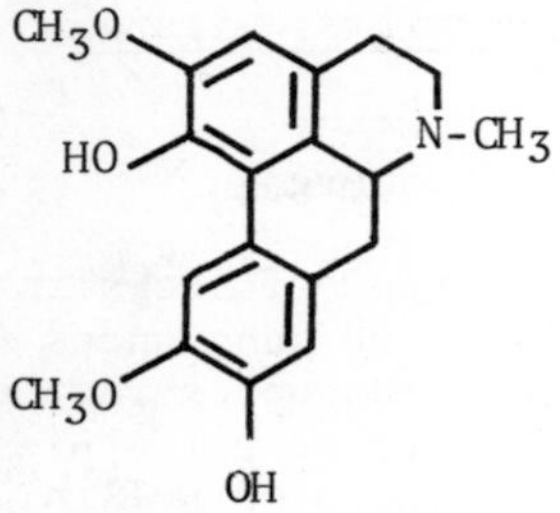

(I) Isoboldine 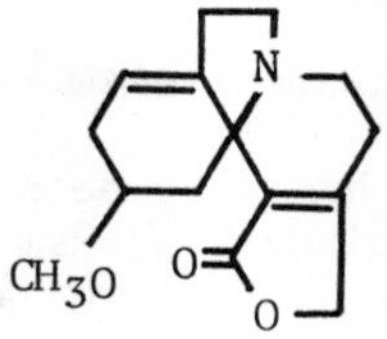 (II) Cocculolidine

The leaf-disc test with Spodoptera litura F. was conducted with 100 and 200 p.p.m. acetone solutions of pure isoboldine. Leaves of sweet potato, one of the many host plants of this insect, were used as feeding discs. This alkaloid showed feeding inhibitory activity at 200 p.p.m. The leaf-disc test with O. excavata eating leaves of C. trilobus was conducted using 500 and 1000 p.p.m. acetone solutions of pure isoboldine, but the insects ate up the treated leaf discs. These facts suggest an interesting hostinsect interrelationship about the constituents in host plants [16].

B. Antifeedants of Parabenzoin trilobum Nakai (=Lindera triloba Blume)

Leaves of P. trilobum ('Shiromoji' in Japanese) are not attacked by the larvae of Spodoptera litura. The crude extract also showed antifeeding activity in the leaf-disc test. The active antifeedant material was benzene-extractable, neutral, and eluted by 30% ether in n-hexane from a silica-gel column. Two active compounds were isolated from the extract and named shiromodial diacetate (III) and shiromodiol monoacetate (IV) and their chemical structures were assigned as (III) and (IV), [17,18]

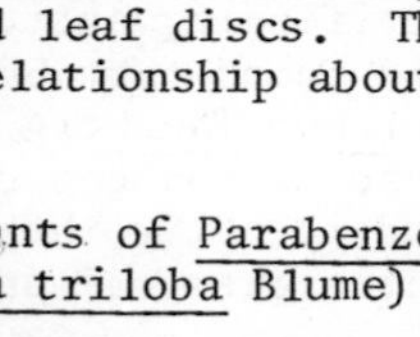

(III) Shiromodiol diacetate. $R_1=R_2=Ac$
(IV) Shiromodiol monoacetate. $R_1=H;R_2=Ac$

The threshold concentration of IV was about 0.125% and 0.033% in acetone when tested with S. litura and Abraxas miranda, respectively [17].

C. Antifeedants from Orixa japonica Thunberg

This plant has long been used to insect-proof books in Japan, and after preliminary assessment of antifeeding substances for tobacco cutworm, Spodoptera litura F., the benzene extract of the leaves showed antifeeding activities. The active principles in the leaves were separated and identified as isopimpinellin, bergapten and kokusagin. These and related compounds were assayed for antifeeding activities, and it was revealed that isopimpinellin and bergapten are rather potent [19].

D. Antifeeding substances in Clerodendron tricotomum Thunberg

In 1962 T. Miyake investigated antifeeding activities of the hot water extracts from the fresh leaves of various plants. Euproctis pseudoconspersa Strand and Cicadella viridis Linn., and particularly the extracts of Clerodendron tricotomun Thunberg (Verbenaceae), showed a strong antifeeding activity [20]. We have reinvestigated the matter and found that benzene extracts from air-dried leaves of C. tricotomum Thunberg showed distinct antifeeding activity for the larvae of tobacco cutworm, Spodoptera litura F. In order to isolate and identify the active principle present in the species, a systematic chemical analysis monitored by leaf-disc test was undertaken.

Purification by column chromatagraphy over alumina and silica-gel gave two crystallines named clerodendrin A and B. Clerodendrin A and B were determined as (VI) [21] and (VII) [22], both having a clerodon skeleton [23], from the chemical and spectroscopic evidence. The absolute configuration of clerodendrin A was confirmed as (VI) by x-ray crystallography of the p-bromobenzoate chlorohydrin derivative [21]. The structure of clerodin (V), a bitter principle of the Indian bhat tree, Clerodendron infortunatum, had been established in 1961 [23].

Clerodendrin A and B, having slightly bitter taste, showed 100% antifeeding activities at 300 and 200 p.p.m. concentrations, respectively [24]. Clerodin exhibited a 100% antifeeding activity at 80 P.P.M.

These diterpene dervatives having a clerodon skeleton showed bitter taste. However, there seems to be no linear relationship between the bitter taste and the antifeeding ability for the insect larvae, since the polyol derivatives with weak activities showed slightly bitter taste, whereas the methanol adduct derivative, which does not manifest a marked bitterness, showed the strongest antifeeding activity at 15 p.p.m. Furthermore, it was revealed that the larvae did not bite the treated leaves, and eventually starved to death, when the concentration of the test solution was raised above twice the concentrations showing the 100% antifeeding

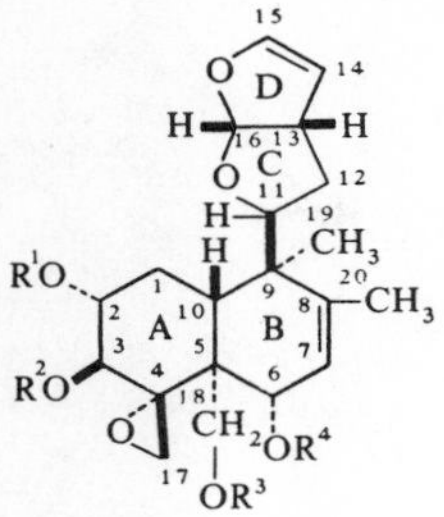

(VI) $R^1 = H$

$R^2 = C_2H_5C(CH_3)(OAc)CO$-

$R^3 = R^4 = Ac$

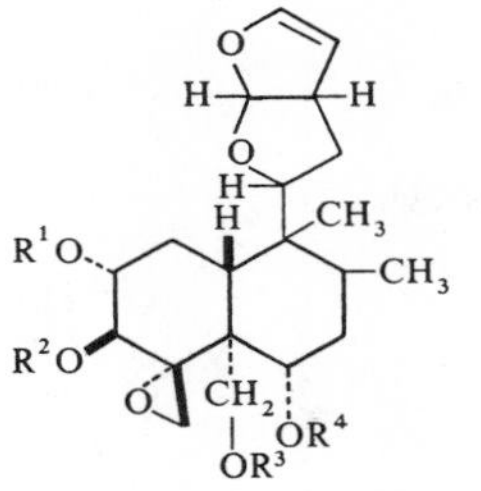

(VII) $R^1 \sim R^4$ = same groups as in (VI)

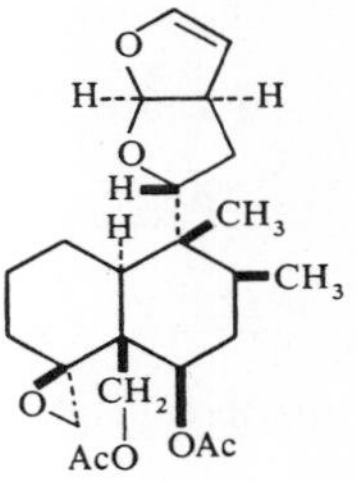

(V)

activities with the exception of the polyol and acetate derivatives. Although the exact mode of action has not been pinpointed, both 'by smell' and 'by taste' are possible.

Clerodendrin A and B did not inhibit feeding of the larvae of *Calospilos miranda* at concentrations below 5000 p.p.m. and the larvae of the European corn borer, *Ostrinia mubilalis* Hubner, showed 100% antifeeding activities at the concentration of 5000 p.p.m. The principle also prevented feeding of the larvae of oriental tussock moth, *Euproctis subflava* Bremer, at the level of 1000 p.p.m. These limited results suggest that these compounds may act as antifeedants for the larvae of polyphagous rather than monophagous insects.

E. Antifeeding substances in *Verbenaceae*

After the discovery of insect antifeeding substances from *Clerodendron* plants, constituents of *Vervenaceae* plants have interested us for screening of insect antifeedants.

Plants of the family *Vervenaceae*, three in Japan, 13 in Taiwan, were collected, and their benzene extracts were subjected to the leaf-disc test of tobacco cutworm. The antifeeding activities of the benzene extracts are shown in Table 2. *Caryopteris divaricata*, *Callicarpa japonica*, *Clerodendron fragrans*, *Clerodendron calamitosum*, *Clerodendron cryptophyllum* showed strong antifeeding activities at One percent concentration is the leaf-disc test.

Table. 2 Antifeeding activity of *Verbenaceae* plants

Species		Activity
Caryopteris divaricata	(leaf)	+ + + +
Callicarpa japonica	(leaf)	+ + + +
Clerodendron fragrans	(leaf)	+ + + +

Table 2. Continued

	(stem)	+
C. inerme	(stem)	+
C. paniculatum	(leaf)	+
	(stem)	+
C. calamitosum	(leaf)	+ + + +
	(stem)	+ + +
C. cryptophyllum	(leaf)	+ + + +
	(stem)	+
C. paniculatum var. *albiflora*	(stem)	-
Callicarpa formosana	(leaf)	+
	(stem)	+
Duranta repens	(leaf)	-
	(stem)	-
Lantana camara	(reaf)	-
Lippia nodiflora	(whole plant)	-
Premna integrifolia	(leaf)	-
	(stem)	-
Vitex negundo	(stem)	-
V. trifolia	(seed)	+ + +
Verbena officinalis	(whole plant)	-

From these plants many clerodon compounds were isolated and their chemical structures were investigated.

The extract of *Caryopteris divaricata* Maxim showed a strong antifeeding activity for the larvae, and eight antifeeding diterpenes having a clerodon skeleton-clerodin (I) [23], caryoptin (II), dihydroclerodin-I(V), dihydrocaryoptin (VI), clerodin hemiacetal (VII), caryoptin hemiacetal (VIII), caryoptionol (IX) and dihydrocaryoptionol (X)-were isolated and identified [25,26].

Clerodendron fragrans gave no strong antifeedants, but a weak antifeedant phytol in a high concentration (0.2% yield on dried basis) was isolated from the active fractions, and identified as phytol. *C. calamitosum* gave a new antifeedant named 3-epicaryoptin

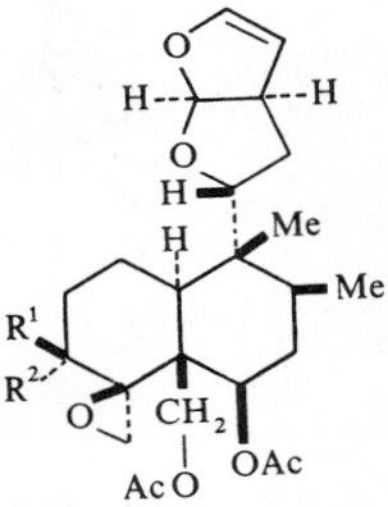

I $R^1 = R^2 = H$
II $R^1 = OAc, R^2 = H$
IX $R^1 = OH, R^2 = H$
XI $R^1 = H, R^2 = OAc$

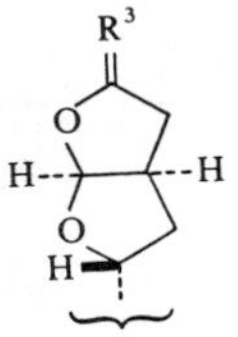

V $R^1 = R^2 = H, R^3 = H_2$
VI $R^1 = OAc, R^2 = H, R^3 = H_2$
VII $R^1 = R^2 = H, R^3 = , OH$
VIII $R^1 = OAc, R^2 = H, R^3 = H,OH$
X $R^1 = OH, R^2 = H, R^3 = H_2$

(XI), and from C. cryptohpyllum clerodendrin-A (XII)[23] was isolated and identified to be the antifeeding substance [27].

It is interesting that the compounds containing the clerodon skeleton are found in both Clerodendron and Caryopteris species. Moreover, it is interesting in view of biogenesis that 3-epi caryoptin rather than caryoptin is observed in C. calamitosum.

The 11 antifeeding substances and one derivative obtained were tested against the larvae of S. litura at different concentrations. The antifeeding activity was embodied as the concentration of sample solution which exhibited the 100% antifeeding avtivity within the defined time (2h) (Table 3).

Table 3. Activities of Verbenaceae antifeedants against S. litura

Compound	p.p.m.[a]	Compound	p.p.m.[a]
Phytol	5000	Clerodin hemiacetal	50
Caryoptin	200	Caryoptinol	200
Dihydrocaryoptin	80	Dihydrocaryoptinol	100
Caryoptin hemiacetal	200	3-Epicaryoptin	200
Clerodin	50	Clerodendrin-A	200
Dihydroclerodin-1	50	3-Epidihydrocaryoptin[b]	100

a: The concentration of the sample solution showing the 100% antifeeding activity within 2h.
b: Catalytic reduction product of X.

The antifeeding activities of a series of clerodin derivatives are much stronger than those of the caryoptin derivatives. The difference in the activities may be attributed to the presence of 3β-acetoxyl group in ring A. It can be concluded that the functional groups in ring A contribute greatly to the antifeeding activity of these diterpenes.

In addition, the feeding test was allowed to continue for 6h and more at the concentration showing the 100% antifeeding activity within 2h. The test leaves treated with the antifeeding diterpenes were not bitten after 24h and more, and the larvae eventually starved to death. Therefore , the term absolute antifeedant is applicable to these diterpene compounds. The term relative antifeedant is used of compounds retarding the feeding of larvae only for the defined time.

F. Antifeedants in Parabenzoin praecox and Piper futokazura

The benzene extracts of Parabenzoin praecox gave two kind of insect antifeeding substances and they were identified as (+)-epieudesmin (I) and (+)-eudesmin (II). Piper futokazura also gave two compounds and they were identified to be isoasaron (III) and a new compound named piperenone (IV), and antifeeding activities are shown in Table 4. [28,29].

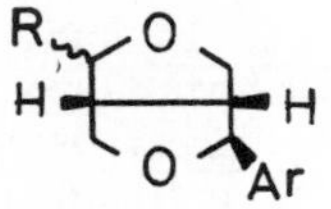

I. R=''''Ar

II. R=◄Ar

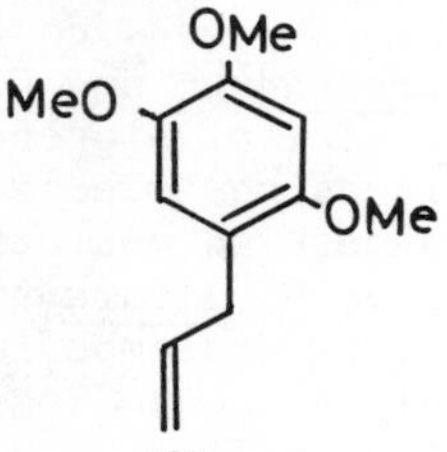

III

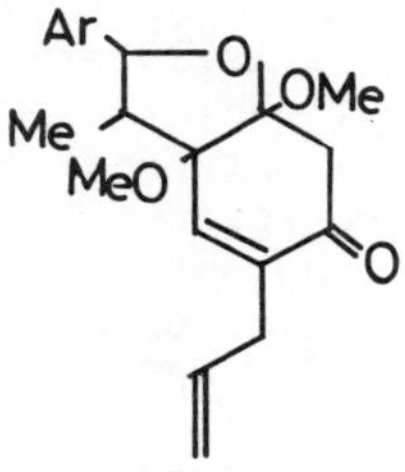

IV

Ar= 3,4-Dimethoxyphenyl

Table 4. Antifeeding Activities of I, II, III and IV

Comp.	Concentration of sample solution (%)						
	1	0.5	0.1	0.05	0.01	0.005	0.001
I	+ + +a)	+ +	+ + +	+ + +	+	+	
II	+ + +	+ +	-				
III	+ + +	+ + +	-				
IV				+ + +	+ + +	+ + +	-

a) Feeding ratio = $\frac{\text{Consumed amount of sample disk}}{\text{Consumed amount of control disk}}$

+ + + (0-10%), + + (10-30%), + (30-50%), - (50%).

I and II, which are epimers, showed 90-100% antifeeding activities at different concentration of 0.05 and 1%, respectively. III showed 90-100% antifeeding activities at 0.5% and IV at 0.005%. These structures are phenyl propanoids. The whole absolute structure of IV was recognized by X-ray crystallography and chemical experiments. [30,31]

IV. Conclusions

Many natural products of plants possess insect antifeeding activity. These compounds probably play a role as resistance factors, protecting the plants against insect attack.

Based on our continuous experience with isolation and identification of insect antifeedants from plants we have identified many new natural products having insect antifeeding activities. We have noted several general characteristics of these compounds:

1. Some compounds show relative and others show absolute antifeeding activity.
2. The bitter taste of a compound may have no relation to its antifeeding activity.
3. Naturally occuring antifeedants may be systemic in plants and decomposable in the environment-both advantageous characteristics.
4. Compounds having insect antifeeding activity can and should be used as genetic index substances in selective breeding programs for insect-resistant plant varieties.

Thus, there are two ways in which insect control may be effected by insect antifeeding substances:

1. The development of powerful synthetic insect antifeedants
2. The development of resistant crop varieties that naturally incorporate the antifeeding substances

When used for insect pest management antifeedants might be especially advantageous because they control the insects indirectly through starvation, and they may not be harmful to parasites, predators, and pollinators. If crops were sprayed with efficient antifeedants, perhaps the pests would turn from the crops to weeds.

REFERENCES

1. A.J. Thorsteinson, Ann. Rev. Entomol. 5, 193 (1960)
2. H.Buhr, R.Toball and K.Schreiber, Ent. Exp. Appl. 1, 209 (1958)
3. E.P. Lichtenstein, R.M. Strong and D.G. Morgan, J.Agric. Fd. Chem. 10, 30 (1962)
4. E.P. Lichtenstein and J.E. Casida, J. Agric. Fd. Chem. 11, 410 (1963).
5. E.E. Smissman, J.B. Lapidus and S.D. Beck, J. Amer. Chem. Soc. 79, 4697 (1957).
6. C.F. Asenjo, L.A. Marin, W, Torres and A, Campillo, Chem. Abstr. 53, 22707 (1959).
7. R. Rudman and F.J. Gay, Hortzforschung, 15, 117 (196)
8. F. Strarchoe and I. Loaw, Ent. Ezp. Appl. 4, 133 (1961)
9. T. Jermy, Ent. Exp. Appl. 9, 1 (1966)
10. D. Lavie, M.K. Jain and S.R. Shpan-Gabrielit Gabrielith, Chem. Comm. 1967, 910
11. J.H. Butterworth, E.D. Morgan and G.R. Percy, J.C.S. Perkin 1, 2445 (1972).
12. P.R. Zanno, I.Miura, K. Nakanishi, and D.L. Elder, J. Am. Chem. Soc., 97 1975 (1975).
13. I. Kubo, Y.N. Lee, M. Pettei, F. Pilkiewicz, and K. Nakanishi, J.C.S. Chem. Comm. 1013 (1976)
14. V.G. Dethier. Chemical Insect Attractants and Repellents, P. 210. Blakiston; Philadelphia (1974).

15. K. Munakata. 1970. Insect antifeedants in plants. In Control of Insect Behaviour by Natural Products. D.L. Wood, R.M. Silverstein, and M. Nakajima, Eds. New York, Academic, pp. 179-187.
16. (a) K. Wada and K. Munakata, Agric. Biol. Chem. 31, 336 (1967)
(b) K. Wada, S. Marumo and K. Munakata, Agric. Biol. Chem. 31, 452 (1967)
(c) K. Wada, S, Marumo and K. Munakata, Agric. Biol. Chem. 32, 1187 (1968).
(d) K. Wada and K. Munakata, J. Agric. Fd. Chem. 16, 471 (1968)
17. K. Wada, Y. Enomoto, K. Matsui and K. Munakata, Tetrahedron Letters, No. 45, 4673 (1968).
(b) K. Wada and K. Munakata, Tetrahedron Letters, No 45, 4677 (1968).
(c) K. Wada, K. Matsui, Y. Enomoto, O. Ogiso and K. Munakata, Agric. Biol. Chem. 34, 942 (1970).
(d) K. Wada, Y. Enomoto and K. Munakata, Agric. Biol. Chem. 34, 947 (1970).
18. R.J. McClure, G.A. Sim. P. Caggon and A.T. Mcphail, J.C.S. Chem. Comm. 128 (1970).
19. T. Yajima, K. Tsuzuki, N. Kato and K. Munakata, Abstract of Papers, Annual Meeting of the Agricultural Chemical Society of Japan, Tokyo, April, 1969, P.145.
20. T. Miyake, Private communication.
21. N. Kato, S. Shibayama, K. Munakata and C. Katayama, Chem. Commun. 1632 (1971).
22. N. Kato and K. Munakata Unpublished results.
23. D.H.R. Barton, H.T. Cheung, A.D. Cross, L.M. Jacman and M. Martin-Smith, Proc. Chem. Soc. 76 (1971); J. Chem. Soc. 5061 (1961); D.A. Sim, T.A. Hamor, I.C. Paul and J.M. Robertson, J. Chem. Soc. 75 (1961); I.C. Paul, D.R. Sim, T.R. Hamor and J.M. Robertson, J. Chem. Soc. 4133 (1962)
24. N. Kato, M. Takahashi, M. Shibayama and K. Munakata, Agric. Biol. Chem. 36. 2579 (1972).
25. S. Hosozawa, N. Kato and K. Munakata, Phytochemistry, 12, 1833 (1973).
26. S. Hosozawa, N. Kata and K. Munakata, Agric. Biol. Chem. 38, 823 (1974).
27. S. Hosozawa, N. Kato, K. Munakata and Yuh-Lin Chen, Agric. Biol. Chem. 38, 1045 (1974)
28. K. Matsui, K. Wada and K. Munakata, Agr. Biol. Chem., 40, 1045 (1976).
29. K. Matsui and K. Munakata, Tetrahedron Letters, 1905 (1975).
30. K. Matsui and K. Munakata, Agr. Biol. Chem., 40, 1113 (1976).
31. K. Matsui, Y. Katsube and K. Munakata, Bull. Chem. Soc. Japan, 49, 62 (1976).

14

Natural Insecticides from Cotton *(Gossypium)*

ROBERT D. STIPANOVIC, ALOIS A. BELL, and MAURICE J. LUKEFAHR

National Cotton Pathology Research Laboratory, U. S. Department of Agriculture, Agricultural Research Service, College Station, TX 77843

Resistance has been defined as the ability of a plant to prevent, restrict or retard partially the injurious effects of a pest (1, 2). The mechanisms of resistance are defined as: a) tolerance -- the plant suffers only slight injury despite a pest population that severely damages a susceptible plant (1, 2); b) nonpreference -- the plant is unattractive for feeding, reproduction, or shelter (1); c) antibiosis -- the plant adversely affects the biology or ecology of the pest because of toxic chemicals in the plant tissue (1); and d) hypersensitivity -- the plant inactivates and localizes the invading pest by rapid morphological and histochemical changes that cause the invaded tissues to die prematurely (2, 3).

A plant breeder may employ any or all of these mechanisms in order to obtain the most desirable agronomic characteristics. Historically, plant lines have been improved by performance and observations in the field. Until recently there was no concerted effort to understand the chemical bases of pest resistance. However, as early as 1906 Cook suggested that "oil glands" in the cotton plant might be involved in resistance to cotton insects (4). Although the term "antibiosis" had not been coined, Cook recognized this mechanism of pest resistance in cotton.

Heliothis Resistance in Cotton.

In the early 1960s, the resistance of the *Heliothis* complex (cotton bollworm, *H. zea* and tobacco budworm, *H. virescens*) to insecticides was recognized. In 1965 research was initiated to discover new sources of host plant resistance to these pests. Wild race stocks and "backyard" plantings from Mexico, Central and South America, and the Caribbean Islands were investigated. Over 1200 different races of *Gossypium hirsutum* were tested of which 78 had some resistance. In general, resistance among races increased with increasing numbers of glands and with increasing gossypol content (5, 6).

Gossypol (I), a dimeric sesquiterpenoid found in cottonseed pigment glands, was usually measured in leaves or flower buds by the spectrophotometric method of Smith (7). This procedure is based on the reaction of aniline with gossypol, which presumably

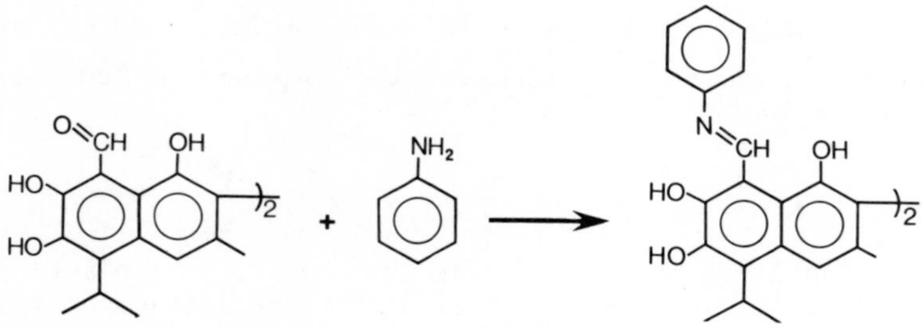

I

occurs by a condensation reaction of the amine group of aniline with the aldehyde group in gossypol. A yellow color is produced and the extinction of light at 445 nm is measured. Aniline is not a specific reagent for gossypol, because it will react with other aromatic aldehydes, giving products which are also yellow in color.

Pigment glands are located in most tissues of the cotton plant, including foliar parts and the seed (8, 9). They are absent from the seed coat xylem. In the foliar parts, the glands are located below the epidermis and hypodermis. The "oil" in these glands is composed of low molecular weight, volatile terpenoids that solubilize higher molecular weight pigments such as gossypol. Certain *Heliothis*-resistant cotton varieties contain more pigment glands than susceptible varieties. The resistant varieties, thus, are called high glanded or high gossypol cottons (5, 6, 10, 11).

The toxicity of gossypol to *Heliothis* spp. was demonstrated (9, 11), and gossypol therefore was considered the active component in the glands of the cotton flower bud and boll. Most commercial varieties of cotton contain 0.5% gossypol in flower buds (dry weight), as measured by the method of Smith (7). In laboratory tests, 1.2% gossypol significantly inhibited larval growth and development of *Heliothis* (12).

Recently, certain wild or primitive cottons were found to exhibit more insecticidal activity than could be accounted for by the gossypol concentration alone. The toxic activity was generally ascribed to "X-factors" (13, 14, 15). The "X-factors" have been identified as the sesquiterpenoids, hemigossypolone (II) and hemigossypolone-7-methyl ether (III), and the derived terpenoids, heliocides H_1 (IV), H_2 (V), H_3 (VI), H_4 (VII), B_1 (VIII), B_2 (IX), B_3 (X) and B_4 (XI). The "H" and "B" designations indicate that these compounds were first found in *G. hirsutum* and *G. barbadense* cottons, respectively.

Detection, Isolation and Location of the Heliocides in *Heliothis* Resistant Cottons.

Freeze dried (0.5 gm) whole cotton flower buds are ground in a blender and extracted with EtOAc:Hex (1:3). This extract is spotted directly on Si gel plates and developed with Et_2O:Hex:HCOOH (15:84:1). The plates are then sprayed with phloroglucinol reagent (equal volumes of 5% phloroglucinol in 95% EtOH and conc. HCl) to detect terpenoid aldehydes. Hemigossypolone, hemigossypolone-7-methyl ether and gossypol form magenta to maroon colors, and the heliocides form orange colors.

Susceptible normally glanded cultivars of *G. hirsutum* contain heliocides H_1, H_2, H_3, H_4, hemigossypolone and gossypol. The resistant high glanded cultivars of *G. hirsutum* contained the same compounds but the concentrations, particularly of H_1 and H_4, are as much as three times higher than that found in susceptible normally glanded cotton (16).

The foliar parts of glandless cottons do not contain terpenoid aldehydes. Thus the terpenoid aldehydes appear to be located specifically in the glands. Unpublished histochemical tests confirm this. It is of interest to note that glandless cottons are highly susceptible to *Heliothis* spp. and many other insects (17).

G. barbadense cottons form methyl ether derivatives of gossypol and its biosynthetic precursors in root bark (18, 19, 20). *G. barbadense* also synthesizes terpenoid methyl ethers in its foliar tissues. For example, the *G. barbadense* cultivar Seabrook Sea Island contained hemigossypolone-7-methyl ether and the methylated heliocides B_1 and B_4 in slightly higher concentrations than hemigossypolone, heliocide H_1 and H_4.

Heliocides were prepared in quantity from young bolls (2-3 days-old) and bracts of field grown plants. Tissues were lyophilized and ground to powder in a blender. The powder was extracted and heliocides were purified by chromatography as previously described (21, 22).

Characterization of Cotton Terpenoids.

Structure Determination. Spectroscopic methods (UV, IR, MS, ^{1}H-NMR and ^{13}C-NMR), X-ray crystal analysis, and synthesis were employed in determining the structures of heliocides from cotton. Heliocide H_2 (V) was the first C_{25}-terpenoid analyzed (23). It was synthesized by a Diels-Alder reaction of hemigossypolone and myrcene. The product which crystallized was identical (UV, IR, MS, m.p., mixed m.p.) to heliocide H_2 from pigment glands. The ^{13}C-NMR data indicated that the side chain was located as shown in V. To prove this, the dibromide was prepared and submitted to X-ray crystal analysis (Fig. 1). This unequivocally established the position of the side chain and proved that the cyclohexene ring was *cis*-fused. Because heliocide H_2 was synthesized from hemigossypolone, the X-ray crystal anlaysis also proved that

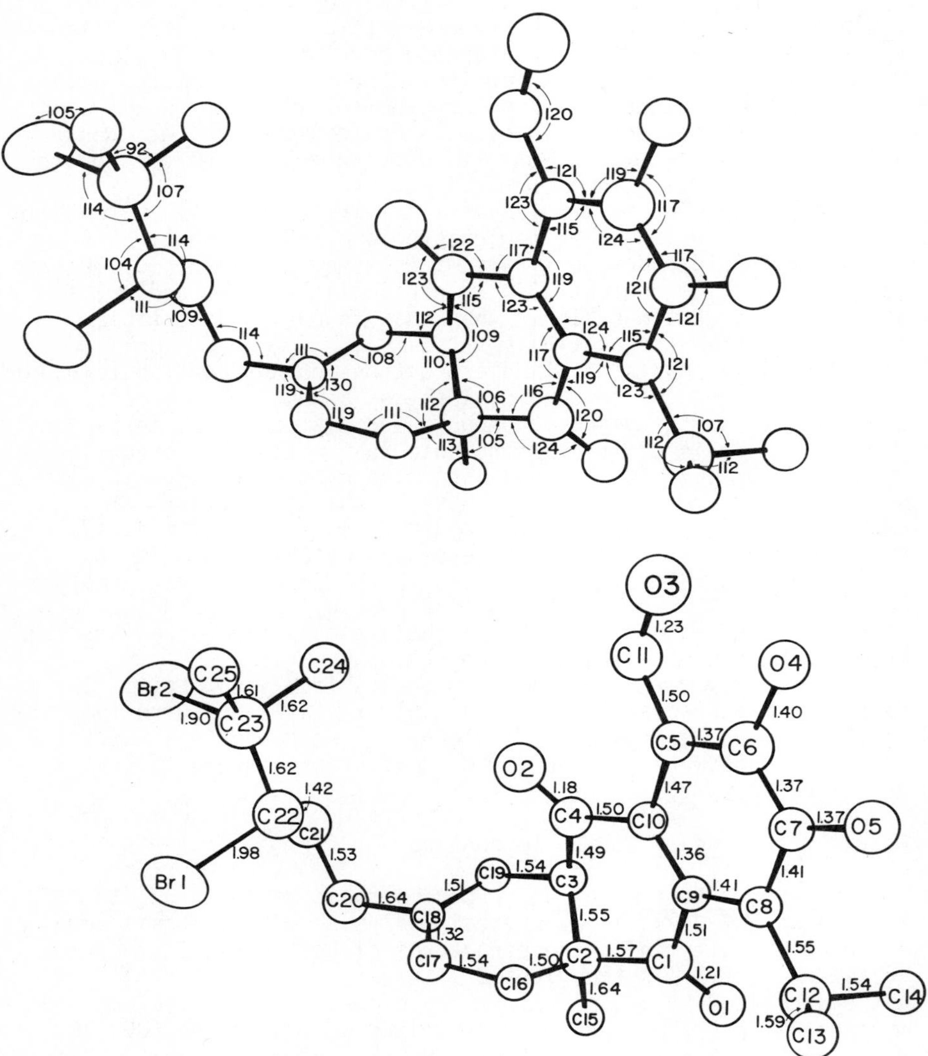

Figure 1. The confirmation of the dibromide derivative of heliocide H_2 as determined by x-ray crystal analysis

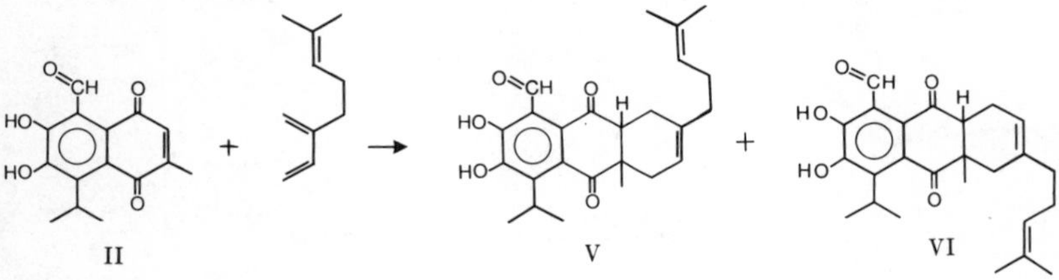

II V VI

hemigossypolone is a *para*-napthoquinone with hydroxyls at C6 and C7 (24, 25).

The mother liquors from the Diels-Alder reaction of hemigossypolone and myrcene yielded another C_{25}-terpenoid which was identical (UV, IR, MS, m.p. and mixed m.p.) to heliocide H_3 from pigment glands (21). The ^{13}C-NMR spectrum of heliocide H_3 was very similar to that of heliocide H_2 but indicated a difference in the location of the side chain. That is, the major chemical shift changes were at the methylene carbons of the cyclohexene ring, and were in a direction commensurate with the side chain located as shown in structure VI.

Heliocides H_1 and H_4 were synthesized by the Diels-Alder reaction of hemigossypolone with *trans*-β-ocimene (22). Heliocides H_1 and H_4 do not revert to starting material at room temperature

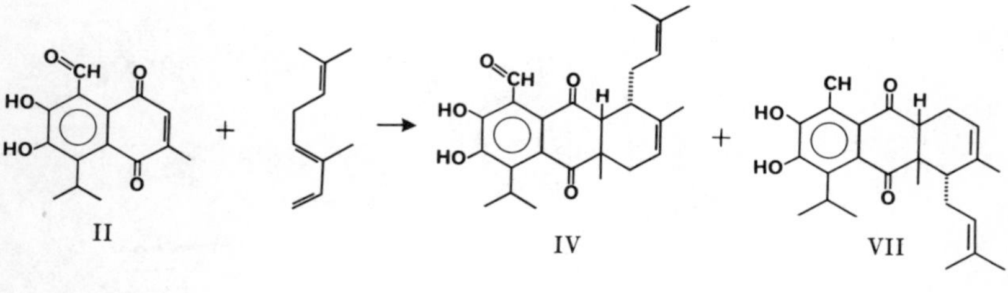

II IV VII

(i.e., they do not undergo the reverse Diels-Alder reaction). Therefore, they are the products of a kinetically controlled reaction that proceeds through an *endo*-transition state giving a *cis*-fused product (26). Further, *trans*-1-substituted butadienes and quinones form adducts in which the diene substituent is *trans* to the bridgehead substituents on the quinone ring (27, 28). Thus heliocides H_1 and H_4 have *cis*-fused ring systems with *transoid* isopentenyl side chains as shown for structures IV and VII.

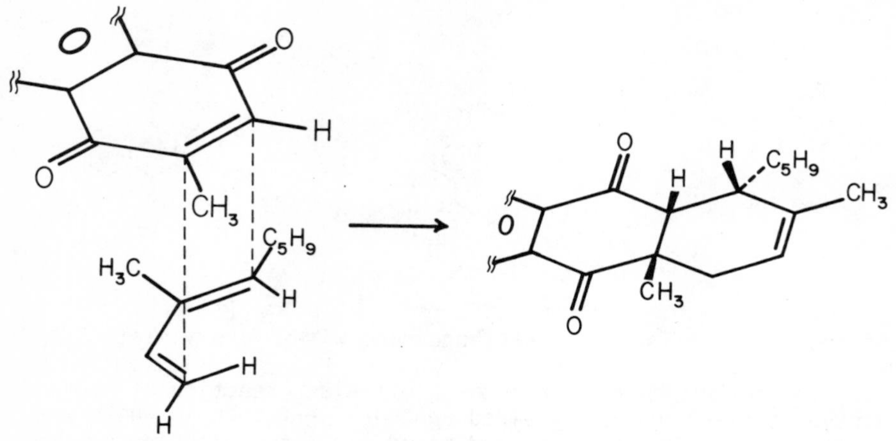

The ^{13}C-NMR spectra of heliocides H_1 and H_4 were compared to those of heliocides H_2 and H_3. On the basis of these comparisons heliocides H_1 and H_4 were assigned structures IV and VII, respectively.

Heliocides B_1 (VIII) and B_4 (XI) are found in *G. barbadense* [Seabrook Sea Island 12B2 (SBSI)]. They were synthesized by the Diels-Alder reaction of hemigossypolone-7-methyl ether (III) and *trans*-β-ocimene (29). The structures were assigned by comparing

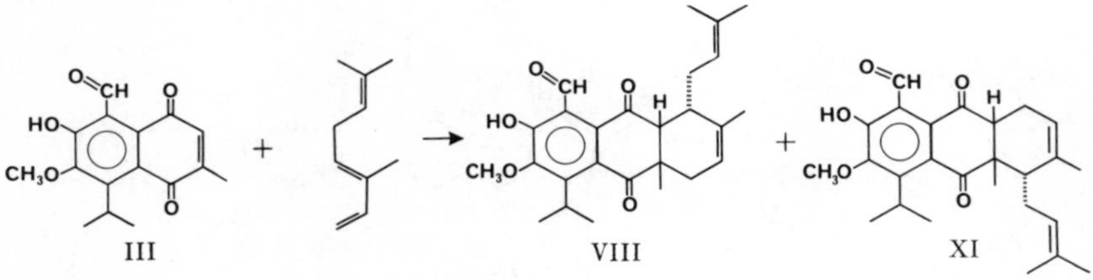

III VIII XI

the ^{13}C-NMR spectra of heliocides B_1 and B_4 with heliocides H_1 and H_4.

Heliocides B_2 (IX) and B_3 (X) have been synthesized from hemigossypolone-7-methyl ether (III) and myrcene. These compounds are apparently not found in SBSI, but they have been tentatively identified in F_2 progeny from crosses of *G. hirsutum* and *G. barbadense* (30).

III + → IX + X

<u>^{1}H-NMR Screening Techniques</u>. To breed plants which contain the most effective mixture of terpenoid aldehydes, a rapid method of screening was desirable. We have developed a ^{1}H-NMR technique that differentiates most of the terpenoid aldehydes and estimates their concentrations in crude extracts. The aldehyde proton of the quinones, heliocides and gossypol appear in the region between δ 10 and 12. Examples of the ^{1}H-NMR (δ 10 to 12) of extracts from a high gossypol, *G. hirsutum* cotton (HG-6-1-N) and a *G. barbadense* cotton (SBSI) are shown in Fig. 2 and Fig. 3, respectively. Acetone was used as the solvent because it gave the best peak resolution. Pure terpenoids were added to the crude extracts to determine which compound(s) corresponded to each peak. Heliocides H_2, H_3, B_2, and B_3 could not be resolved completely from each other by this technique. However, the other heliocides, quinones and gossypol are all resolved, and their concentrations may be estimated. Unfortunately, the chemical shifts of the peak change slightly with increases in concentration; consequently, resolution is inversely proportional to the concentration.

<u>TLC Screening Techniques</u>. The ^{1}H-NMR method of screening plant progeny requires about 3 gm of freeze dried powder. A TLC method that requires a single leaf has been developed (<u>17</u>). This method uses fresh terminal leaves, allowing terpenoid quantitation before flowering. A three-man team can screen 240 progeny/day. An extract from the leaf is spotted on a Si gel plate and developed with Hex:Et_2O:HCOOH (84:15:1). The sequential appearance of the compounds going up the plate are gossypol, hemigossypolone (R_f ca. 0.30), heliocides H_2, H_3 and H_4 (mixed as one spot), hemigossypolone-7-methyl ether mixed with heliocide H_1, heliocides B_2, B_3, and B_4 (mixed as one spot), and heliocide B_1. The individual spots are visualized by spraying with phloroglucinol reagent. Gossypol forms rose spots, hemigossypolone and hemigossypolone-7-methyl ether form magenta to maroon colors, and the heliocides give orange spots.

This method separates heliocides H_2 and H_3 from heliocides

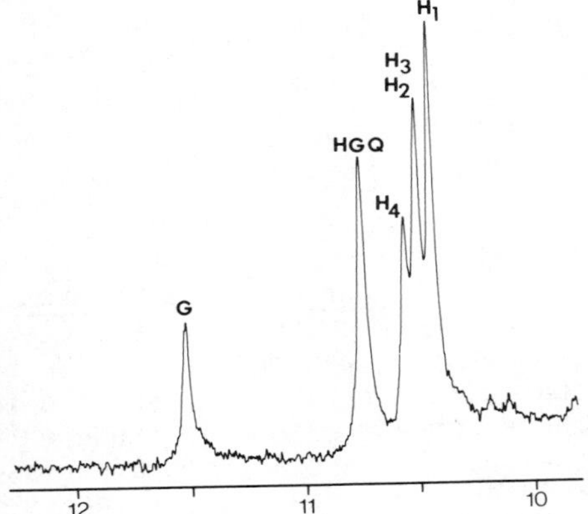

Figure 2. ¹H NMR (δ 10–12) of an ethyl acetate:hexane (1:3) extract of freeze dried 3–4-day-old bolls with bracts from G. hirsutum. *G = gossypol; HGQ = hemigossypolone; H_1, H_2, H_3, H_4 = heliocide H_1, H_2, H_3, and H_4.*

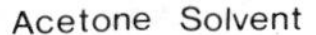

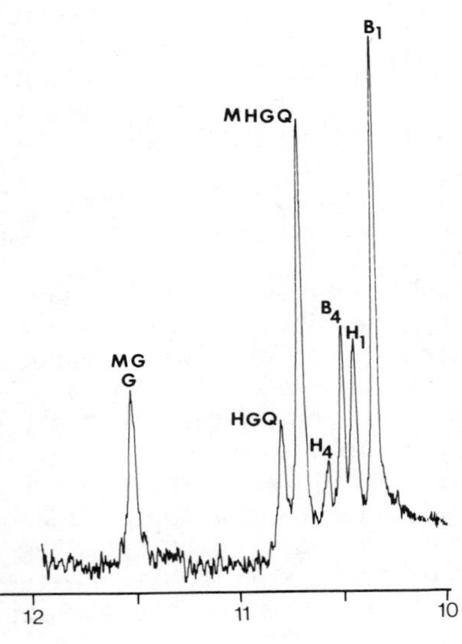

Figure 3. ¹H NMR (δ 10–12) of an ethyl acetate:hexane (1:3) extract of freeze dried 3–4-day-old bolls with bracts from G. barbadense. *MG = 6-methyl ether derivatives of gossypol; G = gossypol; HGQ = hemigossypolone, MHGQ = hemigossypolone-7-methyl ether, B_1, H_1, B_4, H_4 = heliocide B_1, H_1, B_4, H_4.*

B_2 and B_3. Thus, the ^{1}H-NMR and TLC screening techniques, when used together, give good estimations of the heliocides present.

Occurrence. Hemigossypolone and the heliocides H_1, H_2, H_3, and H_4 occur in both upland cottons (*G. hirsutum*) and Egyptian cottons (*G. barbadense*). Most commercial varieties of upland cotton which were screened by the TLC technique above contained larger quantities of heliocides H_2 and H_3 than of heliocides H_1 and H_4. The methyl ether derivatives, hemigossypolone-7-methyl ether and heliocides B_1 and B_4, were found in the *G. barbadense* cotton, SBSI, which did not contain heliocides H_2, H_3, B_2 and B_3. Apparently SBSI does not synthesize myrcene in the foliar glands and therefore heliocides H_2, H_3, B_2 and B_3 are not produced. When F_2 plants from crosses between *G. hirsutum* and *G. barbadense* were screened by the TLC technique, spots corresponding to heliocides H_2, H_3, B_2 and B_3 were observed from some progeny, but the identity of these compounds has not been confirmed. Gossypol is present in both species. Its methyl ether derivatives, 6-methoxygossypol and 6,6'-dimethoxygossypol, are probably present at least in *G. barbadense*, but this has not been confirmed.

Distribution of Heliocides and Related Compounds in Plant Tissue.

Extracts have been prepared from various tissues of glandless and glanded *G. hirsutum* and *G. barbadense*, and quality and quantity of terpenoid aldehydes were compared by thin layer chromatography. Only glanded leaves, stems, and flowers yielded terpenoid aldehydes indicating that the terpenoids are localized in the pigment glands. This has been confirmed by the histochemical tests developed by Mace, et al. (31, 32, 33).

Terpenoid content varies considerably in quantity and quality depending on tissue, age, and whether the plant is healthy or diseased. Other variables, such as temperature, sunlight, and photoperiod, are presently being investigated. The distribution of terpenoids among parts and tissues of healthy, mature plants of *G. hirsutum* and *G. barbadense* is shown in Table I.

In pigment glands, the methylated terpenoids occurred in *G. barbadense* but not *G. hirsutum*. Juvenile plants of *G. hirsutum*, however, contained methylated terpenoids in the epidermis and cortex of healthy roots and in parenchyma cells of diseased stele and hypocotyl (18, 19, 30, 32). Terpenoid quinones, and their heliocide derivatives, were the primary components in glands of stems, leaves, and green parts of flower buds. Neither the quinones nor the heliocides were found in tissues that lacked chlorophyll (e.g. petals and stamens) or were shielded from light (e.g. embryos, stele, and phloem). In healthy plants, appreciable amounts of gossypol or its methyl ether derivatives were found only in embryos, bark (phloem), petals, and stamens. More than 95% of the terpenoids in seeds of *G. hirsutum* were gossypol and its methyl ether derivatives. In healthy glandless plants, terpenoids were

Table I. Major terpenoid aldehyde components in pigment glands in different tissues of cultivated cotton.

Tissue	Upland Cotton (*G. hirsutum*)	Egyptian Cotton (*G. barbadense*)
Seed:		
Embryo	G[a]	G, MG, DMG
Coat	no glands	no glands
Stems:		
Cortex	H_2, H_3, (H_1, H_4)[b]	B_1, B_4, H_1, H_4
Phloem[c]	G	G, MG, DMG
Xylem	no glands	no glands
Leaves:		
Cotyledonary	G	G, MG, DMG
True	HGQ, H_2, H_3, (H_1, H_4)	HGQ, MHGQ, B_1, B_4, H_1, H_4
Petiole	H_2, H_3, (H_1, H_4)	B_1, B_4, H_1, H_4
Flowers:		
Bracts and calyx	H_2, H_3, (H_1, H_4)	B_1, B_4, H_1, H_4
Petals and stamens	G	G, MG, DMG
Pollen	no glands	no glands
Ovary and stigma	HGQ, H_2, H_3, (H_1, H_4)	HGQ, MHGQ, B_1, B_4, H_1, H_4
Roots:		
Cortex[c]	G	G, MG, DMG
Phloem[d]	G	G, MG, DMG
Xylem	no glands	no glands

[a]Abbreviations: G=gossypol; MG=6-methoxygossypol; DMG=6,6'-dimethoxygossypol; H_1, H_2, H_3, H_4, B_1, and B_4=heliocides H_1, H_2, H_3, H_4, B_1, and B_4; HGQ=hemigossypolone; and MHGQ=hemigossypolone-7-methyl ether.

[b]Parentheses indicate that these components were prominent in only a few cultivars.

[c]Most of the terpenoids in this tissue were in the epidermis or phelloderm of the root rather than in glands.

[d]Glands do not occur in the phloem of seedlings, but develop as the plant ages.

totally absent from all tissues, except the outer cells of the root bark.

We compared the glandular terpenoids from leaves of 24 *Gossypium* and three *Cienfuegosia* species. The results for the *Gossypium* species are shown in Table II. The terpenoid quinones and the heliocides were missing in most D genome cottons, and in the three *Cienfuegosia* species. These compounds were present in only trace amounts in African cotton of the E genome. Methylated terpenoids occurred in the B, C, F, AD_2 and AD_3 genomes, but were absent in the A, D, E, and AD_1 genomes. *G. raimondii, G. davidsonii,* and *G. klotzschianum* possessed unique terpenoids missing in the other species.

Biosynthesis of the Terpenoids.

Heinstein, et al. have shown that gossypol is biosynthesized from acetate *via* the isoprenoid pathway (34, 35). They isolated an enzyme system from homogenates of cotton roots that stereospecifically incorporated six molecules of mevalonate-2-^{14}C in each molecule of gossypol. The biosynthesis was *via* a specific cyclization of *cis, cis*-farnesyl pyrophosphate. Veech, et al. showed that peroxidase catalyzes the coupling of hemigossypol to form gossypol (36). Thus, *cis, cis*-farnesyl pyrophosphate is also a biosynthetic precursor to hemigossypol. The proposed steps in terpenoid biosynthesis are shown in Fig. 4. Enzymatic reactions appear to be involved in the formation of desoxy-6-methoxyhemigossypol (dMHG, Fig. 4) from desoxyhemigossypol (dHG), and in the oxidation of hemigossypol (HG) or methoxyhemigossypol (MHG) to hemigossypolone (HGQ) and hemigossypolone-7-methyl ether (MHG), respectively. dHG and dMHG spontaneously oxidize to HG and MHG in the presence of air (20), and myrcene and *trans*-β-ocimene react with the quinones, HGQ and MHGQ, at room temperature without catalysts to form heliocides (21, 22, 29). Thus, an enzyme may not be required for these latter reactions.

Antibiosis.

When cottons without pigment glands (glandless) were discovered and developed in the 1960's (37), it was found that many insects which did not attack glanded cotton almost completely destroyed glandless strains. These observations stimulated studies on the importance of pigment glands in host plant resistance. A review of the importance of pigment glands in host plant resistance has been published (17).

The toxicity of glanded flower buds to insects has been correlated with their gossypol content (6). However, the gossypol content was determined by the aniline method. As mentioned previously, this is a nonspecific reagent for aromatic aldehydes. It is now known that many compounds containing aromatic aldehyde groups are present in the glands, and these give reaction products

Table II. Terpenoid Content of Pigment Glands in Young Leaves of *Gossypium* Species

Genome Type	*Gossypium* Species	Relative Concentration[a] of Terpenoids[b]									
		HG	G[c]	HGQ	H_1	H_2[d]	MHG	MHGQ	B_1	B_2[e]	U[f]
A_1	*G. herbaceum*	tr	+	++	+++	+	-	-	-	-	-
A_2	*G. arboreum*	tr	+	+	tr	++	-	-	-	-	-
B_1	*G. anomalum*	tr	+	++	+	+++	+	+	tr	++	-
C_1	*G. sturtianum*	++	-	+++	tr	tr	++	+++	+	tr	-
C_3	*G. australe*	-	-	+	+	-	tr	-	tr	-	-
C_4	*G. bickii*	+	-	+++	++	tr	tr	+	++	tr	-
D_1	*G. thurberi*	tr	+++	-	-	-	-	-	-	-	-
D_2	*G. harknessii*	tr	++	-	-	-	-	-	-	-	-
D_2-1	*G. armourianum*	tr	++	-	-	-	-	-	-	-	-
D_3-d	*G. davidsonii*	tr	++	-	-	-	-	-	-	-	+
D_3-r	*G. klotzschianum*	+	+++	-	-	-	-	-	-	-	++
D_5	*G. raimondii*	tr	+	-	-	-	-	-	-	-	+++
D_6	*G. gossypioides*	tr	-	++	+++	+++	-	-	-	-	-
D_8	*G. trilobum*	tr	+++	-	-	-	-	-	-	-	-
E_1	*G. stocksii*	tr	++	-	++	+	-	-	-	-	-
E_2	*G. somalense*	tr	tr	-	-	-	-	-	-	-	-
E_4	*G. incanum*	tr	++	tr	-	+	-	-	-	-	-
F_1	*G. longicalyx*	tr	-	+++	++	tr	tr	++	++	tr	-
AD_1	*G. hirsutum*	tr	+	++	+	++	-	-	-	-	-
AD_2	*G. barbadense*	tr	+	++	++	tr	tr	++	++	tr	-
AD_3	*G. tomentosum*	+	-	++	+	tr	+	++	+	tr	-

[a] +++ = large intense spot; + = small but distinct spot; and tr = trace on thin layer chromatograms.
[b] Identities and structures of the abbreviated chemicals are given in Table I.
[c] Spots contain up to three compounds, gossypol, 6-methoxygossypol, 6,6'-dimethoxygossypol.
[d] Spots contain up to three compounds, heliocides H_2, H_3, and H_4.
[e] Spots contain up to three compounds, heliocides B_2, B_3, and B_4.
[f] U = unidentified terpenoids.

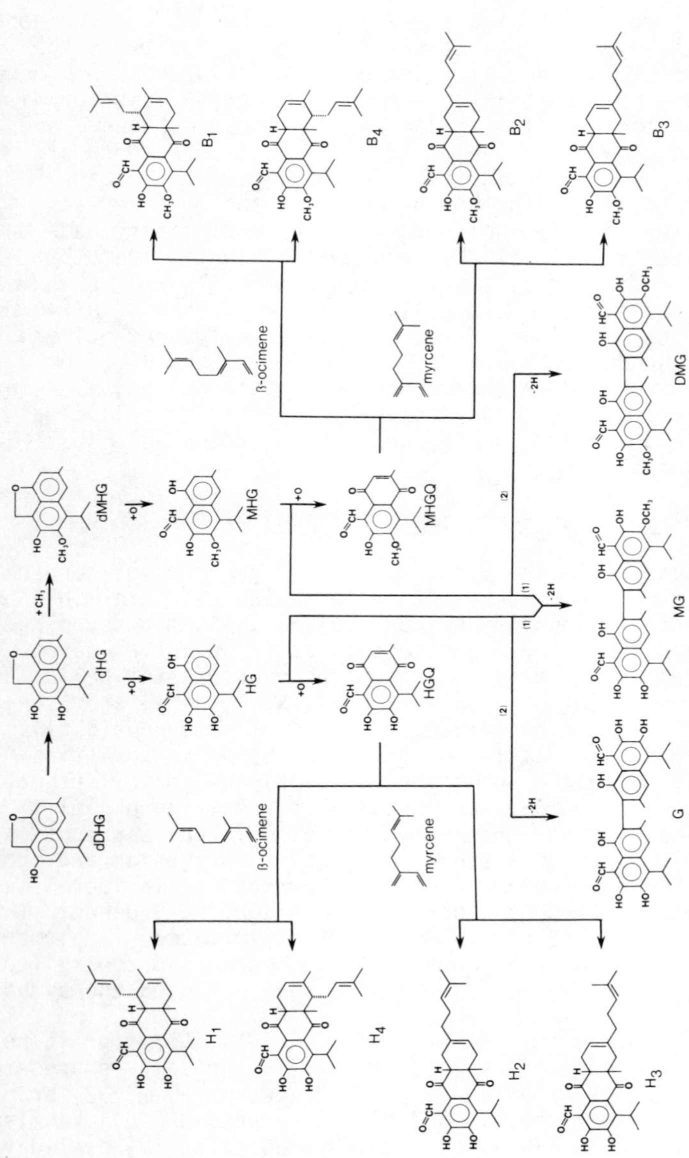

Figure 4. Proposed pathway for biosynthesis of terpenoid aldehydes in cotton

that interfere with the measurement of gossypol. Thus, previous measurements of gossypol, especially in green tissue, are largely erroneous.

We are attempting to bioassay all of the heliocides, quinones, gossypol, and gossypol-methyl ethers for their toxicity to the tobacco budworm (*H. virescens*), pink bollworm (*Pectinophora gossypiella*), and the boll weevil (*Anthonomus grandis*). This work is incomplete, but the results to date for the tobacco budworm and pink bollworm are given in Table III. Synergistic effects of these compounds are being studied. The boll weevil has been insensitive to gossypol, the heliocides, and the quinones at concentrations up to 0.2% (net weight). The preliminary data that we have indicate that all of the terpenoid aldehydes are about equally toxic to the larval stage of the pink bollworm. *G. barbadense* is more resistant to the bollworm (*H. zea*) than *G. hirsutum* (35). Heliocides in *G. barbadense* are formed only from ocimene and are more than 50% methylated. Heliocides H_1 and B_1 (from ocimene) are more toxic to tobacco budworms than Heliocides H_2 or H_3 (from myrcene). Thus, qualitative differences in heliocides between *G. hirsutum* and *G. barbadense* may be responsible for differences in resistance.

Genetics of Methyl Ether Formation in Glands.

We have made preliminary studies of the genetics of terpenoid methyl ether formation. Methyl ether formation was determined by estimating concentrations of hemigossypolone and hemigossypolone-7-methyl ether. These are major terpenoids in juvenile leaves, flower buds, and young bolls, and are precursors to the heliocides.

Three varieties of *G. hirsutum* (Acala SJ-1, CAM-1 and Stoneville 213) were crossed reciprocally with *G. barbadense* (SBSI). One-half expanded terminal leaves of the *G. hirsutum* varieties contained hemigossypolone but no hemigossypolone-7-methyl ether. Similar leaves of *G. barbadense* contained slightly more hemigossypolone-7-methyl ether than hemigossypolone. The segregation of hemigossypolone-7-methyl ether in F_1 and F_2 progenies are shown in Table IV. Hemigossypolone-7-methyl ether and heliocides B_1 and B_4 always occurred together among more than 2000 F_2 progeny. The results indicate that terpenoid methyl ether formation in pigment glands in progeny from *G. hirsutum* X *G. barbadense* is controlled by a single recessive locus for which we have proposed the symbol tm_1 (29).

While terpenoid methyl ether formation in *G. hirsutum* is not expressed in pigment glands (Table IV), it is partially expressed (about 20% of that in *G. barbadense*) in diseased xylem (33) or cambium (18). Further, nearly 50% of all terpenoids in juvenile tissues of *G. hirsutum* [7-day-old healthy radicals (19) or 7-day-old infected hypocotyls (30)] are present as methyl ethers; this is the same percentage as in *G. barbadense*. We conclude that *G. hirusutum* has a structural gene, that gives it the same basic

Table III. ED_{50} Values[a] for *Heliothis virescens* and *Pectinophora gossypiella*.

Compound	ED_{50} *H. virescens*[b] (μmoles/g diet)	ED_{50} *P. gossypiella*[c] (μmoles/g diet)
Heliocide H_1	2.5	0.8
Heliocide H_2	11.2	2.4
Heliocide H_3	3.9	---
Heliocide B_1	4.6	---
Heliocide B_2 & B_3[d]	N.E.[e]	---
Hemigossypolone	10.5	1.4
Methoxyhemigossypolone	N.E.	---
Gossypol	0.8	1.8

[a]Concentration required to reduce larval growth by 50%.

[b]Two-day-old larvae were placed on media (Vanderzant-Adkisson diet) at the beginning of the experiment; after 7 days on amended media, all larvae were returned to regular media.

[c]Amended diet ingested for duration of larval stage.

[d]Added to the diet as a mixture of heliocides B_2 and B_3 (67% B_2, 33% B_3).

[e]Compounds had no appreciable effect on insect growth at a concentration of 2.5μmoles/g diet.

Table IV. Genetics of Terpenoid Methyl Ether Formation[a] in Crosses Between Varieties of *G. hirsutum* (ASJ-1, CAM-1, and S-213) and *G. barbadense* (SBSI).

Cross	F_1		F_2		
	MHGQ	No MHGQ	MHGQ	No MHGQ	Ratio
	Number of Plants				
ASJ-1 X SBSI	0	30	86	263	1/3.06
SBSI X ASJ-1	0	30	124	263	1/2.12
CAM-1 X SBSI	0	30	91	376	1/4.13
SBSI X CAM-1	0	30	82	377	1/4.60
S-213 X SBSI	0	30	71	355	1/5.00
SBSI X S-213	0	30	99	354	1/3.58
Mean	0	30	92	331	1/3.59

[a]The presence of hemigossypolone-7-methyl ether (MHGQ) in one-half expanded terminal leaves was noted as a marker for methyl ether formation. Terpenoid methyl ethers originally were present only in the *G. barbadense* parent.

potential for methyl ether formation as is found in *G. barbadense*. However, methyl ether formation is not expressed in foliar pigment glands of *G. hirsutum* because of the product of the regulator allele $\underline{TM_1}$. The patterns of terpenoids found in leaves (Table II) suggests that the $\underline{tm_1}$ homozygote occurs universally in D, E, and AD_1 genome cottons.

Search for New Compounds Involved in Host Plant Resistance.

Table II indicates new compounds observed in *G. davidsonii*, *G. klotzschianum*, and *G. raimondii*. The structure and biochemical activity of these compounds must be studied. Crosses of these species with *G. hirsutum* or *G. barbadense* might lead to a different series of compounds that could increase host plant resistance.

Conclusion.

Cotton leaves and flower buds contain several toxic terpenoid aldehydes. The concentration of these compounds vary among races and species. We are presently investigating which compound or mixture of compounds is the most effective insecticide for various cotton pests. The biosynthetic pathway leading to these compounds is being studied. Genes that control critical branches in this pathway are known, and we are applying this knowledge in a breeding program aimed at incorporating effective natural insecticides into an acceptable agronomic line.

A specific example is found in heliocides H_1 and H_4. Our present data indicates that these may be the most toxic heliocides in *G. hirsutum*. They are formed from *trans*-β-ocimene and hemigossypolone. Heliocides H_2 and H_3, formed from myrcene, are less effective toxins, and their formation leads to lower levels of insecticidal activity. SBSI produces *trans*-β-ocimene, with little or no myrcene, because it produces only heliocides H_1, H_4, B_1 and B_4. Therefore a specific gene that directs the synthesis of *trans*-β-ocimene rather than myrcene would seem to be operating. If this gene rather than the gene allowing the synthesis of myrcene could be incorporated into an acceptable agronomic line, the resistance to *Heliothis* should be increased. Furthermore, the quinones are not as toxic as heliocides H_1 and H_4. Therefore, a variety of cotton that produced larger quantities of *trans*-β-ocimene also would be desirable because this would increase the concentration of heliocides and decrease the concentrations of quinones.

One aspect that appears to be beyond genetic control is the ratio of heliocides H_1 to H_4. The reaction of the quinone with *trans*-β-ocimene appears to occur in a nonenzymatic fashion and this ratio appears to remain almost constant in all varieties tested.

Our understanding of the structures and biosynthesis of the cotton terpenoids is allowing us to predict what can and can not be accomplished to increase host plant resistance by genetic manipulation of the plant.

Acknowledgements:

We thank J. K. Cornish, G. W. Tribble, M. E. Bearden and J. G. Garcia for excellent technical assistance. We are grateful to Dr. Ron Grigsby for high resolution mass measurements and Dr. Daniel O'Brien for ^{13}C-NMR studies that were invaluable in our structure determinations.

LITERATURE CITED

1. Painter, R. H., "Insect Resistance in Crop Plants," MacMillan, New York, NY. 1951.
2. Bell, A. A., "Biological Control of Plant Insects and Diseases," F. G. Maxwell and F. S. Harris, Ed. 403-461. The University Press of Mississippi, Jackson, Miss. 1974.
3. Muller, K. O., "Plant Pathology," Vol. 1, J. G. Horsfall and A. E. Diamond, Ed. 469-519. Academic Press, New York, NY. 1959.
4. Cook, O. F., U. S. Dept. Agric. Bur. Plant Ind. Bul. No. 88 (1906) 87.
5. Bottger, G. T., Sheehan, E. T., Lukefahr, M. J., J. Econ. Entomol. (1964) 57, 283.
6. Lukefahr, M. J., Bottger, G. T., Maxwell, F., Proc. Beltwide Cotton Prod. Res. Conf., Jan. 11-12, Memphis, Tenn. (1966) 215.
7. Smith, F. H., J. Am. Oil Chem. Soc. (1958) 35, 261.
8. Reeves, R. G., Beasley, J. O., J. Agric. Res. (1935) 51, 935.
9. Stanford, E. E., Viehoever, A., J. Agri. Res. (1918) 13, 419.
10. Lukefahr, M. J., Marlin, D. F., J. Econ. Entomol. (1966) 59, 176.
11. Bottger, G. T., Patana, R., J. Econ. Entomol. (1966) 59, 1166.
12. Lukefahr, M. J., Shaver, T. N., Parrott, W. L., Proc. Beltwide Cotton Prod. Res. Conf., Jan. 7-8, New Orleans, LA (1969) 81.
13. Lukefahr, M. J., Houghtalling, J. E., Cruhm, D. G., J. Econ. Entomol. (1975) 68, 743.
14. Leigh, T. F., Proc. Beltwide Cotton Prod. Res. Conf., Jan. 7-8, New Orleans, LA (1975) 140.
15. Niles, G. A., Proc. Beltwide Cotton Prod. Res. Conf., Jan. 5-7, Las Vegas, Nev. (1976) 168.
16. Bell, A. A., Stipanovic, R. D., Proc. Beltwide Cotton Prod. Res. Conf., Jan. 5-7, Las Vegas, Nev. (1976) 52.
17. Bell, A. A., Stipanovic, R. D., Proc. Beltwide Cotton Prod. Res. Conf., Jan. 10-12, Atlanta, GA (1977) in press.
18. Bell, A. A., Stipanovic, R. D., Howell, C. R., Fryxell, P. A., Phytochem. (1975) 14, 225.
19. Stipanovic, R. D., Bell, A. A., Mace, M. E., Howell, C. R., Phytochem. (1975) 14, 1077.

20. Stipanovic, R. D., Bell, A. A., Howell, C. R., Phytochem. (1975) 14, 1809.
21. Stipanovic, R. D., Bell, A. A., O'Brien, D. H., Lukefahr, M. J., Phytochem. (1977) in press.
22. Stipanovic, R. D., Bell, A. A., O'Brien, D. H., Lukefahr, M. J., J. Agric. Food Chem. (1977) in press.
23. Stipanovic, R. D., Bell, A. A., O'Brien, D. H., Lukefahr, M. J., Tetrahedron Letters (1977) 567.
24. Stipanovic, R. D., Bell, A. A., Lukefahr, M. J., Gray, J. R., Mabry, T. J., Proc. Beltwide Cotton Prod. Res. Conf., Jan. 5-7, Las Vegas, Nev. (1976) 91.
25. Gray, J. R., Mabry, R. J., Bell, A. A., Stipanovic, R. D., Lukefahr, M. J., J. Chem. Soc., Chem. Comm. (1976) 109.
26. Creiger, R., Becher, P., Chem. Ber. (1957) 90, 2516.
27. Bloom, S. M., J. Org. Chem. (1959) 24, 278.
28. Martin, J. G., Hill, R. K., Chem. Rev. (1961) 61, 537.
29. Bell, A. A., Stipanovic, R. D., O'Brien, D. H., Lukefahr, M. J., Phytochem. (1977) in press.
30. Unpublished observations.
31. Mace, M. E., Bell, A. A., Stipanovic, R. D., Proc. Beltwide Cotton Prod. Res. Conf., Jan. 7-9, Dallas, TX, (1974) 46.
32. Mace, M. E., Bell, A. A., Stipanovic, R. D., Phytopathol. (1974) 64, 1297.
33. Mace, M. E., Bell, A. A., Beckman, C. H., Can. J. Bot. (1976) 54, 2095.
34. Heinstein, P. F., Smith, F. H., Tove, S. B., J. Biol. Chem. (1962) 237, 2643.
35. Heinstein, P. F., Herman, D. L., Tove, S. B., Smith, F. H., J. Biol. Chem (1970) 245, 4658.
36. Veech, J. A., Stipanovic, R. D., Bell, A. A., J. Chem. Soc., Chem. Comm. (1976) 144.
37. McMichael, S. C., Agron. J. (1960) 52, 385.

15

Role of Repellents and Deterrents in Feeding of *Scolytus Multistriatus*

DALE M. NORRIS

642 Russell Laboratories, University of Wisconsin, Madison, WI 53706

Bark beetles in the genus: Scolytus (Family: Scolytidae) not only "attack" suitable host trees for production and maturation of progeny, but also feed facultatively (i.e., such feeding is not obligatory) in the twigs of vigorously healthy hosts. Findings reported in this paper relate especially to phytochemicals (allomones) in non-host trees which repel the species, Scolytus multistriatus, from alighting on such healthy plants and/or deter feeding on their tissues. Such repellent and deterrent chemicals of non-host species of trees combined with attractants, arrestant and feeding-stimulant chemicals of healthy hosts (especially Ulmus spp., family: Ulmaceae) are major determinants of the beetle arrival at, and acceptance of, host elm twigs for feeding (1).

Though the complete complement of allomonic allelochemicals in species of plants may be complex and variable among individuals and plant parts (2), our study of certain allomones against S. multistriatus in ten genera of trees in nine families of dicotyledonous angiosperms revealed apparently similar structural and/or electrochemical molecular characteristics among the identified chemicals. It is intriguing and perhaps significant that most of these plant defense chemicals would seem capable of interacting with the lipoprotein neural membrane receptors for the potent allomone, juglone (5-hydroxy-1,4-naphthoquinone, Figure 1), which have been isolated and partially characterized from antennae of S. multistriatus and Periplaneta americana (3-10).

Because of the hundreds of allomonic chemicals of non-host plants which an insect species may encounter, a specific receptor macromolecule for each chemical messenger obviously is genetically untenable. This now obvious situation thus encourages hypotheses of mechanisms for sensory neural receptor-chemical messenger interactions which allow numerous messenger ligands to interact with a given receptor macromolecule in a "quantal" manner. We (11) proposed such an unifying working hypothesis based on our early experimentation with the "juglone-type" receptor and on much supportive data from other areas of receptor neurobiology. Our continued experimental testing of this hypothesis has yielded

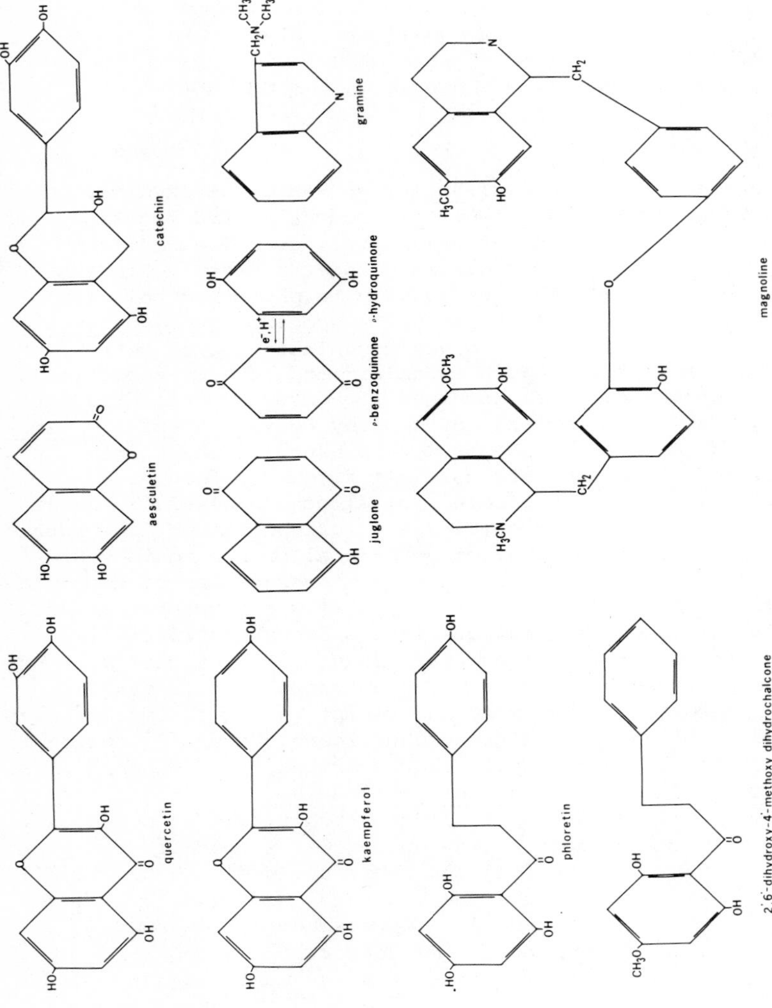

Figure 1. Molecular structures of studied allomones from non-host tree species and kairomones from host Ulmus americana *for* Scolytus multistriatus

encouraging data. The manners in which the allomones reported in this paper seem to support the hypothesis will be discussed.

Table I. Non-Host Trees* Examined For Allomones Against *Scolytus multistriatus* Feeding

Botanical Family	Genus	Species	Common Name
Aceraceae	*Acer*	*negundo*	Boxelder
		saccharinum	Silver maple
Hippocastanaceae	*Aesculus*	*octandra*	Yellow buckeye
Juglandaceae	*Carya*	*cordiformis*	Bitternut hickory
		ovata	Shagbark hickory
	Juglans	*nigra*	Black walnut
Oleaceae	*Fraxinus*	*americana*	White ash
Magnoliaceae	*Magnolia*	*acuminata* var. *acuminata*	Cucumber tree (typical)
Rosaceae	*Malus*	*pumila*	Apple
Salicaceae	*Populus*	*deltoides*	Eastern cottonwood
Fagaceae	*Quercus*	*macrocarpa*	Bur oak
		alba	White oak
Leguminosae	*Robinia*	*pseudoacacia*	Black locust

*Taxonomy taken from (12).

Methods Of Phytochemical Analyses

Procedures used in our initial extraction, purification and identification of chemicals from tissue of non-hosts (Table I) and hosts (i.e., *Ulmus*), and their bioassay with insect species, have been detailed (13-18). More recent investigation also employed chemical methods detailed, or cited, by Harborne (19). Our recent bioassay methods remained the same as in earlier studies. A schematic view of the petri-dish assay chamber used in our bark beetle bioassays termed "Assay #1" and "Assay #2" is shown in Figure 2. Assay #1 was a simple feeding-choice test between two uniform discs of elderberry stem pith (presented on the "floor" of the chamber); one disc usually had been soaked in the solvent and the other in solvent plus candidate chemical or fraction of crude extract from a plant tissue. Assay #2 evaluated the volatile properties of candidate extracts, fractions or pure chemicals. Such a treatment was applied to one of two pith discs positioned on the lid of the petri dish (Figure 2) such that one was located directly above each of the two pith discs on the floor of the chamber as used in Assay #1. The other disc on the lid was treated with solvent. In Assay #2, both lower discs were uniformly treated with 0.1 M sucrose, which is a feeding stimulant for our test bark beetles. The difference in beetle feeding on the lower disc under the treated upper disc versus on the one under the upper solvent-treated disc was measured.

Effects Of Extracts Of Non-Host *Carya* And *Juglans* Trees On *S. multistriatus* Feeding

Our initial studies focused on species of *Carya* (Table I), and their chemical constituents which prevent *S. multistriatus* alightment on, and feeding in, tissues of hickories (14, 15). The aglycone, juglone (5-hydroxy-1,4-naphthoquinone, Figure 1), was shown to keep *S. multistriatus* off *Carya cordiformis* and *Carya ovata* (Figure 3, Table II). This allomone occurs mostly as α-hydrojuglone-4-β-D-glucoside in intact healthy cells in *Carya*. However, some free naphthoquinone is perceptible even by human olfaction in the atmosphere surrounding vigorous hickories. If cells are ruptured, the plant becomes physiologically stressed as during drought, or the plant experiences a microbially caused disease, additional amounts of juglone are released into the atmosphere. It would seem that such additional releases of this potent allomone during initial predator or parasite attack, or other environmentally induced adverse conditions may exemplify the dynamic nature of a plant species' allomonic chemistry. Our field observations and experiments have shown that once a *Carya* tree becomes irreversibly diseased, it no longer can release juglone for defense. At this point, secondary predators and parasites which attack dying trees of several species appear on the plant. It would seem that such secondary herbivores are not confronted by significant amounts of the allomones which defensively serve healthy species of plants, *Carya* spp. in this particular example.

More recent studies of the type described for the *Carya* spp. have been conducted on *Juglans nigra*, black walnut (Table I), another representative of the family: Juglandaceae. Juglone in *Juglans nigra* also proved to be the major allomone against *S. multistriatus* feeding.

Molecular Properties Positively Correlated With Allomonic Activity Of 1,4-Naphthoquinones. To investigate the molecular properties of 1,4-naphthoquinones involved in making them allomonic against *S. multistriatus*, the order of relative deterrency and inhibition of feeding of 2,3-dichloro-; 2-methyl-; 2-hydroxy-; 5-hydroxy-; 5,8-dihydroxy-; and the unsubstituted 1,4-naphthoquinone was determined. Results indicated that the allomonic activity of 5-hydroxyl- > 5,8-dihydroxy- > unsubstituted 1,4- > 2-hydroxy- > 2-methyl- > 2,3-dichloro-. With the sole exception of 2,3-dichloro-1,4-naphthoquinone, this order of relative allomonic action of the tested 1,4-naphthoquinones correlated positively with the order of their relative combined redox potential and

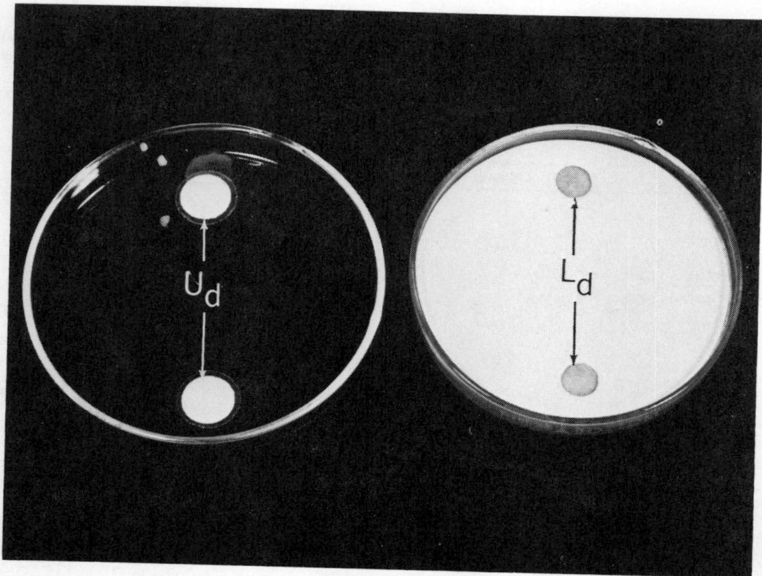

Figure 2. Photograph of top and bottom of petri dish chamber used to bioassay extracts and pure chemicals for effects on Scolytus multistriatus *feeding. Assay No. 1 used only two pith discs on floor of chamber. Assay No. 2 involved four discs: two on the floor* (L_d) *and two on the lid* (U_d).

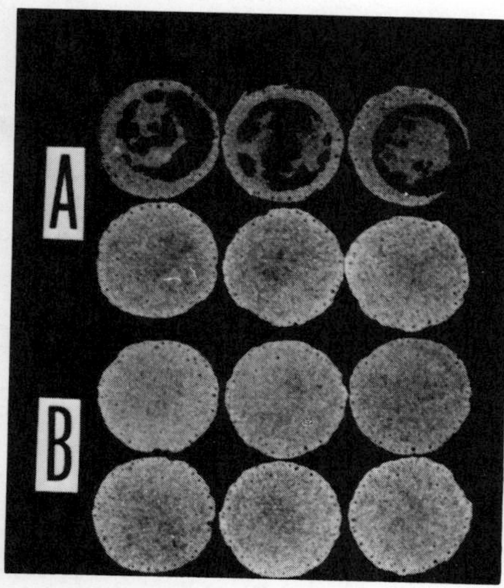

Figure 3. Inhibition of Scolytus multistriatus *feeding on pith discs treated with* Ulmus americana *extract by addition of* 2×10^{-3}M *Juglone. (A) Feeding on* Ulmus *extract; (B) inhibition of feeding by adding Juglone. Solvent-treated control discs are in lower row of "A" and "B."*

Table II. Mean Feeding (Mm^2) By 25 Scolytus multistriatus In 48 Hours on Pith Discs Treated With A Standard Ulmus americana Bark Extract, Fractions* Of Carya ovata Twig Bark Extract, Commercial Juglone Or Combinations

Treatment	Feeding (Mm^2)		Rating
	Treated**	Control	
T.L.C. Band (R_f 0.60-0.80) Of C. ovata Extract (=Juglone) + Ulmus Extract	0.8[a]	0	Band R_f 0.60-0.80 Of C. ovata Extract Was Highly Inhibitory
Ulmus Extract	44.5[b]	0	Highly Stimulatory
T.L.C. Bands (R_f 0.0-0.60, 0.80-1.00) Of C. ovata Extract + Ulmus Extract	40.2[bc]	0	Bands (R_f 0.0-0.60, 0.80-1.00) Of C. ovata Extract Were Non-inhibitory
Distillate (=Juglone) Of C. ovata Extract + Ulmus Extract	0.2[ad]	0	Distillate Of C. ovata Extract Was Highly Inhibitory
Ulmus Extract	39.2[bce]	0.4	Highly Stimulatory
Residue After Distillation Of C. ovata Extract	34.8[ef]	0.0	C. ovata Residue Was Highly Stimulatory
Juglone (2×10^{-3}M) Re-Crystallized From T.L.C. Band R_f 0.60-0.80 + Ulmus Extract	0.3[adg]	0.0	Juglone was Highly Inhibitory
Commercial Juglone (2×10^{-3} M) + Ulmus Extract	0.4[adgh]	0.0	Juglone Was Highly Inhibitory

*Methods previously detailed by Gilbert et al. (14, 15).
**Values not followed by the same superscript letters are significantly different at $P < 0.01$ probability.

hydrogen-bonding capabilities. Hydroxyl substitution(s), with the intra- and inter-molecular hydrogen-bonding capabilities which they bring to the 1,4-naphthoquinone molecule, unfailingly made the naphthoquinone more repellent, or inhibitory to feeding, than the relative redox potential predicted. Two examples are (1) 2-methyl- has a slightly higher (i.e., more negative) redox potential than 2-hydroxy-, but 2-hydroxy- proved more repellent and inhibitory; and (2) the unsubstituted 1,4- has a higher redox potential than either 5,8-dihydroxy- or 5-hydroxy-, but both of the latter chemicals were more repellent and inhibitory to *S. multistriatus*. The 2,3-dichloro-1,4-naphthoquinone showed the least repellency and inhibition of feeding even though it has a higher redox potential than the unsubstituted 1,4-naphthoquinone. Our total studies suggest that this weak repellency and inhibition to *S. multistriatus* are probably attributable to the redox potential of the naphthoquinone being so high that only limited reversible energy exchange between the messenger and receptor could occur. This interpretation gains further support from the experimentally determined fact that 2,3-dichloro-1,4-naphthoquinone is not significantly ($P < 0.05$) repellent or inhibitory to *Periplaneta americana*. All of the other above-mentioned naphthoquinones not only are also repellent and inhibitory to this very different insect *P. americana*, but the order of their relative allomonic activity is the same as against *S. multistriatus*. Thus it would appear that quinones which are allomonic to particular species of insects, or other organisms, must have a redox potential (i.e., energy-sharing or -exchanging capability) within a finite range. Such species-characteristic limits on the redox range within which its chemical sensing system may exchange informational energy with its environment may serve a major role in the chemical ecological ordering of species distributions and densities. Our studies (14, 15, 20) comparing the behavioral effects of juglone on *S. multistriatus*, for which *Carya* spp. are non-hosts; and on the close beetle relative *Scolytus quadrispinosus*, for which *Carya* are major hosts, offer encouraging further evidence that such species-characteristic limits on the redox range of chemical sensing systems operate in helping to determine insect-host plant interactions. *S. quadrispinosus* feeds readily on amounts of the potent allomone, juglone, which kill *S. multistriatus*. Differing capabilities for detoxifying juglone probably exist in these two *Scolytus* species, but the primary chemical sensing systems must differ electrochemically.

Electrochemical Difference Between An Allomone And A Kairomone For Scolytus multistriatus. In conjunction with the aforementioned experimental evaluations of various 1,4-naphthoquinones as allomones against *S. multistriatus*, the simple quinone, ρ-benzoquinone (Figure 1), also was bioassayed. In the same study the classical redox couple of ρ-benzoquinone, ρ-hydroquinone (Figure 1) (i.e., 1,4-dihydroxybenzene), was evaluated. These combined studies (1, 21, 22) showed that ρ-hydroquinone is a significant kairomone (i.e., feeding stimulant) for S. multistriatus, and ρ-benzoquinone is an allomone. Thus, in this insect the electrochemical difference between an allomone and a kairomone is the difference between accepting or donating (or variously sharing) one or two electrons and protons (Figure 1).

Flavonoids As Allomones And Kairomones To Scolytus multistriatus Feeding

The compound (+)-catechin-5-β-D-xylopyranoside was reported (24) as a feeding stimulant in Ulmus for S. multistriatus. Our studies (25) of phytochemical stimulants of S. multistriatus feeding indicated that the flavonoid aglycone, (+)-catechin (Figure 1), from Ulmus americana also promoted feeding. Such a flavan-3-ol involves the most highly reduced C_3 unit (i.e., propane unit) found in flavonoids (23, 26). The different levels of oxidation among the types of flavonoids are expressed in the C_3 unit that joins the "A" and "B" rings. Thus, the flavan-3-ols (catechins) and dihydrochalcones are the most reduced, and the flavonols are the most oxidized (Table III).

Table III. Oxidation States* In The C_3 Units Of Different Types Of Flavonids

Flavonid Name	Structure of C_3 Unit
Flavan-3-ols	A-CH_2-CHOH-CHOH-B
Hydrochalcones	A-CO-CH_2-CH_2-B
Chalcones	A-CO-CH=CH-B
Flavanones	A-CO-CH_2-CHOH-B
Leucoanthocyanidins	A-CHOH-CHOH-CHOH-B
Flavones	A-CO-CH_2-CO-B
Anthocyanidins	A-CH_2-CO-CO-B
Benzalcoumaranones	A-CO-CO-CH_2-B
Flavanonols	A-CO-CHOH-CHOH-B
Flavonols	A-CO-CO-CHOH-B

*Taken from Geissman (23).

In all the structures in Table III, the A ring is assumed to carry an o-hydroxyl group which engages in ring closure to form the chromanol, chromanone, etc., ring. Enolization, loss of the elements of water or flavylium salt formation are processes that do not affect oxidation level. Thus, flavones, anthocyanidins and benzalcoumaranones, of quite different structures, all possess the equivalent of two -CO- and one $-CH_2-$ groups.

Based on data presented in Table IV, the relatively reduced, non-carbonyl-bearing flaval-3-ol, (+)-catechin, significantly stimulated S. multistriatus feeding. When mixed with a standard amount (5 mg/ml) of benzene extractables from U. americana twig bark and phloem, catechin had an additive effect on beetle feeding. Moving to two non-host-derived dihydrochalcones, phloretin from Malus pumila and 2',6'-dihydroxy-4'-methoxy dihydrochalcone from Populus deltoides, neither stimulated a significant ($P < 0.05$) amount of feeding. Upon addition to the Ulmus extract, each compound significantly ($P < 0.01$) reduced the amount of feeding (Table IV). Thus, these two relatively reduced flavonoids, but each with a carbonyl group, clearly inhibited feeding on Ulmus extract; whereas, the comparably reduced (+)-catechin, without the carbonyl, stimulated feeding. These differences in effects of comparably reduced flavonoids, flaval-3-ols and hydrochalcones, may be explained by (1) differences in the distributions of charge in the respective molecules and/or (2) an unique functional group (i.e., carbonyl) on the C_3 (propane) portion of the molecules. Swain (27) commented that the carbonyl group in flavonoids does not show the general reactions of such groups with keto reagents, except Grignard reagents. However, these Grignard reaction properties perhaps make them comparable to the previously discussed simple quinone, ρ-benzoquinone; and certain 1,4-naphthoquinones which are inhibitory to S. multistriatus feeding. The carbonyl at C_4 of the propane portion of these flavonoids is situated as an α, β-unsaturated ketone with regard to ring A of the flavonoid molecule. This apparently is the primary common chemical structural characteristic which makes these flavonoids and the discussed quinones deterrents and inhibitors of S. multistriatus. The hydroxyl situated at C_5 on the A ring also is sufficiently close to the C_4 carbonyl to allow intramolecular hydrogen bonding between the hydrogen of the hydroxyl and the C_4 carbonyl. Our studies of the effects of various substituent groups on the allomonic activity of 1,4-naphthoquinones have clearly shown that such adjacent hydroxyls enhance the activity of such quinones. Thus, there would seem to be this important hydroxyl situation in certain allomonic naphthoquinones and flavonoids.

If we next consider two flavonoids, kaempferol as isolated from Robinia pseudoacacia and quercetin as isolated from Quercus macrocarpa, which are among the most highly oxidized flavonols (Figure 1, Table IV), they showed levels of inhibition of S. multistriatus feeding, when mixed with the Ulmus extract, which were significantly ($P < 0.05$ or 0.01) greater than those caused by the two tested hydrochalcones. The additional levels of oxidation in the flavonols plus the apparently important C_4 carbonyl and adjacent hydroxyls probably explain their greater allomonic effects under these test conditions.

Table IV. Mean Feeding (Mm^2) By 25 Scolytus multistriatus Adults In 48 Hours On Pith Discs Treated With A Standard Extract Of Ulmus americana Twig Bark, A Known Flavonid Or Combination Of Both

Flavonid	Molar Concentration	Plant Source	Feeding (Mm^2) Treated*	Solvent
(+)-Catechin	2×10^{-3}	Ulmus americana	24.3^{a}	0.0
(+)-Catechin + Ulmus Extract	2×10^{-3}	Ulmus americana	52.4^{b}	0.0
Ulmus Extract	--	--	39.3^{c}	0.0
Phloretin	2×10^{-3}	Malus pumila	4.3^{d}	0.0
Phloretin + Ulmus Extract	2×10^{-3}	Malus pumila	22.6^{ae}	0.0
Ulmus Extract	--	--	36.4^{cf}	0.0
2',6'-Dihydroxy-4' -methoxy dihydro-chalcone	2×10^{-3}	Populus deltoides	3.1^{dg}	0.0
2',6'-Dihydroxy-4' -methoxy dihydro-chalcone + Ulmus Extract	2×10^{-3}	Populus deltoides	20.6^{aeh}	0.0
Ulmus Extract	--	--	34.8^{cfi}	0.0
Kaempferol	2×10^{-3}	Robinia pseudoacacia	0.0^{gj}	0.0
Kaempferol + Ulmus Extract	2×10^{-3}	Robinia pseudoacacia	14.3^{k}	0.0
Ulmus Extract	--	--	37.6^{cfil}	0.0
Quercetin	2×10^{-3}	Quercus macrocarpa	0.0^{gjm}	0.0
Quercetin + Ulmus Extract	2×10^{-3}	Quercus macrocarpa	16.1^{kn}	0.8
Ulmus Extract	--	--	35.7^{cfilo}	0.0

*Values not bearing the same superscript letter are significantly different from each other ($P < 0.05$).

Effects Of Certain Coumarins And Phenolic Acids On *S. multistriatus* Feeding

Compounds containing the C_6-C_3 phenylpropane unit are abundant in phytochemistry. Coumarins are phenylpropanoid in origin, and are formed by ring closure of a C_6-C_3 compound such as o-hydroxy cinnamic acid. The coumarin, aesculetin, from the non-host *Aesculus octandra* proved to be a potent antifeedant for *S. multistriatus* (Figure 1, Table V). Fraxetin, a coumarin isolated from the non-host *Fraxinus americana*, also yielded significant antifeedant effects on *S. multistriatus* (Figure 1, Table V). It is interesting and apparently chemically significant that the phenolic acids o-hydroxy- and ρ-hydroxy- cinnamic acid also were very inhibitory to *S. multistriatus* feeding (Table V). Because *trans*-cinnamic acid (Table V) significantly stimulated *S. multistriatus* feeding, the presence of a hydroxyl on the basic cinnamic acid structure changed the effects of the molecule on beetle feeding from stimulatory to inhibitory. Thus, among the tested coumarins and cinnamic acids, the hydroxyl and its effect on the overall charge distribution in the molecule may determine feeding activity with this specific insect.

In view of our previously presented evidence that the effects of various flavonoids upon *S. multistriatus* feeding correlate with (1) the relative degree of oxidation, especially in the C_3, propane, unit; and (2) the presence or absence of particularly a carbonyl and/or one or more hydroxyls, common functional group - and electrochemical properties seemingly exist among these studied allomonic chemicals including the simple quinone, ρ-benzoquinone, and the 1,4-naphthoquinones.

Based on our best understanding of the functional group - and electrochemical properties of the chemosensory neural membrane receptor in *S. multistriatus* and *Periplaneta americana* (5, 11) for 1,4-naphthoquinones, this receptor mechanism should operate *in vivo* and *in vitro* with the other allomonic chemicals discussed thus far in this chapter.

To illustrate further the apparent importance of unsaturation in a carbon sidechain and of the charge distribution in phenolics for determining whether a molecule stimulates or inhibits *S. multistriatus* feeding, it is significant that protocatechuic acid (Table V) was stimulatory. Previously published data (1, 18, 22) also showed that several phenolics similar to protocatechuic acid (e.g., procatechuic aldehyde, vanillin, syringaldehyde and ρ-hydroxybenzaldehyde) stimulate *S. multistriatus* feeding. Likewise the C_{18} lignan, α conidendrin, from *Ulmus* stimulated feeding. It is a dimer of C_6-C_3 units with a lower degree of unsaturation in the C_3, propane, portions than is present in that unit of the inhibitory ρ-hydroxycinnamic acid.

Table V. Mean Feeding (Mm^2) By 25 Scolytus multistriatus Adults In 48 Hours On Pith Discs Treated With A Standard Extract Of Ulmus americana Twig Bark, A Coumarin, A Phenolic Acid, A Lignan Or Combinations

Chemical	Molar Concentration	Plant Source	Feeding (Mm^2) Treated*	Solvent
Aesculetin	2×10^{-3}	Aesculus octandra	0.0[a]	0.0
Aesculetin + Ulmus Extract	2×10^{-3}	Aesculus octandra	12.6[b]	0.0
Ulmus Extract	--	--	34.9[c]	0.0
Fraxetin	2×10^{-3}	Fraxinus americana	4.1[ad]	0.0
Fraxetin + Ulmus Extract	2×10^{-3}	Fraxinus americana	19.3[e]	0.0
Ulmus Extract	--	--	36.0[cf]	0.0
o-Hydroxy cinnamic acid	2×10^{-3}	Commercial	0.0[adg]	0.0
o-Hydroxy cinnamic acid + Ulmus Extract	2×10^{-3}	Commercial	10.3[bh]	0.0
Ulmus Extract	--	--	38.1[cfi]	0.0
ρ-Hydroxy cinnamic acid	2×10^{-3}	Commercial	0.0[adgj]	0.0
ρ-Hydroxy cinnamic acid + Ulmus Extract	2×10^{-3}	Commercial	14.8[bk]	0.0
Ulmus Extract	--	--	35.2[efil]	0.4
trans-cinnamic acid	2×10^{-3}	Commercial	24.8[em]	0.6
Protocatechuic acid	2×10^{-3}	Commerical	27.1[mp]	0.9
α-Conidendrin**	2×10^{-3}	Ulmus	16.8[eknq]	0.3

*Values not followed by the same superscript letter are significantly different ($P < 0.05$).

**A lignan, a dimer of C_6-C_3 units.

Alkaloids In Non-Host Species As Allomones For *Scolytus multistriatus*

About 15-20% of all vascular plants contain alkaloids (28). Our primary definition of an alkaloid is a basic, nitrogen-containing compound. Further descriptive characteristics which we ascribe to "alkaloids" are (1) by-products of protein metabolism which are (2) methylated on either nitrogen or, when present, on hydroxyl groups and so removed from general metabolism. They are more or less toxic substances which act primarily on the nervous system. These physiological properties make them prime candidates as allomones against insects such as *S. multistriatus*.

Our studies of major allomones in non-host *Acer negundo*, *Acer saccharinum* and *Magnolia acuminata* var. *acuminata* against *S. multistriatus* feeding revealed two types of alkaloidal deterrents and antifeedants, the indole substance gramine (Figure 1, Table VI), from the *Acer* spp; and the benzlisoquinoline derivatives, such as magnoline (Figure 1), from the *Magnolia* sp. (Table VI). It is believed that the quinoline alkaloids are formed from indole precursors (29).

Table VI. Mean Feeding (Mm^2) By 25 *Scolytus multistriatus* Adults In 48 Hours On Pith Discs Treated With A Standard Extract Of *Ulmus americana*, A Known Alkaloid Or Combination Of Both

Alkaloid	Molar Concentration	Source Plant Genus	Source Plant Species	Feeding (Mm^2) Treated*	Feeding (Mm^2) Control
Gramine	2×10^{-3}	*Acer*	*negundo*	0.0^{a}	0.0
Gramine + *Ulmus* Extract	2×10^{-3}	*Acer*	*negundo*	21.4^{b}	0.0
Ulmus Extract	--	--	--	39.8^{c}	0.0
Magnoline	2×10^{-3}	*Magnolia*	*acuminata* var. *acuminata*	5.2^{ad}	0.4
Magnoline + *Ulmus* Extract	2×10^{-3}	*Magnolia*	*acuminata* var. *acuminata*	30.4^{e}	0.0
Ulmus Extract	--	--	--	38.6^{cf}	0.0

*Values not followed by the same superscript letter are significantly different ($P < 0.05$ or 0.01).

Our limited studies of alkaloids as inhibitors of feeding by *S. multistriatus* do not allow significant speculation or a hypothesis about their structural chemical properties responsible for the observed inhibition of *S. multistriatus* feeding; however, Horowitz (30) did report the interesting observation that forming the nitrogen oxime derivative (i.e., =N-OH) of the carbonyl group in aglycones of certain flavonoids did not abolish the bitter taste of these compounds. Thus, some inhibitory alkaloids, flavonoids, coumarins, naphthoquinones and simple ρ-benzoquinones may have rather similar structural chemical characteristics which are contributory to their feeding deterrency and inhibition against S. multistriatus.

General Discussion

Our reported research findings on some apparent common molecular characteristics found among a rather diverse group of phytochemicals which repel and/or inhibit feeding by S. multistriatus suggest that they all may interact with our demonstrated "quinone" receptor in the chemosensory primary neurons in the insect. This view seems to be supported significantly by the findings of Horowitz (30).

Horowitz (30) discussed the roles of free disaccharides, aglycones and the corresponding glycosides in the elicitation of taste sensations in humans. His evidence indicated that the dissacharides, neohesperidose or rutinose, need to be attached to an aglycone to elicit intense taste. The next consideration, in view of our results reported from assays with S. multistriatus, is the role(s) of the aglycone. Regarding the structure of the aglycone, the carbonyl and its association with α,β-unsaturation and an adjacent hydroxyl seemed strongly correlated with significant inhibition of S. multistriatus feeding. Phloretin, which has the carbonyl and an adjacent hydroxyl, was the least inhibitory of tested flavonoids. Adding α,β-unsaturation unfailingly increased the effects on gustation. Removal of the carbonyl (e.g., catechin) resulted in significant stimulation rather than inhibition of insect feeding. Horowitz (30) concluded that there was scarcely any information regarding the role of the carbonyl group of flavonoids in eliciting the taste response in humans. However, he reported that removal of the carbonyl from the aglycone (i.e., reducing it to hydroxyl) seemed to cause the taste of bitterness to disappear.

These previously reported findings with humans thus would seemingly agree well with our demonstrated importance of the carbonyl group or a functionally comparable nitrogenous group, with appropriate associated molecular unsaturation and hydroxyls, in aglycones which promote repellency of, or feeding inhibition in, S. multistriatus. There seems to be an encouraging amount of evidence that the "quinone" receptors demonstrated in our studied species of insects interact as chemoreceptors with many, if not most, of the other studied allomones.

Acknowledgments

This research was supported by the College of Agricultural and Life Sciences, the Wisconsin Department of Natural Resources;

and by research grants Nos. GB-6580, 8756, 41868 and BNS 74-00953 from the National Science Foundation.

Literature Cited

1. Norris, D.M., Baker, J.E., Borg, T.K., Ferkovich, S.M., Rozental, J.M., Contrib. Boyce Thompson Inst., (1970), 24, 263-274.
2. Rice, E.L., "Allelopathy", p. 353, Academic Press, New York, 1974.
3. Norris, D.M., Ferkovich, S.M., Rozental, J.M., Baker, J.E., Borg, T.K., Science, (1970), 170, 754-755.
4. Ferkovich, S.M., Norris, D.M., Experientia, (1972), 28, 978-979.
5. Rozental, J.M., Norris, D.M., Nature, (1973), 244, 370-371.
6. Rozental, J.M., Norris, D.M., Life Sciences, (1975), 17, 105-110.
7. Singer, G., Rozental, J.M., Norris, D.M., (1975), Nature, 256, 222-223.
8. Norris, D.M., Bull. Entomol. Soc. Amer., (1976), 22, 27-30.
9. Norris, D.M., Symp. Biol. Hung., (1976), 16, 197-201.
10. Norris, D.M., Rozental, J.M., Samberg, G., Singer, G., Comp. Biochem. Physiol., (1977), in press.
11. Norris, D.M., Experientia, (1971), 27, 531-532.
12. Little, E.L., "Check List of Native and Naturalized Trees of the United States (Including Alaska)", Agric. Handbook, 41, p. 472, U.S. Govt. Printing Office, Washington, D.C., 1953.
13. Norris, D.M., Baker, J.E., J. Insect Physiol., (1967), 13, 955-962.
14. Gilbert, B.L., Baker, J.E., Norris, D.M., J. Insect Physiol., (1967), 13, 1453-1459.
15. Gilbert, B.L., Norris, D.M., J. Insect Physiol., (1968), 14, 1963-1068.
16. Baker, J.E., Norris, D.M., Ann. Entomol. Soc. Amer., (1968), 61, 1248-1255.
17. Baker, J.E., Norris, D.M., Ent. exp. appl., (1968), 11, 464-469.
18. Meyer, H.J., Norris, D.M., J. Insect Physiol., (1974), 20, 2015-2021.
19. Harborne, J.B., "Phytochemical Methods", p. 278, Chapman and Hall, London, 1973.
20. Goeden, R.D., Norris, D.M., Ann. Entomol. Soc. Amer., (1964), 57, 743-749.
21. Norris, D.M., Nature, (1969), 222, 1263-1264.
22. Norris, D.M., Ann. Entomol. Soc. Amer., (1970), 63, 476-478.
23. Geissman, T.A., pp. 213-250, "Comprehensive Biochemistry", 9, Florkin, M. and Stotz, E.A., Eds., p. 265, Elsevier, Amsterdam, 1963.
24. Hoskotch, R.W., Chatterji, S.K., Peacock, J.W., Science, (1970), 167, 380-382.

25. Meyer, H.J., "Ph.D. Thesis", p. 246, University of Wisconsin, Madison, 1974.
26. Swain, T., "Chemical Plant Taxonomy", p. 543, Academic Press, London, 1963.
27. Swain, T., pp. 211-245, "Chemistry and Biochemistry of Plant Pigments", Goodwin, T.W., Eds., p. 583, Academic Press, London, 1965.
28. Hegnauer, R., pp. 389-427, "Chemical Plant Taxonomy", Swain, T., Ed., p. 543, Academic Press, London, 1965.
29. Relijk, J., Pharm. Weekbl., (1958), 93, 625.
30. Horowitz, R.M., pp. 545-571, "Biochemistry of Phenolic Compounds", Harborne, J.B., Ed., p. 618, Academic Press, London, 1964.

16

Behavioral and Developmental Factors Affecting Host Plant Resistance to Insects

P. A. HEDIN and J. N. JENKINS

Boll Weevil Research Laboratory, Agricultural Research Service, U.S. Department of Agriculture, P. O. Box 5367, Mississippi State, MS 39762

F. G. MAXWELL

Department of Entomology and Nematology, University of Florida, Gainesville, FL 32604

Plants that are inherently less severely damaged or less infested by a phytophagous pest under comparable environments in the field are termed "resistant" (1, 2). Painter(2); Beck (3); Maxwell et al. (4); and Maxwell (5) have reviewed the development of resistant host plant varieties. While considerable success has been achieved in breeding for resistance to certain key insect pests, often little is known about the chemistry of resistance. There are several reasons for this paucity of knowledge about the chemical bases. First, plant breeders have been able to make selections successfully without chemical analyses for guidance. Second, micro-techniques for identifying chemical resistance factors have only been fairly recent developments. Third, resistance as expressed in the field most often involves not only physical and biochemical factors, but also frequently complex interrelationships among the insect, the plant and the environment. Fourth, we have not had the necessary basic behavioral information on the insect pest to devise adequate chemical and biological assay techniques to properly assess the response of insects to various chemical fractions and compounds derived from plants.

Recently, the interest in elucidating some of the chemical aspects of resistance has increased because of several factors: (1) greater interest in the field of host plant resistance as an alternative to our current dependence upon pesticides, and the accompanying recognition that resistance can play an important role in the development of effective integrated pest management programs; (2) degree of success registered through recent programs; (3) information derived from basic insect-plant interactions from host plant resistance programs that contributes materially to the field of insect behavior and control methodology; (4) greater interest by private and public agencies in supplying the support necessary to conduct more basic chemical studies; (5) a greater awareness and interest on the part of natural product chemists of the opportunities for research contributions in the field; (6) advent of better methodology and technology in microchemical techniques and (7), probably most important for the future, are the new Food and Drug Administration regulations that will necessitate chemical research to document that changes (toxins, nutritional, etc.) made in the plant through development of resistance to a pest will not constitute a hazard to the health or nutrition of humans and other animals that might consume the resistant crop. These regulations will have a tremendous impact on research of insect-plant interactions, and more specifically insect behavior, chemical ecology, and applied insect control.

We have approached the subject of the chemical basis of insect resistance in plants in the following areas: (1) a discussion of those chemicals affecting the preference or nonpreference toward the plant for feeding and oviposition, i.e. those that elicit an attractive, repellent, feeding, oviposition, or deterrent response; (2) an examination on a selected basis of some toxins, growth inhibitors and other antibiotic factors in certain plants that affect the survival of insects (antibiosis); and (3) an examination of some important nutritional factors in plants that affect survival and development of insects (antibiosis).

This review is not intended to be comprehensive, but to present selected current work in the field. For additional information attention is directed to reports of insect antifeedants and toxic agents by other contributors to this book.

Biologically Active Compounds in Plants Affecting Insect Behavior

Components in plants that direct the initial selection by insects such as attractants, feeding and oviposition stimulants, repellents, and feeding and oviposition deterrents affect the mechanism of resistance defined as preference or nonpreference by Painter (1). Although they are not included in this report, other insect behavior and development agents such as sex pheromones, kairomones, and hormones may be directly or indirectly obtained from plants. We have surveyed the literature and attempted to categorize the classes of chemicals that are involved in the various insect responses. The results, which have been tabulated in Tables I - VI, include the type of response, the insect, the host, the compounds or classes of compounds if known, and the investigators. The information in Table 1 demonstrates that in most cases, feeding stimulants are comprised of chemicals that fall into what have been grouped as secondary plant substances (6), that is, these substances are largely thought to possess no primary functions (as nutrient, or energy source) in the plant or in the insect.

When the known feeding stimulants were classified by chemical structure, 21 were glycosides, 15 were acids, 9 were flavonoid aglycones, 6 were carbonyls, 5 each were phospholipids and terpenoids, and 21 were miscellaneous classes. When these feeding stimulant classes were further subdivided by insect order, somewhat more specific preferences were suggested. Among the Lepidoptera, 8 glycosidic feeding stimulants were reported, along with 4 aglycones that could become glycosidated; there were 11 miscellaneous classes. Among the Coleoptera, 10 acids, 7 glycosides, 4 flavonoid aglycones, and 4 terpenes were reported. There were also 10 miscellaneous. In the order Homoptera, both compounds reported were glycosides. In the order Orthoptera, acids, and their esters and salts, were most prevalent. These compounds accounted for 8 of the 14 listed classes.

In reports on insect feeding stimulants, no study has provided clear evidence that any compound acts individually. Furthermore, when a "glycoside," "sugar," or "purine" has been implicated, it has often been questionable whether the activity could have been maintained after rigorous purification. In addition, when pure compounds have been demonstrated to give activity (7, 8), several compounds might have been isolated from the same source, each of which could have elicited some activity or synergized the activity of the compound. The concept of host-plant specificity implies that insects can discriminate flavors. Therefore, though insects almost certainly do sometimes respond to a single dominant compound, probably in a much larger number of situations, no dominant compound exists. Consequently, an adequate response more likely requires a complicated profile of compounds.

One insect that feeds as a result of a complicated, and probably sequential, presentation of stimuli is the silkworm. (*Bombyx mori* L.). Volatile substances including citral, linalyl acetate, linalool and terpinyl acetate attract the larvae to the mulberry leaves. Biting factors such as β-sitosterol-β-glucoside, lupeol, isoquercetrin, morin, and 2', 2,4'5,7-pentahydroxyflavone cause initiation of plant consumption. Swallowing factors such as sitosterol, sugar,silica, cellulose, and potassium phosphate were demonstrated. From these results, the substances controlling the feeding behavior of silkworm larvae were described, but these substances are not found exclusively in mulberry leaves, and are in fact rather common in green leaves. The preference for mulberry leaves may depend on the amounts and proportions of these compounds and on the absence of repellents. A repellent effect could be demonstrated by adding raw soybean cake or powdered milk to a mulberry leaf preparation. Extraction of these additives with methanol removed the repellent components (9).

Work conducted at the Boll Weevil Research Laboratory illustrates some of the complexities of feeding stimulant research. Feeding-stimulant components extractable with petroleum ether, chloroform, acetone, and chloroform-methanol from cotton buds (*Gossypium hirsutum* L.) appear to belong to several major groups that cause feeding activity in the boll weevil, *Anthonomus grandis* Boheman, (10, 11, 12, 13, 14). However, isolational efforts frequently result in a dissipation of feeding activity that cannot be fully regenerated by recombination, indicating a breakdown in the chemical structure of some components during purification.

Table I. Feeding stimulants.

Insect				
Scientific name	Common name	Plant	Compound	Reference
Pieris rapae (L.)	Imported cabbageworm	Cruciferae	Sinigrin	(126)
Pieris brassica (L.)	Cabbageworm	Cruciferae	Sinigrin & related glucosides, glucocapparin, glucoiberin	(126) (127)
Papilio polyxenes asterius Stoll	Black swallowtail	Umbelliferae	Carvone, methyl chavicol, coriandrol	(128)
Serrodes partita (F.)	Plum moth	Pappae capensis (wild plum)	Quebrachitol	(129)
Leptinotarsa decemlineata (Say)	Colorado potato beetle	Potato leaves	Phospholipid, chlorogenic acid, glycoside $C_{17}H_{29}O_{10}N$	(123), (130), (131), (131)
Bombyx mori (L.)	Silkworm	Mulberry	glycoside: $C_{30}H_{62}O$(?)	(6)
Diabrotica virgifera LeConte	Western corn rootworm	Corn, Zea mays L.	Unknown	(132)
Diabrotica longicornus (Say)	Northern corn rootworm	Corn, *Zea mays*	Unknown	(132)
Musca domestica L.	House fly	Yeast	Guanine, monophosphate	(8)
Anthonomus grandis Boheman	Boll weevil	Cotton	glycoside, flavonoids	(133), (134), (135)
Anthonomus grandis	Boll weevil	Cotton	Gossypol, α-ketoglutaric acid, malonic acid, vanillin, formic acid, lactic acid, ℓ-malic acid, quercetin, β-sitosterol, succinic acid, valine, quercetin-7-glucoside, quercetin-	(13)

Insect		Plant	Compound	Reference
Scientific name	Common name			
contd.			3'-glucoside, cyanidin-3-glucoside	
Anthonomus grandis Boheman	Boll weevil	Cotton Gossypium sp.	Pheophytin a, Pheophytin b	(14)
Galla Pyrrhalta xanaena luteola (Shiller) (Muller)	Elm leaf beetle	Ulmus americana L., American elm	Unknown	(136, 135)
Gastrophysa cyanea (Melsheimer)	Dock beetle	Dock, Rumex obtusifolius L.	Unknown	(137)
Danaus plexippus (L.)	Monarch butterfly	Milkweed, Ascelpius syriaca L.	Unknown	(138)
Anasa tristis (DeGeer)	Squash bug	Pumpkin, Curbita pepo L.	Glycoside	(137)
Heliothis zea (Boddie)	Corn earworm (Bollworm)	Corn, Zea mays L.	Unknown	(139)
Heliothis zea	Corn earworm	Corn, Zea mays L.	Sugars, amino acids	(140)
Choristoneura fumiferana (Clemens)	Spruce budworm	White spruce, Picea glauca (Moench)	Unknown	(141)
Laspeyresia caryana	Hickory shuckworm	Pecan, Carya illinoensis (Wang) K. Koch	Unknown	(142)
Scolytus multistri-atus (Marsham)	Smaller European elm bark beetle	Elm	Unknown	(143)
Scolytus multistri-atus (Marsham)	Smaller European elm bark beetle	Elm	(+)-catechin-T-B-D-xylopyranoside Lupeyl cerotate	(144)
Manduca sexta (L.)	Tobacco hornworm	Tomato	Glycoside	(40)
Plutella xylostella (L.) (Curt.)	Diamondback moth	Cruciferae	Sinigrin, Sinalbin, Glucocheirolin	(145)

Insect		Plant	Compound	Reference
Scientific name	Common name			
Epilachna varivestis Mulsant	Mexican bean beetle	*Phasaeolus* sp.	Phaseolunatin,	(*50*)
Epilachna fulvosignata suahelorum (Weise)		*Solanum campylacanthum* L.	essential oil	(*146*)
Ceratomia catalpae (Boisduval)	Catalpa sphinx	Catalpa, *Catalpa bignonoides*	Unknown, Catalposide	(*131*), (*47*)
Melanoplus bivittatus (Say)	Twostriped grasshopper	Corn, soybeans, other plants	Lecithin, phosphatidyl inositol	(*147*)
Camnula pellucida (Scudder)	Clearwinged grasshopper	Corn, soybeans, other plants	Lecithin, phosphatidyl inositol, amyl acetate	(*147*), (*148*)
Operophtera brumata (L.)	Winter moth	*Rosaceae*, *Ericaceae Salicinae*	Tannins	(*149*)
Nygmia phaeorrhoea (Donovan)	Browntailed moth	Oak	Tannins	(*150*)
Malacosoma neustria (L.)	A Tent caterpillar	Oak	Tannins	(*150*)
Euproctis chrysorrhoea (L.)		Oak	Tannins	(*150*)
Periophorus padi		*Rosaceae*	Glycoside	(126)
Malacosoma americanum (F.)	Eastern tent caterpillar	*Rosaceae*	Amygdalin	(*126*)
Gastrophysa viridula (DeGeer)		*Polygonaeae rumix* sp.	Oxalic acid	(*126*)
Hylemya antiqua (Meigen)	Onion maggot	Onion	Allyl sulfide	(*151*)
Gonioctena vitallinae (L.)	a Willow beetle	Willow var.	Salicin	(*152*)
Diabrotica undecimpunctata howardi (Barber)	Spotted cucumber beetle	*Cucurbitaceae* sp.	Curcurbitacins	(*153*), (*154*)

Insect				
Scientific name	Common name	Plant	Compound	Reference
Trichoplusia ni (Hübner)	Cabbage looper	Laboratory diets	proteins, sugars, wheat germ, oil, salts, phospholipids, lecithins	(_122_)
Poekilocerus bufonius (Klug)	a Grasshopper	Milkweed, _Asclepius syrica_ (L.)	Calotropin	(_155_)
Brevicoryne brassicae (L.)	Cabbage aphid		Mustard oil, glycosides	(_156_)
Oulema melanopus (L.)	Cereal leaf beetle		Sucrose	(_157_)
Agasicles hygrophilia (Nov.) Selmar and Vogt	a Chrysomelid beetle	Alligatorweed _Alternanthera phylloxeroides_, Amaranthaceae	7-alpha-L-rhamnosyl-6-methoxyluteolin	(_158_)
Scolytus multistriatus (Marsham)	Smaller European elm bark beetle	Elm, _Ulmus americana_	_p_-hydroxyacetophenone, _o_-hydroxybenzyl alcohol, _p_-hydroxybenzaldehyde	(_159_)
Aplocera plagiata (L.)	a Geometrid moth	_Hypericum_	Cuticular waxes	(_160_)
Calophasia lunula (Hufnagel)	a noctuid moth	Linaria	Unknown	(_161_)
Heliothis virescens (F.)	Tobacco budworm	Corn kernels & silks, cotton buds & flowers	Unknown	(_162_)
Hypera postica (Gyllenhal)	Alfalfa weevil	Alfalfa	Adenine salts, nucleotides	(_163_)
Scolytus multistriatus (Marsham)	Smaller European elm bark beetle	_Ulmus americana_ (L.)	(+)-catechin-5-alpha-D-xylopyranoside, Lupeyl cerotate	(_164_)

Insect				
Scientific name	Common name	Plant	Compound	Reference
Papilio polyxenes asterius (Stoll)	Black swallowtail	Anise, Coriander, celery, angelica, citrus	Anethole, anisic aldehyde	(165)
Spodoptera frugiperda (J. E. Smith)	Fall armyworm	Corn, cotton, tomato, sorghum	Unknown	(166)
Spodoptera litoralis (Boisdural)	a noctuid moth	Cotton	Terpineol, citronellol pinene	(167)
Heliothis zea (Boddie)	Corn earworm (bollworm)	Corn, cotton, tomato, sorghum	Unknown	(166)
Heliothis virescens (F.)	Tobacco budworm	Corn, cotton, tomato, sorghum	Unknown	(166)
Tribolium confusum (J. DuVac)	Confused flour beetle	Wheat germ, yeast	Unknown, Palmitic acid	(143)
Solenopsis richteri Forel	Black imported fire ant	Boll weevil, cabbage looper, ground beef	Linoleic acid, lino- Trilinolein	(168)
Sitona cylindricollis Fåhraeus	Sweetclover weevil	Sweetclover	Adenosine	(169)
Chorthippus curtipennis (Harris)	a Grasshopper	General	Ascorbic acid, thiamine, betaine, monosodium glutamate	(170)
Melanoplus bivittatus (Say)	Twostriped grasshopper	General	Amides, anisic acid, benzoic acid, ammonium salts, pentyl acetate, plant phospholipids, wheat germ oil	(120)
Schistocerca gregaria Forskål	a Desert locus	Bran, olives, peanuts, miaouli	Unknown	(171)
Acrididae	Grasshopper, several species	Host plants	Unknown	(172)

Insect				
Scientific name	Common name	Plant	Compound	Reference
Pieris brassicae (L.)	Cabbageworm	*Brassica* sp., *Capparidaceae*, *Nasturtium trogacolum majus* (L.)	glucotropacolin, glucoapparin, glucoiberin, glucocheirolin, progortrin, glucosinalbin, sinigrin, glucoerucin	(127)
Clenicera aeripennis destructor (Brown)	Prairie grain wireworm	Germinating rye seed	Water extract, triolein	(121, 173)
Calliphora spp.		*Hypericum* sp. (St. Johns wort)	Hypericin	(174)
Phrydiuchus topiarius (Germar)		*Salvia* sp.	Unknown	(175)
Gastrophysa cyanea Melsheimer		*Rumex crispis*	Water extract	(176)
Calophasia lunula (Hufnagel)		Toadflax, *Linaria vulgaris*	Pyridine extract	(161)
Pectinophora gossypiella (Saunders)	Pink bollworm	*Gossypium* sp.	volatiles	(177)
Dsydercus koenigii (F.)	Red cotton bug	Cottonseed	sucrose, glucose, raffinose	(178)
Heliothis zea (Boddie)	Corn earworm	16 host plants	water extracts	(179), (180)
Diatraea grandiosella (Dyar)	Southwestern corn borer	corn, other	β-sitosterol, Stigmasterol	(181)
Chalcodermus aeneus Boheman)	Cowpea curculio	Southern pea	Eicosenoic acid	(182)

This inability to regenerate full feeding activity by recombination is especially true for many of the TLC systems studied.

Since recombination or fortification of fractions by sugars and buffers often rejuvenated part of the activity, efforts were directed to formulating an active feeding mixture from known cotton constituents, common metabolites, and compounds inducing primary mammalian sensations of taste and odor (13). Of 286 compounds bioassayed individually, 52 elicited substantial activity, and 14 of these previously had been reported in cotton. They include gossypol, α-ketoglutaric acid, malonic acid, vanillin, formic acid, lactic acid, ℓ-malic acid, quercetin, β-sitosterol, succinic acid, valine, quercimeritrin, quercetin-3'-glucoside, and cyanidin-3-glucoside. The insect was found to express preference for sweet, sour, and cooling taste properties, but odor preferences were difficult to establish. Sixteen carboxylic acids, 8 alcohols, 8 carbonyls, 8 phenols, and 10 amides or amines were among the most stimulatory. When the active components were factored on the bases of taste and molecular weight, it became apparent that sweet substances having molecular weights above 200 were consistently well accepted. Since most of these compounds were di- or triterpenoids or steroids, hydroxylated, and much less sweet than the sugars, their activity may be associated with their predicted low rate of desorption. The most favored molecular weight for sour and salty compounds was below 150. Bitter deterrent compounds were concentrated in the 100-200 range. Pungent compounds of 150-200 were most deterrent.

From a review of biologically active substances in host plants affecting insect behavior, a frequency table was prepared relating feeding stimulation and/or attractants and deterrents and/or repellents to molecular structure. The leading attractive compounds were: 8 acids attractive to 1 unattractive; monoterpene hydrocarbons, 9 to 2; di- or triterpenoids and steroids, 23 to 1; esters and alcohols, each 7 to 0; and nucleotides and tannins, each 5 to 0. The leading repellent compounds were: 14 alkaloid glycosides unattractive to 1 attractive; and lactones, 7 to 1. Flavonoid, cyanogenic, and other unclassified glycosides were normally stimulatory, 17 to 6. The tastes of the stimulatory substances were sour, cooling, semisweet, and salty, and the stimulatory odors appeared to be floral, musky, pepperminty, and camphoric. Repellent substances such as the lactones had bitter tastes and strongly pungent odors.

Plant Attractants. When the known attractants were classified by chemical structure (Table 2), 22 were terpenes, 8 were alcohols, 4 each were esters, acids, and sulfur containing, and 2 were phenolics. There were 3 that possessed little volatility and probably were misclassified as attractants. When these attractants were further subdivided by insect order, somewhat more specific preferences were suggested. Among the Coleoptera, 17 terpenes, 3 ketones, 2 alcohols, 2 acids, and 5 miscellaneous

Table II Attractants.

Insect				
Scientific name	Common name	Plant	Compound	Reference
Pieris rapae L.	Imported cabbageworm	Cruciferae	Allyl isothiocyanate	(183)
Papilio polyxenes asterius	Black swallowtail	Umbelliferae Stoll	Carvone, methyl chavi- Coriandrol	(184)
Bombyx mori (L.)	Silkworm	Mulberry, Morus alba L.	2-hexenol, 3-hexenol, β-gamma-hexenol, α-β-hexenal	(185) (186)
Bombyx mori (L.)	Silkworm	Mulberry, Morus alba L.	Citral, terpinyl acetate, linalyl acetate linalool	(187) (7) (188), (189)
Bombyx mori (L.)	Silkworm	Mulberry, Morus alba L.	Isobornyl acetate, butyl alcohol, isoamyl alcohol, α,β-hexenal, cis and trans-β-γ-hexenol, linalool	(190)
Anthonomus grandis Boheman	Boll weevil	Cotton, Gossypium sp.	Unknown	(191)
Anthonomus grandis Boheman	Boll weevil	Cotton, Gossypium sp.	β-bisabolol, β-carophyllene oxide, ℓ-pinene, limonene, β-carophyllene, trimethylamine, ammonia	(16) (192)
Chilo suppressalis (Walker)	Asiatic rice stem borer	Rice, Oryza sativa L.	Oryzanone	(193)
Listroderes costirostris obliquus (Klug)	Vegetable weevil	Crucifera and Umbelliferae	leaf alcohol 3-hexenol	(194)

Insect				
Scientific name	Common name	Plant	Compound	Reference
Listroderes costirostris obliquus (Klug)	Vegetable weevil		mustard oil, isothiocyanates; methyl, ethyl, allyl, isobutyl, n-butyl, phenyl, benzyl, β-phenylethyl, α-naphthyl	
?	?	Sweet clover, Melilotus officianalis L.	Coumarin	(195)
Laspeyresia caryana (Fitch)	Hickory shuckworm	Pecan, Carya illinoensis (Wang) K. Koch	Unknown	(142)
Manduca sexta (L.)	Tobacco hornworm	Jimson weed, Datura stromonium	Amyl salicyiate	(196)
Plutella xylostella (L.)	Diamondback moth	Cruciferae	Progoitrin, glucocouringlin, glucoerucin, glucotropacolin, gluconastertium, gluconapin	(125)
Oncopeltus fasciatus (Dallas)	Large milkweed bug	Milkweed seed	Unknown	(197)
Cosmopolites sordidus (Germar)	Banana root borer	Banana	Unknown	(198)
Gastrophysa viridula (DeGeer)		Polygonaeae rumix sp.	Oxalic acid	(126)
Malacosoma americanum (F.)	Eastern tent caterpillar	Rosaceae	Amygdalin	(126)
Hylemya antiqua (Meigen)	Onion maggot	Onion	Allyl sulfide, n-propyl disulfide, n-propyl mercaptan, methyl	(151)

Insect: Scientific name	Insect: Common name	Plant	Compound	Reference
contd.			disulfide, Isopropyl mercaptan	
Drosophila melanogaster (Meigen)	a fruit fly	Banana	Volatiles	(199)
Dacus diversus (Coquillett)	a fruit fly	Fruit	Oil of citronella	(200)
Dacus zonatus (Saunders)	a fruit fly	Fruit	Oil of citronella	(200)
Dacus dorsalis (Hendel)	Oriental fruit fly	Fruit	Methyl eugenol	(201)
Phyllotreta cruciferae (Goeze)	a flea beetle	*Cruciferae* spp.	Allyl isothiocyanate mustard oils, glycosides	(202)
P. striolata (F.)	Striped flea beetle			
Reticulitermes flavipes (Kollar)	Eastern subterranean termite	Decayed wood by *Lenzites trabea*	Essential oil	(203)
Dendroctonus pseudosugae Hopkins	Douglas fir beetle	Douglas fir, *Pseudotsuga menziezei* Mirb. Franco	alpha-Pinene	(204)
Musca domestica L.	House fly	Mushroom, *Amarita muscaria* L.	1,3-Diolein	(205)
Adris tyrannus amurensis (Staudinger)		Grapes	Neutral volatile	(206)
Chrysopa septempunctata (Wesmael)	a lacewing	Matatabi, *Actinidia polygama* (Miq.)	Iridodiol, metatabiol, 5-hydroxymatatabiether, 7-hydroxydehydromatatabiether, allomatatabiol	(207)
Blastophagus piniperda (L.)	Pine beetle	*Pinus densiflora*, *Pinus silvestris*	Benzoic acid α-terpineol	(208)

Insect				
Scientific name	Common name	Plant	Compound	Reference
Gnathotrichus sulcatus (LeConte)	Ambrosia beetle	Western hemlock *Tsuga heterophylla* (R.) Sargent	Ethanol	(209)
Hylemya brassicae (Wiedemann)	Cabbage maggot	Cabbage	Volatiles	(36)
Schistocerca gregaria (Forskål)	a desert locust	Grasses	Ammonium dihydrogen phosphate, Diammonium hydrogen phosphate, ammonium sulfide	(210), (211)
Hypera postica (Gyllenhal)	Alfalfa weevil	Alfalfa	Unknown	(212)
Bruchophagus roddi (Gussakovsky)	Alfalfa seed	Alfalfa	Seed pod extract	(213)
Hylobius pales (Herbst)	Pales weevil	Loblolly pine stem	Monoterpenoid(s), eugenol, anethole, alpha-pinene	(214)
Ips pini (Say)	Pine engraver beetle	*Pinus resinosa*	Terpenes	(215)
Dendroctonus pseudosugae Hopkins	Douglas fir beetle	Douglas fir, *Pseudotsuga menziezei* Mirb. Franco	Terpenes	(216)
Dendroctonus brevicomis LeConte	Western pine beetle	*Pinus ponderosa* (Laws)	"Oleoresin," α-pinene	(217)
Scolytus multistriatus (Marsham)	Smaller European elm bark beetle	Decaying hardwood	Syringaldehyde,	(218)
Popillia japonica (Newman)	Japanese beetle		Geraniol, Citronellol	(219)

Insect		Plant	Compound	Reference
Scientific name	Common name			
Enoclerus sphegeus (Fab.)	a checkered beetle	Douglas fir, Ponderosa pine, grand fir	α-pinene, β-pinene, limonene	(220)
Thanasimus undatulus (Say)	a checkered beetle	Douglas fir, Ponderosa pine, grand fir	α-pinene, β-pinene, limonene	(220)
Laspeyresia pomonella (L.)	Codling moth		esters, oil of cloves, oil of citronella, alcohols, acids, propionates, pine tar oil, Anethole, terpene, alcohols, Terpinyl acetate	(221), (128) (222)
Grapholitha (*Laspeyresia*) *molesta* Busck	Oriental fruit moth		Terpinyl acetate	(223)
Diabrotica undecimpunctata howardi Barber	Spotted cucumber beetle	*Cucumis melo* (L.) *Cucurbitae foetidissima* (HBK.)	Cucurbitacins	(104)
Kalotermes flavicollis (Fab.); *Zootermopsis nevadensis* (Hagen); *Heterotermes indicola* (Wasmann) *Reticulitermes*	Termites	Wood infected Basidomycetes	Vanillic acid, *p*-hydroxy-benzoic acid, *p*-coumaric acid, protocatechuic acid, ferulic acid	(224)
Leptinotarsa decemlineata (Say)	Colorado potato beetle	Potato leaves	Alcoholic extract	(123)
Bruchophagus roddi Gussakovsky	Alfalfa seed chalcid	Alfalfa	β-Carotene, niacin, Vitamin D_2, cholesterol, diethylstilbesterol, DL-	(33)

Insect		Plant	Compound	Reference
Scientific name	Common name			
			aspartic acid, L-proline, histidine, Pangamic acid	
Dysdercus koenigii (F.)	a red cotton bug	Cotton	Ether extract	(178)
Hylotrupes bajulus (L.) *H. ater* Paykull	House beetles Bark beetles	*Pinus* sp. and others	α-pinene	(225)
Sitophilus oryzae (L.)	Rice weevil	Rice grains	Ether extract, methyl ketones	(226)
Costelytra zealandica (White)	a grass grub beetle	Elder, *Sambucus niger* (L.)	Essential oil	(227)
Cerambycid sp.	Wood boring beetles	Wood	Turpentine, smoke	(228)
Ostrinia nubilalis	European corn borer	Corn	Essential oil	(229)
Apis mellifera L.	Honey bee	General	Geraniol	(230)
Trichoplusia ni (Hubner)	Cabbage looper		Phenylacetaldehyde	(231)
Hypsipyla grandella (Zeller)	a mahogany shoot	*Cedrela odorata*	leaf acetone extract	(232)
Ithomia iphianassa Doubleday	a butterfly	*Heliotropium indicum* L.	Pyrrolizidine alkaloids	(233)
Chrysopa carnea Stephens	common green lacewing		Tryptophan	(234)

were reported. A parallel with the compounds reported as sex attractants for insects of this order is therefore suggested. Among the Lepidoptera, 6 esters, 2 glycosides (probably improperly identified as attractants), 2 acids, and 5 alcohols were reported.

In our laboratory, Minyard et al. (15) found β-bisabolol to be present in the highest concentration of any polar compound in the cotton bud essential oil. It was not previously reported in nature, but has subsequently been found by our group to be present in several other malvaceous oils. However, β-bisabolol alone is only about 50% as attractive in the laboratory olfactometer bioassay as a hot water extract of cotton. By fractionation and bioassay of the cotton bud essential oil, several other components were identified that were attractive in their own right and that improved the activity of β-bisabolol; they included β-carophyllene, ℓ-limonene, α-D-pinene, and β-caryophyllene oxide (16). Subsequently, α-bisabolol (17) and bisabolene oxide (18) have been identified as other contributing attractants.

Feeding Deterrents. Alkaloid glycosides have been among the most frequent compounds reported as feeding deterrents. Kuhn and Low (19) found several of these compounds in Solanaceae and implicated them as deterrents against the Colorado potato beetle *Leptinotarsa decemlineata* (Say). More recently, Harley and Thorsteinson (20) have shown the deterrency of this class against two-striped grasshopper *Melanoplus bivittalus* (Say) nymphs. Lichtenstein et al. (21) isolated 2-phenylethylisothiocyanate from the turnip as an antifeedant of vinegar flies. Two resistance factors in corn plants, 6-methoxybenzoxazolinone (22) and 2,4-dihydroxy-7-methoxy-1,4-benzoxazine-3-one (23) were shown to be effective against the European corn borer *Ostrinia nubialis* (Hubner). Rudman et al (24) showed that 2 anthraquinones present in teak heartwood inhibited termite activity. Several other interesting compounds include juglone (25), azadirachten (26), nepetalactone (27) cocculolidine and isoboldine (28), and 2 shriomodial acetates (29). When the known feeding deterrents were classified by chemical structure (Table 3), 19 related alkaloids or alkaloid glycosides affecting 5 insects were reported. The other classes and their frequencies were 4 lactones, 4 quinones affecting 2 insects, 5 heterocyclic ring compounds, 1 isothiocyanate, and 1 acid. While 7 orders of insects were studied, all of the reports except 12 involved Coleoptera.

Although Beck (3) and Munakata (30) stress the concept that feeding deterrents do not directly kill the insect, most of those reported are in fact biological poisons. Two representative compounds for which LD_{50} values were given in the Merck Index were the alkaloid glycoside tomatin (25 mg/kg i.p. and 500 mg/kg oral in mice) and nornicotine (23.5 mg/kg i.p. in rats). The alkaloid aglycones would be expected to be more toxic. Most of these compounds are bitter to humans, and some similar perception appar-

Table III. Feeding deterrents.

Insect				
Scientific name	Common name	Plant	Compound	Reference
Leptinotarsa decemlineata (Say)	Colorado potato beetle	Potato, *S. demisum* L.	Demissine	(*235*), (*236*)
Leptinotarsa decemlineata (Say)	Colorado potato beetle	Potato, *S. tuberosum*	dihydro-alpha-solanin	(*19*, *237*)
		S. caulescens	Solacaulin	(*19*)
		S. dulcamare	Soladulcin	(*235*)
		S. polyademium	Tomatin	(*237*), (*235*)
		S. schreiteri	Solanin	(*237*), (*235*)
		S. punae	Solanin	(*237*), (*235*)
		S. acaulia	Solacaulin	(*237*), (*235*)
Sitona cylindricollis Fåhraeus	Sweetclover weevil	Sweet clover	Ammonium nitrate	(*238*)
Anthonomus grandis Boheman	Boll weevil	Rose of Sharon (calyx)	Unknown	(*239*)
Listroderes costirostris obliquus (Klug)	Vegetable weevil	Sweet clover, *Meliolotus officianalis* L.	Coumarin	(*195*)
Ostrinia nubilalis (Hubner)	European corn borer	Corn	6-Methoxybenzoxazolinone, 2,4-dihydroxy-2-methoxy-1,4-benzoxazine-3-one	(*240*) (*22*, *56*) (*23*)
Oraesia excavata (Butler)	a fruit-piercing moth	*Cocculus trilobus* (DC)	Cocculolidine, isoboldine	(*28*)
Oraesia emarginata (F.)	a fruit-piercing moth	*Cocculus trilobus* (DC)	Cocculolidine, isoboldine	(*28*)
Abraxas miranda (Butler)		*Cocculus trilobus* (DC)	Cocculolidine, isoboldine	(*28*)

Insect				
Scientific name	Common name	Plant	Compound	Reference
Spodoptera litura (F.)	a fruit piercing moth	*Cocculus trilobus* (DC)	Cocculolidine, isobol-dine	(28)
Spodoptera litura (F.)	a fruit piercing	*Clerodendron tricotomum* (Thunb.)	Clerodendrin A	(30)
		Parabenzoin trilobum Nakoe	Shiromodiol diacetate, Shiromodiol monoacetate	(29)
Spodoptera eridania (Cramer	Southern armyworm	*Mucana* sp. seeds	L-Dopa	(241)
Diatraea gradiosella Dyar	Southwestern corn borer	Corn, other	Cholesteryl acetate Cholesteryl myristate Cholesteryl oleate	(181)
Musca domestica L.	House fly	*Podocarpus hallii*	Hallactones A & B (Norditerpenelactones)	(242)
Trichoplusia ni (Hubner)	Cabbage looper	Laboratory diets	Ascorbic acid	(122)
Two ant species (Formicadae)	Phytophagous insects	Catnip, *Nepeta cataria*	Nepetalactone	(27)
Acyrthosiphon pisum (Harris)	Pea aphid	Alfalfa	Coumesterol	(243)
Therioaphis maculata (Buckton)	Spotted alfalfa aphid	Alfalfa	Coumesterol	(243)
Melanoplus bivittatus (Say)	Twostriped grasshopper	Catnip, *Nepeta cataria*	Nornicotine dipicrate solanine, tomatine, digitonin, saponin, santonin, indicane dioe-gin, hecogenein, lucenine	(20)
Papillo polyxenes asterius Stoll	Black swallowtail butterfly	*Cruciferae* spp.	Sinigrin	(244)
Scolytus multi-striatus (Marsham)	Smaller European elm bark beetle	*Carya ovata* bark	Juglone	(25)

Insect		Plant	Compound	Reference
Scientific name	Common name			
Schistocerca gregaria (Forskål)	a desert locust	Neem tree, *Azadirachta indica* *melia*, *Azerdarach* *scilla maritima*	Azadirachtin,	(26) (245) (171)
Heliothis zea (Boddie)	Corn earworm	Corn silks	Unknown	(246)
Schizaphis graminum (Rondani)	Greenbug	Small grains	Benzyl alcohol	(247)
Drosophila pachea Patterson and Wheeler		*Senita cacti*	Pilocereine Tophocereine	(248)
Epicauta fabricii (LeConte)	Ashgray blister beetle	Sweet clover	Coumarin, *cis*-Hydroxy-coumaric acid glucoside	(249)
Epicauta vittata (F.)	Striped blister beetle	Sweet clover	Coumarin, *cis*-*o*-Hydroxy-coumaric acid glucoside	
Epicauta pestifera Werner	Margined blister beetle	Sweet clover	Coumarin, *cis*-*o*-Hy-droxy-coumaric acid glucoside	(249)
Epicauta spp.	Blister beetles	(Blood of blister beetles)	Cantharidin	(250)
Drosophila melanogaster Meigen	a vinegar fly	Turnip (*Brassica rapa* L.	2-phenylethyl isothio-cyanate	(21)
Chrysomelidae pieridae		Various	Tomatine Demissine Capsaicin N-Atropine Scopalamine Morphine Sparteine	(251)

ently occurs with insects.

Repellents. Beck (3) defines a repellent as a substance that elicits an oriented response away from the apparent source. The property of volatility is thus inferred for the candidate compound. Another group of repellents includes the defense secretions. However, they are biosynthesized by insects rather than by plants, so they have not been included. Some of the compounds reported (Table 4) were myrcene and limonene (31) and butyric acid and coumarin (33). Several others that were reported as repellents have no appreciable vapor pressure and therefore should have been reported as feeding deterrents. Any repellency probably should have been attributed to impurities in the isolated fraction. They include anacardic acid (34), tannic acid (35), and shikimic acid, pyridoxine, succinic acid, malic acid, betaine, and xanthophyll (33).

Twelve appreciably volatile compounds were reported; they include: 4 hydrocarbons, 3 acids, 2 phenols, 2 terpenes and 1 alcohol (Table 4). The paucity of work on naturally occurring insect repellents is apparent. However, since many plants that are resistant to insects have characteristic odors, it is possible that a number of others exist. The vastly improved capability for identification of volatile compounds, i.e., integrated gas chromatography-mass spectrometry and highly sensitivity nuclear magnetic resonance spectrometry, could facilitate expanded research on repellents in the next few years.

Oviposition Stimulants. Many of the compounds reported as oviposition stimulants (Table 5) were also identified as feeding stimulants or attractants (Tables 1 and 2), but not necessarily for the same insects. More oviposition stimulants were reported for Diptera than for any other order. This may reflect the relative simplicity of the bioassay as compared with that for other orders. The cabbage maggot *Hylemya brassicae* (Wiedemann) was stimulated by sinagrin and β-phenylethylamine (36). Matsumoto and Thorsteinson (37) reported that the onion maggot was stimulated by n-propyl disulfide, methyl disulfide, methyl disulfide, n-propyl mercaptan, and n-propanol. (Sulfides and mercaptans are found in high concentrations in onions.) The carrot rust fly *Psila rosae* (F) was stimulated by methyl isoeugenol and (*E*)-*trans*-1,2-dimethoxy-4-propenylbenzene (38). The mosquito was reported by Perry and Fay (39) to be stimulated by 3 esters; ethyl acetate, methyl propionate, and methyl butyrate.

In Lepidoptera, a partially characterized glycoside for the tobacco hornworm *Manduca sexta* (L.) (40), allylisothiocyanate for the diamond back moth *Plutella zylostella* (L.) (41) and two partially characterized growth regulators for the European corn borer (42) were reported.

Table IV. Repellents.

Insect				
Scientific name	Common name	Plant	Compound	Reference
Leptinotarsa decemlineata (Say)	Colorado potato beetle	Tomato, S. esculentum Mill.	Tomatin	(235)
Leptinotarsa decemlineata (Say)	Colorado potato beetle	Pepper, S. capsicum	Capsaicin	(235)
Leptinotarsa decemlineata (Say)	Colorado potato beetle	Tobacco, Nicotiana tobacum L.	Nicotine	(252) (253)
Heliothis zea (Boddie)	Corn earworm (Bollworm)	Corn, Zea mays L.	Essential oil	(139)
Manduca sexta (Johna.) (L.)	Tobacco hornworm	Nicandria sp.	Alcohol $C_{22}H_{27}O$	(131)
Manduca sexta (Johna.) (L.)	Tobacco hornworm	Nicandria sp.	NIC-2, $C_{28}H_{40}O_7$	(254)
Reticulitermes sp.	a termite	Cashew, Anacardium occidentale L.	Anacardic acid	(34)
Dendroctonus brevicomis LeConte	Western pine beetle	Ponderosa pine	Myrcene, limonene	(31)
Schistocerca gregaria (Forskål)	a desert locust	General, corn	n-valeric acid isovaleric acid, Unknown	(211) (32) (172)
Hypera postica (Gyllenhal)	Alfalfa weevil	Alfalfa	Tannic acid	(35)
Bruchophagus roddi (Gussakovsky)	Alfalfa seed chalcid	Alfalfa	Butyric acid, coumarin, shikimic acid, pyridoxine, succinic acid, xanthophyll, malic acid, betaine	(33)

Insect				
Scientific name	Common name	Plant	Compound	Reference
	Assorted	*Clerodendrum tricotomum* Thunb.	Clerodendrin A (diterpene)	(255)
	Redwood insects	Coast Redwood	Phenolic compounds	(256)

Table V. Oviposition stimulants.

Insect		Plant	Compound	Reference
Scientific name	Common name			
Leptinotarsa decemlineata (Say)	Colorado potato beetle	*S. tuberosum*	Lecithin	(43)
Manduca sexta (L.)	Tobacco hornworm	Tomato	glycoside, $C_{17}H_{29}O_{10}N$	(40), (131)
Plutella xylostella (Curt)	Diamondback moth	*Brassica nigra* L. & other Cruciferae	Allyl isothiocyanate	(41)
Hylemya brassicae (Wiedemann)	Cabbage maggot	Swede, *Brassicae napus* L.	Sinigrin, betaphenyl-ethylamine, allyl isothiocyanate	(36)
Hylemya antiqua (Meigen)	Onion maggot	Onion	n-propyl disulfide,, methyl disulfide, n-propyl mercaptan, n-propyl alcohol	(37)
Schistocerca gregaria (Forskal)	a desert locust	*Commiphora myrrha*	alpha-Pinene, beta-pinene, limonene, eugenol	(44)
Psila rosae (F.)	Carrot rust fly	Carrot leaves	Methyl isoeugenol trans-1,2-dimethoxy-4-propenyl-benzene	(38), (38)
Ostrinia nubilalis (Hubner)	European corn borer	*Zea mays*	"Gibberellin-like" acytokinin, other	(42)
Aedes aegypti (L.)	Yellowfever mosquito	--	ethyl acetate, methyl propionate, methyl butyrate	(39)
Heliothis virescens (F.)	Tobacco budworm	Tobacco	Unknown	(257)
Cryptorhynchus lapathi L.	Poplar and willow borer	Poplar	Unknown	(258)

Insect				
Scientific name	Common name	Plant	Compound	Reference
Callosobruchus chinensis (L.)	a bruchid	Soybeans	Saponins, SBSE, Urease	(45)

In Coleoptera, lecithin was reported for the Colorado potato beetle (43). In Orthoptera, 4 terpenes were reported for the so called desert locust Schistocerca gregaria (Forskal) (44). In Coleoptera, saponins were reported for a multivoltine bruchid Callosobruchus chinensis (L.) (45). All data on the oviposition stimulants are summarized in Table 5.

There were also reports on feeding incitants, flight termination stimulants, swallowing factors, mating factors, growth factors, and an oviposition deterrent; these data are included in Table 6. β-Sitosterol, β-sitosterol glucoside, lupeol, isoquercitrin and 2',3,4',5,7-pentahydroxyflavone were reported as feeding incitants for the silkworm larvae(7, 46, 47, 48, 49). Phaseolunatin and lotaustrin, cyanogenic glycosides, were reported as as feeding incitants for the Mexican bean beetle Epilachna varivestis (Mulsant) (50). Coumarin was reported as a flight termination agent for the sweet clover weevil Sitona cylindricollis (Fahraeus) (51). Cellulose, silica and potassium phosphate were reported as swallowing factors for the silkworm larvae (7), and protein, leucine, and guanosine-5'-monophosphate as swallowing factors for the mosquito (52).

trans-2-Hexenal was reported as a mating factor for the polyphemus moth Antheraea polyphemus (Cramer) (53, 54). Chlorogenic acid was described as a growth factor for the silkworm larva (55), and some partially characterized sapogenic aglycones were reported as oviposition deterrents for the bruchid. The compounds reported for these miscellaneous categories are the same as, or similar to, those reported earlier as attractants, feeding stimulants, and deterrents.

Toxins and Growth

Not all of the chemical constituents of plant tissue are conducive to insect growth and well-being. Plant defense against utilization by insects includes an array of alkaloids, phenols, flavonoids, and other substances, many of which may also have feeding deterrent or repellent properties as discussed in the previous sections. Analogs of insect hormones that have been found in plant tissue also may contribute to host plant resistance; however, they will not be included in this discussion.

No attempt will be made to cite every reported plant toxin, or other growth inhibiting chemical. Rather, a few relatively well understood plant-insect interactions have been selected for discussion. They include the benzoxazolinone compounds with European corn borer in maize; Gossypol with Heliothis zea (Boddie) and Heliothis virescens (F.) in cotton and saponins and protease inhibitors affecting several insects in legumes and stored legume seeds.

European Corn Borer. The resistance of maize varieties to the growth and survival of European corn borer larvae was shown

Table VI. Miscellaneous

Insect				
Scientific name	Common name	Plant	Compound	Reference
		Feeding Incitants		
Bombyx mori (L.)	Silkworm	Mulberry	Beta-sitosterol, beta-sitosterol glucoside, lupeol	(*46*) (*47*)
Epilachna varivestis Mulsant	Mexican bean beetle	*Phaseolus* sp.	Phaseolunatin	(*50*)
Bombyx mori (L.)	Silkworm	Mulberry	isoquercitrin	(*7*)
Bombyx mori (L.)	Silkworm	Mulberry	2',3,4',5,7-pentahydroxyflavone	(*49*)
Leptinotarsa decemlineata (Say)	Colorado potato beetle	Potato leaves	Alcoholic extract	(*123*)
Dysdercus koenigii (F.)	Red cotton bug	Cottonseed	Lipids	(*178*)
		Flight Termination		
Sitona cylindricollis Fahraeus	Sweet clover weevil	*Melilotus officianalis* L.	Coumarin	(*51*)
Bombyx mori (L.)	Silkworm	Mulberry	Cellulose, water extract, silica, potassium phosphate	(*7*)
Musca domestica (L.)	House fly	Laboratory diet	Protein, ℓ-leucine, Guanosine-5'-monophosphate	(*52*)

Insect				
Scientific name	Common name	Plant	Compound	Reference
		Mating Factor		
Antheraea polyphemus (Cramer)	polyphemus moth	Oak leaves	*trans*-2-hexenal	(*53*)
		Growth Factor		
Bombyx mori (L.)	Silkworm		chlorogenic acid	(*55*)
?	an alfalfa aphid	Alfalfa	Unknown	(*117*)
		Oviposition Deterrent		
Callosobruchus chinensis (L.)	a bruchid	Soybeans	Sapogenin	(*45*)

to be the result of feeding deterrency and/or growth inhibition exerted by benzoxazolinone compounds in certain tissues of young plants (3, 22, 56). Beck (3) covers in some detail the research that led to the discovery of resistant factors A, B, and C. Resistant factor A was identified as 6-MBOA (6-methoxy-benzoxazolinone) (22). Klun (57) found a strong correlation between the amount of 6-MBOA produced by 11 maize inbreds at the whorl stage of development and the field rating of resistance of the inbreds to the first-brood European corn borer. Highly resistant inbreds yielded 10 times more 6-MBOA than highly susceptible inbreds. He also suggested that a precursor of 6-MBOA may be more biologically active than 6-MBOA. Klun et al. (23) in a followup study evaluated 2,4-dihydroxy-7-methoxy-2*H*-1,4-benzoxazin-3-one (DIMBOA) which Virtanen (58) and Wahlroos and Virtanen (59) had reported as the precursor of 6-MBOA. When DIMBOA was incorporated in the diet, it inhibited larval development and caused 25% mortality. It was concluded that DIMBOA is a chemical factor in the resistance of corn to first-brood European corn borer.

The adequacy of DIMBOA in accounting for differences in levels of resistance was studied in 11 inbreds and their diallel combinations by Klun et al. (60). Leaf-feeding ratings were obtained under artificial infestation and whorl-leaf samples analyzed for DIMBOA. The correlation between DIMBOA and resistance to leaf feeding was 0.89 at the inbred level and 0.74 at the hybrid level. The variance for general combining ability accounted for 84% of the variation in resistance ratings and 91% of the variation in concentration in DIMBOA.

Efforts to find resistance to second-brood borers is in progress. Pesho et al. (61) identified several inbred lines possessing an acceptable level of resistance. Scott and Guthrie (62) reported a split-plot study contrasting chemicals and resistance as control measures for second-brood borers. The resistant X resistant crosses exhibited a 4% yield reduction, the yield of the susceptible X susceptible cross was reduced by 12%. The chemical and genetic bases of second-brood borer resistance in corn remain unknown. Feeding studies have not implicated DIMBOA in second brood resistance.

Heliothis spp. *Heliothis zea*, the bollworm, and *Heliothis virescens*, the tobacco budworm, are generally referred to as the bollworm complex. Both species feed on a wide range of host plants (63, 64). A free exchange of hosts occurs between cotton and other host plants as reported in a study in Mississippi (63). In all species of the genus *Gossypium*, lysigenous gossypol glands in the plant parts above-ground have contents known to be toxic to nonruminant animals including insects.

Three cotton gland pigments, gossypol, quercetin, and rutin, were incorporated into the standard bollworm diet and found by Lukefahr and Martin (65) to decrease larval growth. They suggested that plant breeders might select for cotton plants with higher pigment content as a mechanism of resistance. High gossypol cotton lines (gossypol content 1.7% in X-G-15) have been found in the dooryard cottons (66). Shaver and Lukefahr (67) found in dietary studies that naturally occurring flavonoid pigments have potential as sources of resistance to *Heliothis* spp. larvae.

Oliver et al. (68) found that glanded cotton diets caused a 21-31% reduction in feeding by all ages of larvae and that the efficiency of food conversion was decreased. Shaver et al. (69) also found that food utilization by *H. zea*, but not *H. virescens*, was reduced by a diet containing high gossypol cotton. Larval weight gains of both species were also reduced by feeding the high gossypol diet.

There are two recessive genes, gl_2 and gl_3 that produce cotton plants devoid of gossypol glands in the plant and seed (70). In a study of larval feeding by *Heliothis* on 14 pairs of glanded and glandless cotton lines, weight gains were greater by those feeding on the glandless member of each pair (71, 72). In field studies, larvae feeding on glandless lines were larger, but the damage to the plant was not significantly greater (68, 73). However, the glandless lines in some variety backgrounds tended to receive more damage than the glanded counterpart in both these studies. Thus, these authors suggest a careful consideration of the variety-glandless interaction. Wilson (74, 75) found that *H. virescens* larvae at 6 days of age could discriminate among the 9 genotypes of the two glandless genes. Their preference was inversely related to gossypol content of the 9 genotypes. He also suggests that it might be possible to select seedlings by counting pigment glands in the petiole and cotyledon. Through cooperative efforts of several scientists across the U. S. Cotton Belt, breeding lines with adequate gossypol levels to confer good field resistance to *Heliothis* have been developed.

Gossypol and related terpenoids are also related to disease resistance in cotton. Gossypol and four related terpenoid aldehydes, 6-methoxygossypol, 6,6'-dimethoxygossypol, hemigossypol and 6-methoxyhemigossypol, were identified in roots of 1-wk-old Acala 4-42 cotton seedlings (76). Histochemical procedures revealed the localization of these terpenoids in the epidermis and in scattered cortical parenchyma cells of the healthy tap root.

These compounds may be related to resistance of cotton to *Verticillium dahliae*. Since gossypol and related terpenoids are absent from the first 3 cm of healthy root tissue and since the root cap zone appears to be the primary point of the fungus infection, the fungus probably does not contact these terpenoids when it first penetrates the root. Minton (77) reported that the root knot nematode (*Meloidogyne incognita acrita*) penetrated the roots of Auburn 56 and Rowden cotton seedlings principally within the apical 2 cm of the tap root. This area of root in 1-wk-old seedlings of these varieties is free of gossypol and related terpenoids (78). This root tip zone may be causally related to the susceptibility to nematode penetration.

Legume Insects. Applebaum and Birk (79), reviewing the natural mechanisms of resistance to insects in legume seeds, cited the isolation of a *Tribolium* larval protease inhibitor from the soybean. They noted that *Tribolium* protease inhibitors are present in legume species that are normally regarded as susceptible to damage by *Tribolium* spp. and that the activities of these inhibitors are invariably specific on *Tribolium* *in vivo* and *in vitro*. They suggest that selection and breeding of seed varieties (e.g., groundnuts or chickpeas) containing higher concentrations of the specific *Tribolium* protease inhibitors would afford at least partial resistance to damage in storage, without being detrimental to human or animal nutrition.

Applebaum and Guez (80) found a specific heteropolysaccharide comprising about 1% of the haricot bean that was responsible for resistance to the bruchid beetle. Higher concentrations of this factor also impart resistance to a second bruchid beetle, *Acanthoscelides obtectus* (Say), a major pest of haricot beans (80).

Saponins are triterpenoid glycosides that occur in legume seeds and plants including soybean. Five aglycones have been isoisolated and their chemical structures have been determined (81). These and related saponins also occur in checkpeas, garden peas, broad beans, haricot beans, lentils and groundnuts (82). A correlation was also reported between the toxicity of certain legume saponin fractions to *Callosobruchus* spp. larvae, and the relative resistance of these different legume seeds to insect damage (82).

Saponins have also been implicated as possible factors for resistance in alfalfa to white grubs (83), pea aphid, potato leafhopper and certain other alfalfa insects (84, 85). Hanson et al. (86) showed that most alfalfa (*Medicago sativa* L.) populations selected for high saponin concentration had more pea aphid (*Acyrthosiphon pisum* (Harris) resistance than those with low saponin concentration. Pedersen et al. (87), however, found that the foliage and root saponin concentrations of six pea aphid resistant and five pea aphid-susceptible alfalfa varieties were not significantly correlated with insect damage.

There was significant negative correlation of pea aphid resistance with root saponin content, but there was a significant positive correlation between clover root curculio (*Sitona hispidulus* (F.)) damage and foliage saponin content. There was no evidence that breeding for resistance to pea aphids had changed the foliage saponin concentration, but it appeared to reduce the amount of saponin in the roots.

The apparent conflict between the results of Hanson et al. (86) and Pedersen et al. (87) might be explained by the work of Horber et al. (84) who studied saponin components from Dupuits and Lahontan alfalfa. Lahontan contained more soyasapogenol A than Dupuits, but medicagenic acid predominated in Dupuits. Medicagenic acid was found to be responsible for toxic properties in all assays. Presence or absence of medicagenic acid seemed to explain most of the differences in biological activity of extracts of Dupuits and Lahontan cultivars. Increased mortality of potato leaf hopper *Empoasca gabae* (Harris) and pea aphid nymphs was related to not only increasing saponin concentration, but also to the kind and origin of saponins' host cultivar.

One of the best known cases of saponins being implicated in disease resistance is resistance of oats (*Avena sativa* L.) to take-all disease (*Gaeumannomyces graminis* (Sacc.) Arx and Olivier) in which the triterpenoid saponin called "Avenacin" has been identified as the causative agent (88).

Nutritional Factors in Plants Affecting Survival and Development of Insects

The insect is dependent on the plant for much more than nutrients: Chemostimulation, physical factors, and a satisfactory microenvironment are all interacting factors which play a role in determining whether a plant may be susceptible or resistant (3). A resistant plant is not necessarily nutritionally inadequate, and more frequently than not, nutrition does not appear to be a primary factor. Painter (2) suggested that some instances of resistance may be attributed to the complete absence of specific nutrients required by the insect; however, to our knowledge, there is no absolute documented evidence of this form of resistance. The opposite opinion has been expressed by Fraenkel (6). This view was based on the unproven assumption that all plants are capable of meeting any insect's nutritional needs. We know now from subsequent research that the role of nutritional factors in plant resistance is far too complex to fit either of these views (5, 89).

Although resistance mechanisms involving host nutritional status may be quite prevalent, they are most difficult to prove because the resistance may be due to the amount of some factor that is present or the requirement for some obscure trace factor that is absent. A potential disadvantage of genetic manipulation of the nutritional quality of the insect's host is that a plant developed so that it would be nutritionally inadequate for the insect might also be equally inadequate for man or the animals that would utilize the crop. Additionally, the plant may not be acceptable to FDA regulations restricting major changes in nutritional factors.

For the purposes of this discussion, the relationship of nutritional factors to resistance will be reviewed in relationship to two mechanisms of resistance--antibiosis and preference.

Antibiosis. Antibiosis is the adverse effect of a plant (in this context) on some aspect of the insect's biology (1, 2, 90). More often this type of resistance stems from toxins or other such antibiotic agents (see previous section). From a nutritional standpoint, although there are few documented examples, antibiosis may occur for one or more of the following reasons (90): (1) the absence or deficiency in the plant of some nutritional materials such as vitamins, essential amino acids or specific sterols; and (2) the imbalance in available nutrients, especially sugar-protein or sugar-fat ratios.

Sugar content of plants may be very important to insect pests, usually because of feeding stimulation, but it may also be limiting for proper growth and survival. For example, low levels of soluble sugar content in the host plant of the cabbage aphid *Brevicoryne brassicae* (L.) limits reproduction and development of winged forms in the aphid (91). Akeson et al (92) demonstrated the importance of glucose, fructose and sucrose levels in sweetclover (*Melilotus*) leaves to successful feeding and nutrition of the sweetclover weevil (*Sitona cylindricollis* Fahraeus). Certain pentoses in the bean, *Phaseolus vulgaris*, have been shown to inhibit the growth of the bruchid *Callosobruchus chinensis* (L.) (93).

Some insects' requirements for sugars may vary with age. Reduction in sugar content of the plant at the more critical stages may cause resistance. For example, larvae of European corn borer, *Ostrinia nubilalis*, require glucose until the fourth instar and are capable of minutely differentiating between different concentrations (94). Knapp et al. (95) also showed that the bollworm, *Heliothis zea*, could discriminate between concentrations of sugars and that the sugar balance was important in the resistance of certain lines of corn to this insect.

Most phytophagous insects studied thus far have shown the same qualitative requirements for amino acids as the rat. In pea aphid resistant plants of *Pisum sativum* L., lower concentrations of amino acids occur in resistant than in susceptible lines (96, 97). Similarly, Colorado potato beetle larvae were shown to grow faster with higher survival rates on young potato foliage than on leaf foliage, presumably because of a more favorable amino acid content in the younger tissue (98). The silks of certain corn lines resistant to the corn earworm (bollworm) also have lower concentrations of amino acids than those of susceptible lines (99). It has not been shown, however, that lowered amounts of certain amino acids were the principal factors in resistance in either of the previous cases. Van Emden and Bashford (100) found that the green peach aphid *Myzus persicae* (Sulzer) selects for important amino acids in its host, especially in relation to tissue age. Van Emden (101) further defined the role of certain amino acids in host determination and selection by aphids.

Parrott et al. (102) conducted an extensive test for qualitative and quantitative differences in amino acids of the various species of Gossypium in the hopes of finding a species that would be deficient or void of an essential amino acid required for development of the boll weevil. Unfortunately, no qualitative and no great quantitative differences were found. Benepal and Hall (103) studied the free amino acids in plants of Brassica oleracea var. capitata L. as related to resistance to the cabbage looper Trichoplusia ni (Hubner) and the imported cabbageworm Pieris rapae L. and found pipecolic acid and tyrosine present in susceptible varieties though they were not detected in resistant varieties. In a similar study with Cucurbita pepo L. varieties and squash bugs Anasa tristis (DeGeer), total free amino acids and total soluble N were higher in susceptible than in resistant plants (104).

Most insects utilize dietary fat or fatty acids for energy, for a source of metabolic water, and for building reserves of depot fat and glycogen. The body fat of insects is affected quantitatively and qualitatively by the host on which they feed. Plants that do not provide sufficient amounts of fats in the insects' diet may have a detrimental effect on the insects' biology. Mature larvae and pupae of bollworm (Heliothis zea), contained more fat when the larvae were reared on a high-fat, dough-stage corn than on lower-fat, milk-stage corn although the iodine and saponification numbers of the corn lipids at both stages remained almost constant (93). Cholesterol has been used in most chemical diets developed for plant feeding insects (105), although it is not the characteristic sterol of the higher plants. Grison (106) observed that egg production rates of the Colorado potato beetle, (Leptinotarsa decemlineata), were positively correlated with the phospholipid content of the potato foliage; senescent foliage was deficient in phospholipids required for egg production. Carnitine, lysine, linoleic acid, and inositol are other examples of substances reported to affect the biology of particular insects, when a deficient amount is present (90). Dadd (107) speculated that a derivative common to both carotene and vitamin A constituted a growth factor for Schistocerca and Locusta spp. Waites and Gothilf (108) stated that when thiamine, riboflavin, pyridoxine, calcium pantothenate, folacin, biotin, and niacin were individually omitted from the diet, high mortality of the almond moth larvae, Cadra cautella (Walker), resulted. There are many other similar studies that have been done that show the importance of vitamins and related compounds in the diets of insects. Unfortunately, few studies have been done that involve the analysis of the insect host plant. One of the few studies conducted in recent years is that of Hudspeth et al. (109) who surveyed a cross-section of cotton lines for vitamin C content. Vitamin C is necessary for boll weevil development. Tremendous quantitative differences were evident among the investigated cotton lines, but lines containing very small quantities of Vitamin C were found to support excellent boll weevil growth and development. Whether in a host plant, synthetic diet, or other nutritional substrate, the importance of the dietary proportion of required nutrients may be of greater importance than their absolute quantities (110, 111). Such nutrient ratios may also influence to some extent feeding behavior, of the insect, and they are almost certain to be important factors in the efficiency of conversion of ingested food into energy and insect tissues.

There is much evidence in the literature on the importance of minerals in plants as they relate to resistance. Plants deficient in minerals needed by insects may in addition contain atypical concentrations of organic compounds that can affect the growth or reproductive capacity of the insects feeding on them. Barker and Tauber (112) reported that the pea aphid exhibits lower reproductive capacity on plants deficient in Ca, Mg, N, P, and K. Allen and Selman (113) found that the number of eggs laid by the twospotted spider mite, *Tetranychus urticae* (Koch), was not affected for at least 7 days on leaf discs from plants that received excess amounts of Fe, Mn, Zn, or Co; also a wide range of Fe in the leaves appeared to be tolerated by females without affecting the quantity of eggs laid. The lack of Fe or Zn in diets for the green peach aphid, *Myzus persicae*, increased the percentage of alates formed, whereas larvae deprived of P developed mainly to apterae (115). The imported cabbageworm, *Pieris rapae*, has been observed to be affected in different ways on leaves deficient in N, P, K, and Fe (113).

Branson and Simpson (116) allowed corn leaf aphid, *Rhopalosiphum maidis* (Fitch), to feed on sorghum plants grown under high and low levels of nitrogen. Results showed that more than twice as many aphids were to be found on plants receiving nitrogen than on nitrogen-deficient plants. Kindler and Staples (117) studied the effect of excess, medium, and deficient amounts of Ca, Mg, N, K, P, and S on spotted alfalfa aphid, *Therioaphis maculata* (Buckton), feeding on alfalfa clones resistant and susceptiable to the aphid. None of the treatments made the susceptible clone more resistant. Resistance was significantly decreased but not eliminated when the resistant clone had deficient levels of Ca or K or excess levels of Mg or N, but resistance was significantly increased in plants with deficient levels of P. Sulfur did not affect resistance.

Similar studies conducted by Rodrignez et al. (118) with susceptible and resistant strawberry clones with different N levels demonstrated significant correlation between foliage N and mite injury. Studies of the survival of European corn borer larvae on resistant and susceptible corn plants supplied different levels of N and P showed that increased N did not affect degree of resistance, but amount of N in the soil influenced survival in the field.

In general, parallel tests of resistant and susceptible plant varieties under different fertility conditions show that resistant and susceptible plants tend to be affected in the same way. Under no conditions of soil fertility has a resistant plant been susceptible or a susceptible one become highly resistant. Because each species of insect, each host plant species, and often each type of soil constitutes a separate problem, no general conclusion regarding the importance of particular nutrient elements can be made. However, it is well-known that certain plant species and varieties have the ability to extract and utilize various chemicals from the soil with greater efficiency than other species and varieties. If the increased absorption and use of the element contributes to resistance, the ability may be a basis for resistance.

Preference. Preference or nonpreference is defined as the effect of those chemical or morphological host plant characteristics and insect responses that lead away from the selection and use of a particular plant for food, oviposition, shelter, or a combination of the three (90).

The morphology of the host plant may affect the nutrition of the insect in the following ways; (a) it may limit the amount of feeding because of texture, shape, or color, which would reduce the amount of nutrients being ingested, and (b) it may limit the digestibility and utilization of the food by the insect. Some plant substances may play a dual role in that they stimulate as well as being essential nutritionally (119, 120, 121, 122, 123, 124, 125). Proteins, sugars, phospholipids, inorganic salts, minerals, vitamins, etc., are examples of materials that many times function in dual roles by acting as feeding stimulants or suppressants and also being essential for proper development. Essential oils, glycosides and other secondary substances are usually involved more in a single role--attraction and feeding stimulation.

Literature Cited

(1) Painter, R. H. "Insect Resistance in Crop Plants." 520 pp. Macmillin, New York (1951).

(2) Painter, R. H. Ann. Rev. Entomol. (1958) 3, 267-290.

(3) Beck, S. D. Ann. Rev. Entomol. (1965) 10, 207-232.

(4) Maxwell, F. G., Jenkins, J. N., Parrott, W. L. Advances in Agronomy (1972) 24, 187-265.

(5) Maxwell, F. G. "Insect and Mite Nutrition" 702 pp. (ed. J. G. Rodriguez). North Holland Publ. Co., Amsterdam. (1972).

(6) Fraenkel, G. Science (1959) 129, 1466-1470.

(7) Hamamura, Y., Hayashiya, K., Naito, K., Matsuura, K., Nishida, J. Nature (1962) 194, 754-755.

(8) Robbins, W. E., Thompson, M. J., Yamamoto, R. T., Shortino, T. J. Science (1965) 147, 628-630.

(9) Hamamura, Y. "Control of Insect Behavior by Natural Products", pp. 55-80 ed. by D. L. Wood, R. M. Silverstein and

M. Nakajima. Academic Press, New York (1970).

(10) Hedin, P. A., Thompson, A. C., Minyard, J. P. J. Econ. Entomol. (1966) 59, 181-185.

(11) Struck, R. F., Frye, J., Shealy, Y. F., Hedin, P. A., Thompson, A. C., Minyard, J. P. J. Econ. Entomol. (1968) 61, 270-274.

(12) Struck, R. F., Frye, J., Shealy, Y. F., Hedin, P. A., Thompson, A. C., Minyard, J. P. J. Econ. Entomol. (1968) 61, 664-667.

(13) Hedin, P. A., Miles, L. R., Thompson, A. C., Minyard, J. P., J. Agr. Food Chem. (1968) 16, 505-513.

(14) Temple, C., Roberts, E. C., Frye, J., Struck, R. F., Shealy, Y. F., Thompson, A. C., Minyard, J. P., Hedin, P. A. J. Econ. Entomol. (1968) 61, 1388-1393.

(15) Minyard, J. P., Thompson, A. C., Hedin, P. A., J. Org. Chem. (1968) 33, 909-11.

(16) Minyard, J. P., Hardee, D. D., Gueldner, R. C., Thompson, A. C., Wiygul, G., Hedin, P. A. J. Agr. Food Chem. (1969) 17, 1093.

(17) Hedin, P. A., Thompson, A. C., Gueldner, R. C., Minyard, J. P. Phytochemistry (1971) 10, 1693-1694.

(18) Hedin, P. A., Thompson, A. C., Gueldner, R. C., Ruth, J. M. Phytochemistry. (1972) 11, 2118-2119.

(19) Kuhn, R., Low, I. Angew. Chem. (1957) 69, 236.

(20) Harley, K., Thorsteinson, A. J. Canadian J. Zool. (1967) 45, 305-319.

(21) Lichtenstein, E. P., Strong, F. M., Morgan, D. G. J. Agr. Food Chem. (1962) 10, 30-33.

(22) Smissman, E. E., Lapidus, J. P., Beck, S. D. J. Amer. Chem. Soc. (1957) 79, 4697.

(23) Klun, J., Tipton, C., Brindley, T. J. Econ. Entomol. (1967) 60, 1529-1533.

(24) Rudman, P., Gay, F. J. Hotzforschung (1961) 15, 50-53.

(25) Gilbert, B., Baker, J., Norris, D. J. Insect Physiol. (1967) 13, 1453-1459.

(26) Gill, J., Lewis, C. Nature (1971) 232, 402-403.

(27) Eisner, T. Science (1964) 146, 1318-1320.

(28) Wada, K., Munakata, K. J. Agr. Food Chem. (1968) 16), 471-474.

(29) Wada, K., Enomoto, Y., Matsui, K., Munakata, K. Tetrahedron lettrs. (1968) 45, 4673-4676.

(30) Munakata, K. "In Control of Insect Behavior by Natural Products", pp. 179-87. ed. by Wood, L., Silverstein, R. M., and Nakajima., N. Academic Press, New York (1970).

(31) Smith, R. H. Science (1966) 12, 63-68.

(32) Goodhue, D. Nature (1963) 197, 405-406.

(33) Kamon, J. A., Frank, W. D. Univ. of Wyoming Agr. Exp. Sta. Bull. 413. 35 pp. (1964).

(34) Wolcott, G. M. J. Econ. Entomol. (1949) 42, 273-275.

(35) Bennett, S. E. J. Econ. Entomol. (1965) 58, 372.

(36) Traynier, R. M. M. Nature (1965) (207, 201-208.
(37) Matsumoto, Y., Thorsteinson, A. J. Appl. Entomol. Zool. (1968) 3, 5-12.
(38) Berentes, J., Staedler, E. Z. Naturforsch. B. (1971) 26, 339-340.
(39) Perry, A. S., Fay, R. W. Mosquito News (1967) 27, 175-183.
(40) Yamamoto, R. G., Fraenkel, G. Ann. Entomol. Soc. Amer. (1960) 53, 499-503.
(41) Gupta, P. D., Thorsteinson, A. J. Entomol. Exptl. & Appl. (1969) 3, 305-314.
(42) Schurr, K. Adv. Front & Plant Sci. (1970) 24, 147.
(43) Grison, P. Academse aes Sciences Compt. Rend. (1958) 227, 1172-1174.
(44) Carlisle, D. B., Ellis, P. E., Betts, E. J. Insect Physiol. (1965) 11, 1541-1558.
(45) Applebaum, S. W., Gestetner, B., Bik, Y. J. Insect Physiol. (1965) 11, 611-616.
(46) Hamamura, Y., Hayashiya, K., Naito, K. Nature (1961) 190, 880-881.
(47) Nayar, J. K., Fraenkel, G. Ann. Entomol. Soc. Amer. (1962) 56, 174-178.
(48) Goto, M., Kamada, M., Imai, S., Murata, S., Fejrta, E., Hamamura, Y., Takeda Ken. Nempo (1965) 24, 55-65.
(49) Hayashiya, K., Naito, K., Nishida, J., Hamamura, Y. Nippon Nogei Kagaku Kaishi (1963) 37, 735-737.
(50) Klingenburg, M., Bucher, J. Ann. Rev. Biochem. (1960) 29, 669-708.
(51) Heidweg, H., Thorsteinson, A. J. Entomol. Exptl. & Appl. (1961) 4, 165-177.
(52) Yamamoto, R. T., Jensen, E. J. Insect Physiol. (1967) 13, 91-98.
(53) Reddiford, L. Science (1967) 158 , 139-140.
(54) Reddiford, L. M., Williams, C. M. Science (1967) 155, 589-590.
(55) Kato, M., Yamada, H. Life Sci. (1966) 5, 717-722.
(56) Smissman, E. E., Lapidus, J. P., Beck, S. D. J. Org. Chem. (1957) 22, 220.
(57) Klun, J. A. Diss. Abstr. (1965) 26, 3544.
(58) Virtanen, A. I. Soum. Kemistilehti (1961) B34, 29-31.
(59) Wahlroos, O., Virtanen, A. I. Acta Chem. Scand. (1959) 13, 1906-1908.
(60) Klun, J. A., Guthrie, W. D., Hallauer, A. R., Russell, W. A. Crop Sci. (1970) 10, 87.
(61) Pesho, G. R., Dicke, F. F., Russell, W. A. Iowa State J. Sci. (1965) 40, 85-98.
(62) Scott, G. E., Guthrie, W. D. Crop Sci. (1967) 7, 233-235.
(63) Snow, J. W., Brazzel, J. R. J. Econ. Entomol. (1965) 58, 525-526.

(64) Snow, J. W., Brazzel, J. R. Mississippi Farm Res. (1966) 29, 6-7.

(65) Lukefahr, J. J., Martin, D. F. J. Econ. Entomol. (1966) 59, 176-179.

(66) Lukefahr, M. J., Houghtaling, J. E. J. Econ. Entomol. (1969) 62, 588-591.

(67) Shaver, T. N., Lukefahr, M. J. J. Econ. Entomol. (1969) 62, 643-646.

(68) Oliver, B. F., Maxwell, F. G., Jenkins, J. N. J. Econ. Entomol. (1970) 63, 1965-1966.

(69) Shaver, T. N., Lukefahr, M. J., Garcia, J. A. J. Econ. Entomol. (1971) 63, 1544=1546.

(70) McMichael, S. C. Agron. J. (1960) 52, 385-387.

(71) Lukefahr, M. J., Nobel, L. W., Houghtaling, J. E. J. Econ. Entomol. (1966) 59, 817-820.

(72) Oliver, B. F., Maxwell, F. G., Jenkins, J. N. J. Econ. Entomol. (1971) 64, 396-398.

(73) Jenkins, J. N., Maxwell, F. G., Lafever, H. N. J. Econ. Entomol. (1966) 59, 352-356.

(74) Wilson, F. D. Crop Sci. (1971) 11, 268.

(75) Wilson, F. D. Crop Sci. (1971) 11, 419.

(76) Stipanovic, R. D., Bell, A. A., Howell, C. R. Phytochem. (1975) 14, 1809-1811.

(77) Minton, N. A. Phytopathology (1962) 52, 272-279.

(78) Mace, M. E., Bell, A. A., Stipanovic, R. D. Phytopathology (1974) 64, 1297-1302.

(79) Applebaum, S. W., Birk, Y. "Insect and Mite Nutrition". 702 pp. (J. G. Rodriguez, ed.) North Holland Publ. Co., Amsterdam (1972).

(80) Applebaum, S. W., Guez, M. Entomol. Exptl. et Appl. (1972) 15, 203-207.

(81) Birk, Y. "Toxic Constituents of Plant Foodstuffs". (I. E. Liener, ed.) pp. 169-210. Academic Press, N. Y. (1969).

(82) Applebaum, S. W., Marco, S. Birk, Y. J. Agr. Food Chem. (1969) 17, 618-622.

(83) Horber, E. Intl. Cong. Entomol. XII Proc. pp. 540-541. (1965).

(84) Horber, E., Leath, K. T., Berang, B., Marcarian, V., Hanson, C. H. Entomol. Exptl. & Appl. (1974) 17, 410-424.

(85) Roof, M., Horber, E., Sorensen, E. L. Central Branch Entomol. Soc. Amer. (1972) 27, 140-143.

(86) Hanson, C. H., Pedersen, M. W., Berrang, B., Wall, M. E., Davis, K. H. "Antiquality Components of Forages". 140 pp. Special Publ. No. 4. Amer. Soc. of Agron., Madison, Wisconsin. (1973).

(87) Pedersen, M. W., Sorensen, E. L., Anderson, M. J. Crop Sci. (1975) 15, 254-260.

(88) Maizel, J. V., Burkhardt, H. J., Michell, H. K. Biochem. (1964) 3, 424-426.

(89) Beck, S. D. "Proceedings of the Summer Institute on

Biological Control of Plant Insects and Diseases". 647 pp. (F. G. Maxwell and F. A. Harris, eds.) University Press, Jackson, Miss. (1972).
(90) Painter, R. H. "Principles of Plant and Animal Pest Control". pp. 64-99. Natl. Acad. Sci. Publ., Washington, D. C. (1969).
(91) Evans, A. C. Ann. Appl. Biol. (1938) 25, 558-572.
(92) Akeson, W. R., Gorz, H. J., Haskins, F. A., Manglitz, G. R. J. Econ. Entomol. (1968) 61, 1111-1112.
(93) Friend, W. G. Ann. Rev. Entomol. (1958) 3, 57-74.
(94) Beck, S. D. J. Insect Physiol. (1957) 1, 158-177.
(95) Knapp, J. L., Hedin, P. A., Douglas, W. A. J. Econ. Entomol. (1966) 59, 1062-1064.
(96) Auclair, J. L. Ann. Rept. Entomol. Soc. Ontario (1957) 88, 7-17.
(97) Srivastana, P. N., Auclair, J. L. Canadian Entomol. (1974) 106, 149-156.
(98) Cibulz, A. B., Davidson, R. H., Fisk, F. W., Lapidus, J. B. Ann. Entomol. Soc. Amer. (1967) 60, 626-631.
(99) Knapp, J. L. Diss. Abs. (1966) 26, 4918.
(100) Van Emden, H. F., Bashford, M. A. Entomol. Exptl. & App. (1971) 14
(101) Van Emden, H. F. (ed.) "Insect/Plant Relationships." Halsted (Wiley), New York. 216 pp. Symposia of the Royal Entomol. Soc. of London, No. 6 (1973).
(102) Parrott, W. L., Maxwell, F. G., Jenkins, J. N., Hardee, D. D. Ann. Entomol. Soc. Amer. (1969) 62, 261-264.
(103) Benepal, P. S., Hall, C. V. Ann. Soc. Hort. Sci. Proc. (1967) 91, 425-430.
(104) Benepal, P. S., Hall, C. V. Amer. Soc. Hort. Sci. (1967) 91, 353-359.
(105) Vanderzant, E. S., Reiser, R. J. Econ. Entomol. (1956) 49, 4-10, 454-458.
(106) Grison, P. A. Entomol. Exptl. & Appl. (1958) 1, 73-93.
(107) Dadd, R. H. Nature (1957) 179, 427-428.
(108) Waites, R. E., Gothilf, S. J. Econ. Entomol. (1969) 62, 301-305.
(109) Hudspeth, W. N., Jenkins, J. N., Maxwell, F. G. J. Econ. Entomol. (1969) 62, 583-584.
(110) House, H. L. Entomol. Exptl. & Appl. (1969) 12, 651-659.
(111) House, H. L. J. Insect Physiol. (1971) 17, 1125-1238.
(112) Barker, J. S., Tauber, O. E. J. Econ. Entomol. (1954) 47, 113-116.
(113) Allen, M. D., Selman, I. W. Bull. Entomol. Res. (1957) 48, 229-242.
(114) Cannon, W. N., Terriere, J., Terriere, L. C. J. Econ. Entomol. (1966) 59, 89-93.
(115) Raccah, B., Tahori, A. S., Applebaum, S. W. J. Insect Physiol. (1971) 17, 1385-1390.
(116) Branson, T. F., Simpson, R. G. J. Econ. Entomol. (1966)

59, 290-293.
(117) Kindler, S. D., Staples, R. J. Econ. Entomol. (1970) 63, 938-940.
(118) Rodriguez, J. G., Chaplin, E. E., Stolitz, L. P., Lasheen, A. M. J. Econ. Entomol. (1970) 63, 1855-1858.
(119) Beck, S. D., Hanec, W. J. Insect Physiol. (1958) 2, 85-96.
(120) Thorsteinson, A. J. Ann. Rev. Entomol. (1960) 5, 193-219.
(121) Davis, G. R. F. Arch. Intern. Physiol. Biochem. (1965) 73, 177-187.
(122) Gothilf, S., Beck, S. D. J. Insect Physiol. (1967) 13, 1039-1953.
(123) Hsiao, T. H., Fraenkel, G. Ann. Entomol. Soc. Amer. (1968) 61, 485-493.
(124) Ishikawa, S., Hirao, T., Arai, N. Entomol. Exptl. & Appl. (1969) 12, 544-554.
(125) Nayar, J. K., Thorsteinson, A. J. Canadian J. Zool. (1963) 41, 923-920.
(126) Verschaeffelt, E., Proc. Acad. Sci. Amsterdam (1910) 13, 536-542.
(127) David, W. A. L., Gardiner, B. O. C. Entomol. Exptl. & Appl. (1966) 9, 247-255.
(128) Detheir, V. G. "Chemical Insect Attractants and Repellents." 289 pp. The Blakiston Co., Philadelphia. (1947).
(129) Hewitt, P. H., Whitehead, V. B., Read, J. S. J. Insect Physiol. (1969) 15, 1929-1934.
(130) Cauvin, R. Annales de L.I.N.R.A. (1952) 3, 303-308.
(131) Yamamoto, R. T., Fraenkel, G. Nature (1959) 184, 206-207.
(132) Derr, R. F., Randall, D. D., Kieckhefer, R. W. J. Econ. Entomol. (1964) 57, 963-965.
(133) Keller, J. C., Maxwell, F. G., Jenkins, J. N. J. Econ. Entomol. (1962) 55, 800-801.
(134) Maxwell, F. G., Jenkins, J. N., Keller, J. C., Parrott, W. L. J. Econ. Entomol. (1963) 56, 449-454.
(135) Jenkins, J. N., Maxwell, F. G., Keller, J. C., Parrott, W. L. Crop Sci. (1963) 3, 215-219.
(136) Keller, J. C., Davich, T. B. Personal Communication (1964).
(137) Keller, J. C., Davich, T. B. J. Econ. Entomol. (1965) 58, 165.
(138) Detheir, V. G. Biol. Bull. (1937) 72, 7-23.
(139) Starks, K. J., McMillian, W. W., Sekul, A. A., Cox, H. C. Ann. Entomol. Soc. Amer. (1965) 58, 74-76.
(140) Jones, R. L., McMillian, W. W., Wiseman, B. R. Ann. Entomol. Soc. Amer. (1972) 65, 821-824.
(141) Heron, R. J. Canadian J. Zool. (1965) 43, 247-269.
(142) Howell, G. S., Jr., Maxwell, F. G., Nevins, R. B. Ann. Entomol. Soc. Amer. (1969) 62, 240.
(143) Loschiavo, S. R. Ann. Entomol. Soc. Amer. (1965) 58, 526-588.

(144) Doskotch, P. M., Mikhail, A. A. Lloydia (Cinci) (1973) 35, 473.

(145) Thorsteinson, A. A. Canadian J. Zool. (1953) 31, 52-72.

(146) Stride, G. O. J. Insect Physiol. (1965) 11, 21-22.

(147) Thorsteinson, A. J., Nayar, J. K. Canadian J. Zool. (1963) 41, 931-935.

(148) Parker, J. R. Minnesota Agr. Exp. Sta. Bull. 214 (1924).

(149) Lagerheim, G. Entomol. Tidskv. (1900) 21, 209-232.

(150) Grevillius, A. Y. Bot. Centralbl. Beihefte 18 Abt. II. (1905) 22.

(151) Peterson, A. J. Econ. Entomol. (1924) 17, 87-94.

(152) Kearns, H. G. H. Ann. Rept. Agr. Hort. Sta., Long Ashton. 199 (1931).

(153) Chambliss, O. Y., Jones, C. M. Science (1966) 153, 1392-1393.

(154) DaCosta, C. P., Jones, C. M. Science (1971) 172, 1145-1146.

(155) Euw, J., Fishelson, L., Parson, J., Rechstein, T., Rothschild, M. Nature (1967) 214, 35-38.

(156) Wensler, R. Nature (1962) 195, 830-1.

(157) Panella, J. S., Webster, J. A., Zabik, M. J. J. Kansas Entomol. Soc. (1974) 47 , 348-357.

(158) Zielske, A., Simons, J., Silverstein, R. Phytochemistry (1972) 11, 393-396.

(159) Baker, J., Norris, D. Entomol. Exptl. & Appl. (1968) 11, 464-469.

(160) Harris, P. Canadian Entomol. (1967) 99, 1304-1310.

(161) Harris, P. Canadian Entomol. (1963) 95, 105.

(162) Guerra, A., Shaver, T. J. Econ. Entomol. (1968) 61, 1393-1399.

(163) Hsiao, T. J. Insect Physiol. (1969) 15, 1785-1790.

(164) Dosthotch, R., Chatterji, S., Peakcock, J. Science (1970) 167, 380-382.

(165) Ehrlich, P. R., Raven, P. H. Scien. Amer. (1967) 216, 104-113.

(166) McMillian, W. W., Starks, K. J. Ann. Entomol. Soc. Amer. (1966) 59, 516-519.

(167) Khalifa, A., Rizk, A., Salama, H. S., El-Sharaby, A. F. J. Insect Physiol. (1973) 19, 1501-1509.

(168) Vinson, S. B., Thompson, J. L., Green, H. B. J. Insect Physiol. (1967) 13, 1729-1736.

(169) Beland, C. L., Haskins, F. A., Manglitz, G. R., Gorz, H. J. J. Econ. Entomol. (1973) 66, 1037-1039.

(170) Thorsteinson, A. J. Canadian Entomol. (1955) 87, 49-57.

(171) Chauvin, R., Mentzer, C. Bull. Offic. Natl. Anti-Acrid. (1951) 1, 5-14.

(172) Mulkern, G. B. Ann. Rev. Entomol. (1967) 12, 59-78.

(173) Davis, G. R. F. Canadian, J. Zool. (1961) 39, 299-303.

(174) Rees, C. J. C. Ph.D. dissertation, Univ. of Oxford, England. (1966).

(175) Andrea, L. USDA Spec. Rept. (1964).

(176) Force, D. C. Ann. Entomol. Soc. Amer. (1966) 59, 1119-1125.

(177) Parrott, W. L., Shaver, T. N., Keller, J. C. J. Econ. Entomol. (1968) 61, 1766-1767.

(178) Saxena, K. N. Proc. XII Intl. Cong. Entomol., London. p. 24. (1964).

(179) Wiseman, B. R., McMillian, W. W., Burton, R. L. J. Georgia Entomol. (1969) 4, 15-22.

(180) Allen, G. E., Pate, T. L. J. Invertebr. Pathol. (1966) 8, 129-131.

(181) Chippendale, G. M., Reddy, G. P. V. Experientia (1974) 28, 485-486.

(182) Rymal, K. S., Chambliss, O. L. J. Am. Soc. Hort. Sci. (1976) 101, 722-724.

(183) Hovanitz, W., Chang, V. C. S. J. Res. on the Lepidoptera (1963) 2, 281-288.

(184) Detheir, V. G. Amer. Naturalist (1941) 75, 61-73.

(185) Guenther, E. "The Essential Oils" Vol. 2, D. Van Nostrand Co., New York. (1949).

(186) Watambe, T., Nature (1958) 182, 325-326.

(187) Hamamura, Y. Nature (1959) 183, 1746-1747.

(188) Horie, Y. J. Sericult. Sci. Japan (1962) 31, 258-264.

(189) Ito, T. Bull. Sericult. Exptl. Sta. (Tokyo) (1961) 17, 119-136.

(190) Iwanari, Y. Sanshigakj Zasshi (1974) 42, 403-405.

(191) Keller, J. C., Maxwell, F. G., Jenkins, J. N., Davich, T. B. J. Econ. Entomol. (1963) 56, 110=111.

(192) Folsom, J. W. J. Econ. Entomol. (1931) 24, 827-833.

(193) Munakata, K., Saito, T., Ogawa, S., Ishii, S. Bull. Agr. Chem. Soc. Japan (1959) 23, 65-67.

(194) Matsumoto, Y., Sugiyama, S., Ber. Ohara Inst. Land. Biol. (1969) 11, 359-364.

(195) Matsumoto, Y. A. Japan J. Appl. Entomol. Zool. (1962) 6, 141-149.

(196) Morgan, A. C., Lyon, S. C. J. Econ. Entomol. (1928) 21, 189-191.

(197) Feir, D., Beck, S. D. Ann. Entomol. Soc. Amer. (1963) 56, 224-229.

(198) Cuille, J. Inst. Fruits et Agrumes Coloniaux, Ser. Tech. (1950) 4.

(199) Wright, R. H. Nature (1965) 207, 103-104.

(200) Howlett, F. M. Trans. Entomol. Soc. London (1912) pp. 412-418.

(201) Howlett, F. M. Bull. Entomol. Res (1915) 6, 297-305.

(202) Feeny, P. P., Paauwe, K. L., Demong, N. J. Ann. Entomol. Soc. Amer. (1970) 63, 832-841.

(203) Watanabe, T., Casida, J. E. J. Econ. Entomol. (1963) 56, 300-301.

(204) Heikkenen, H. J., Hruitfiord, B. F. Science (1965) 150.

1457-1459.

(205) Muto, T., Sugawara, R. Fr. Agr. Biol. Chem. (Japan) (1965) 29, 949-955.

(206) Saeto, T., Munakata, K. "Control of Insect Behavior by Natural Products", pp. 225-235, ed. by D. L. Wood, R. M. Silverstein and N. Nakajima. Academic Press, New York (1970).

(207) Sakan, T., Issac, S., Hyeon, S. "Control of Insect Behavior by Natural Products", pp. 237-247, ed. by D. L. Wood, R. M. Silverstein and N. Nakajima. Academic Press, New York. (1970).

(208) Kangas, E., Pertunnen, V., Oksanen, H., Rinne, M. Ann. Entomol. Fenn. (1965) 31, 61-73.

(209) Cade, S., Hruitfiord, B., Gara, R. J. Econ. Entomol. (1970) 63, 1010-1015.

(210) Kennedy, J., Moorehouse, J. Entomol. Exptl. and Appl. (1969) 12, 487-503.

(211) Haskell, P. T., Paskin, W. J., Moorehouse, J. E. J. Insect Physiol. (1961) 8, 53-78.

(212) Byrne, H. D., Steinauer, A. L. Entomol. Soc. Amer. (1966) 59, 303-309.

(213) Tingey, W. M., Nielson, M. W. J. Econ. Entomol. (1974) 67, 219-221.

(214) Thomas, H., Hertell, G. J. Econ. Entomol. (1969) 62, 383-386.

(215) Seybert, J., Gara, R. Ann. Entomol. Soc. Amer. (1970) 63, 947-950.

(216) Johnson, N., Belluschi, P. J. Forestry (1969) 67, 290-295.

(217) Vite, J., Pitman, G. Canad. Entomol. (1969) 101, 113-117.

(218) Meyer, H. J., Norris, P. M. Ann. Entomol. Soc. Amer. (1967) 60, 858-859.

(219) de Wilde, J. Ghent. Land. Hog. Med. (1957) 22, 335-347.

(220) Harwood, W. G., Rudinsky, J. A. Tech. Bull. 95, Agr. Exp. Sta., Corvallis, Oregon. 36 pp. (1966).

(221) Eyer, J. R., Rhodes. J. Econ. Entomol. (1931) 24, 702-711.

(222) Madsen, H. F., Falcon, L. A. J. Econ. Entomol. (1960) 53, 1083-1085.

(223) Brunson, M. H. J. Econ. Entomol. (1955) 48, 390-392.

(224) Becker, von. Holzforschung (1964) 18, 168-172.

(225) Perttunen, V. Ann. Entomol. Fennici (1957) 23, 101-110.

(226) Honda, H., Yamamoto, I., Yamamoto, R. App. Entzool. (1969) 4, 23-31.

(227) Osborne, G. O., Hoyt, C. P. N. Zealand J. Sci. (1968) 11, 137-139.

(228) Gardiner, L. M. Canada Dept. Agr. Forest Biol. Bimonthly Prog. Rept. 13 (1957) 1-2.

(229) Moore, R. H. Proc. Oklahoma Acad. Sci. (1928) 8, 16-18.

(230) Free, J. B. J. Apic. Res. (1962) 1, 52-54.
(231) Creighton, C. S., McFadden, T. L., Cuthbert, E. R. J. Econ. Entomol. (1973) 66, 114-115.
(232) Gara, R. I., Allan, G. G., Wilkins, R. M., Whitmore, J. L. Z. Angew Entomol. (1973) 72, 259-266.
(233) Pliske, T. E., Edgar, J. A., Culvenor, C. C. J. Chem. Ecol. (1976) 2, 255-262.
(234) Van Emden, H. F., Hagen, K. S. Environ. Entomol. (1976) 5, 469-473.
(235) Schreiber, K. Entomol. Exptl. & Appl. (1958) 1, 28-37.
(236) Kuhn, R., Gauhe, A. Z. Naturforsch (1947) 2, 407-409.
(237) Kuhn, R., Low, I. "Origins of Resistance to Toxic Agents" pp. 122-132. Academic Press, New Oork (1955).
(238) Akeson, W. R., Haskins, F. A., Gorz, H. J. Science (1969) 163, 293-294.
(239) Maxwell, F. G., Parrott, W. L., Jenkins, J. N., Lafever, H. N. J. Econ. Entomol. (1965) 58, 985-988.
(240) Beck, S. D., Smissman, E. E. Ann. Entomol. Soc. Amer. (1960) 53, 755-762.
(241) Rehr, S. S., Janzen, D. H., Feeny, P. P. Science (1973) 181, 81-82.
(242) Russell, G. B., Fenemore, P. G., Singh, P. Chem. Commun.
(243) Loper, G. M. Crop Sci. (1968) 8, 104-105.
(244) Erickson, J. M., Feeny, P. Ecology (1974) 55, 103-111.
(245) Butterworth, J. H., Morgan, E. D. J. Insect Physiol. (1971) 17, 969-977.
(246) Straub, R., Fairchild, M. J. Econ. Entomol. (1970) 63, 1901-1903.
(247) Juneja, P. S., Gholson, R. K., Burton, R. L., Starks, K. J. Ann. Entomol. Soc. Amer. (1972) 65, 961-964.
(248) Kircher, H. W., Heed, W. B., Russell, J. S., Grove, J. J. Insect Physiol. (1967) 13, 1869-1874.
(249) Gorz, H. J., Haskins, F. A., Manglitz, G. R. J. Econ. Entomol. (1972) 65, 1632-1635.
(250) Carrel, J., Eisner, T. Science (1974) 188, 755-757.
(251) Levinson, H. Z. Experientia (1976) 32, 408-411.
(252) Buhr, H. Zuchter (1954) 24(7/8), 185-193.
(253) Trouvelot, B. Entomol. Exptl. & Appl. (1958) 1, 9-13.
(254) Bates, P. B., Morehead, S. R. Chem. Commun. (1974) 4, 125-126.
(255) Kato, N., Shibayama, M., Munakata, K. J. Chem. Soc. Perkin Trans. (1973) i(7), 712-719.
(256) Ediz, S. H. Ph. D. Thesis, Univ. Microfilm. 74-14, 358. (1974).
(257) Deutsch, R. G. M.S. Thesis, Texas A&M Univ., 35 pp. (1968).
(258) Cadahia, A. Boln. Serv. Plagas (1965) 8(16), 115-125.

INDEX

A

Abies balsamea 159
Abraxas miranda 188
Acanthoscelides obtectus 261
Acer negundo 227
Acer saccharinum 227
Acids, nucleic 107
Activities of plant extracts 186
Acyrthosiphon pisum 261
Aesculetin 216
Aesculus octandra 225
African armyworm 166, 167
African plants 165
Agricultural crops 117
Agrobacterium radiobacter 79
Agrobacterium tumefaciens 40, 79
Agrotis ipsilon 133
Ajuga remota 171
Ajugarins 171
Aldehyde, terpenoid 206, 209
Alfalfa 40, 124
Alkaloids 155, 227
  glycosides 247
Allelochemics 129
  effects of on assimilation of food 133
  effects of on black cutworm survival 136
  interactions with nutrients 131
  plant 137
Allomone(s) 215
  electrochemical difference between kairomone and 221
  non-host trees examined for 217
Alternaria mali 36
Alternaria tenuis 36
Amanita muscaria 157
Amanita pantherina 157
Amino acids 263
Anabasis aphylla 155
Anacyclus pyrethrum 156
Analyses, methods of phytochemical 217
Anasa tristis 264
Anhydro-β-rotunol 62
Antheraea polyphemus 256
Anthonomus grandis 239
Antibiosis 116, 207, 262
Antifeedant(s) 166
  activities of plant leaf extracts 187
  from African plants, insect 165
  from *Clerodendron tricotomum* 190
  from *Cocculus trilobus* 188
Antifeedant(s) (*continued*)
  from *Orixa japonica* 190
  in *Parabenzoin praecox* 193
  of *Parabenzoin trilobum* 189
  in *Piper futokazura* 193
  sesquiterpene lactones 179
  against *S. litura, verbenaceae* 193
  of *Spodoptera litura* in plants, insect 185
  tests, insect 179, 181
  tests, mammalian 183
  from verbenaceae 191
Antifungal phytoalexins 3, 36
Antifungal properties 12
Arbutus unedo 174
Armyworm, African 166, 167
Aspergillus flavus 158
Assimilation of food, effects of allelochemics on 133
Attractants, plant 240–246
Aubergenone 63
Azadirachta indica 156, 169
Azadirachtin 169, 170

B

Bacillus subtilis 28, 172
Bark beetle 219
Bark, *Carya ovata* twig 220
Bark, *Ulmus americana* twig 220, 224, 226
Barley 43, 124
Bean, *Colletotrichum lindemuthianum* 80, 81
Bean phytoalexins 3
Beet 43
Beetle, bark 219
Behavioral factors affecting host plant resistance to insects 231
ρ-Benzoquinone 216
Bicyclic sesquiterpenes, stereochemistry of 67
Binding activity of trichloroacetate-solubilized fractions, α-galactoside 43
Biochemical marker of toxin resistance 42
Biochemistry of mitochondria from N and T cytoplasms, comparative 103
Biologically active compounds in plants affecting insect behavior 232
Biosynthetic relationships of sesquiterpenoidal stress compounds from the solanaceae 61

Bitter substances 156
Blight
disease
mechanism of southern corn leaf 110
southern corn leaf 90
nature of late 47
Boll weevil 211
Bollworm 211, 259
Bombyx mori L. 239
Borer, corn 121
Botrytis cinerea 8
Brassica oleracea 264
Brassica rapa 185
Bremia lactucae 55
Brevicornyne brassicae 263
Brewer's yeast 32
Budworm, resistance to cotton to tobacco 117
Budworm, tobacco 211, 259

C

Cadra cautella 264
Calcium 105
Callicarpa japonica 191
Callosobruchus 261
chinensis 256, 263
Callus 93
Calospilos miranda 191
Canthium euroides 174
Capsicum frutescens stress metabolites 64
Capsidiol 63, 64
^{13}C NMR spectrum of
2,3-Dihydroxy germacrene 70
enriched phytuberin 74
farnesol 69
lubimin 73
rishitin 71
Carya
cordiformis 218
ovata 218
twig bark 220
trees, non-host 218
Caryopteris divaricata 191
Catechin 216
Celerio euphorbiae 144
Ceratocystis fimbriata 16
Chemicals that elicit their production in plants 1
Chemicals on N and T mitochondria, action of 103
Chestnut tannins 132
Chloroform-extractable HmT toxin on mitochondria 101
Chromatography
high-performance liquid 160
liquid 174
thin-layer 160
Chronic effects hypothesis 130
Chrysanthemum cinerariifolium 154
Cicadella viridis 190
Citrus trees 40
Clerodendron
calamitosum 191
cryptophyllum 191
fragrans 191
infortunatum 190
tricotomun 190
Cocculolidine 189
Cocculus trilobus, antifeedants from 188
Colletotrichum lagenarium 79
protection of cucurbits against 83
Colletotrichum
lindemuthianum 16, 33, 39, 79
elicitation of resistance of bean to 80, 81
Compatible hypocotyls, inoculated 14
Complementarity, gene-for-gene 9, 10
Compositae 179
Composition of N and T mitochondria, chemical 107
Conversion
assimilated food, efficiency of 134
dietary moisture effects on growth and efficiency of 147
ingested food, efficiency of 134
Corn 43, 124
borer 121, 256, 265
resistance of corn to 121
ear worm 122
resistance of corn to 121
leaf blight disease, southern 37
biochemical and ultrastructural aspects of 90
mechanism of 110
silk 122
Corynebacterium insidiosum 40
Cotton 118, 124
G. barbadense 204
G. hirsutum 204
heliothis resistance in 197, 199
natural insecticides from 197
tannin 118, 119
terpenoids 199
aldehyde components of cultivated 206
tobacco budworm, resistance to 117
toxicity in 120
Coumarin(s) 225, 226
Coumestrol 2, 11
Crops, HPR in agricultural 117
Crops that exhibit resistance to insect attack 124
Crotepoxide 173
Croton macrostachys 173
Cucumber leaf 85
Cucurbita pepo L. 264

Cucurbits against *Colletotrichum lagenarium*, protection of ... 83
Cutworm, black ... 135, 140
larvae, black ... 137, 145, 148, 149
survival, effects of some allelochemics on ... 136
Cytochromes ... 108
Cytoplasm, comparative biochemistry of mitochondria from N & T ... 103
Cytoplasm, effects of *H. maydis* Race T toxin on T ... 91, 99

D

Daidzein ... 11
Damage to T mitochondria, mechanism of HmT toxin-induced ... 102
Datura stramonium ... 66
stress metabolites ... 63
Decenylsuccinic acid ... 105
Defense system ... 87
against disease in plants, chemical 78
Dehydrogenase, NADH ... 102
Densities, population ... 129
Depolarization, membrane ... 53
Derris ... 154
Deterrents, feeding ... 247–250
of *Scolytus multistriatus* ... 215
Diacrisia virginica ... 179
Dibromide derivative of heliocide ... 200
Diels–Alder reaction ... 199
Diet(s) ... 147, 181–183
insect ... 130
moisture levels ... 144
effects on growth and efficiency of conversion ... 147
interrelationships nutritional indices and ... 139
Digitonin ... 105
15-Dihydrolubimin ... 62
2,3-Dihydroxy germacrene, ^{13}C NMR spectrum of ... 70
2,4-Dihydroxy-7-methoxy-1,4-benzoxazin-3-one ... 121
11,13-Dihydroxysolavetivone ... 65
3,4-Dimethoxyphenyl ... 194
2,4-Dinitrophenol (DNP) ... 105
Disease
chemical defense against ... 78
headblight ... 39
mechanism of southern corn leaf blight ... 110
resistance ... 2, 78
elicitors of phytoalexin accumulation in plant ... 27
phytoalexins as a mechanism for .. 7
and susceptibility, plant ... 35
southern corn leaf blight ... 90
Disk, leaf ... 166, 168

E

Earworm, resistance in corn ... 122
Echinacea angustifolia ... 156, 159
Edothia parasitica ... 79
Electrochemical difference between an allomone and a kairomone ... 221
Electrophysiology ... 166
Elicitor(s) ... 53
abiotic ... 15
biotic ... 15
chemical nature of the *Phytophthora megasperma* var. *sojae* ... 29
non-specific ... 15
phytoalexin ... 1, 15
accumulation in plant disease resistance ... 27
Pms ... 28, 33
receptor hypothesis specific ... 19
species specificity ... 31
specific ... 15
Empoasca gabe ... 262
Enzyme(s) ... 39
inhibitors ... 132
Epilachna varivestis ... 166, 256
10-Epilubimin ... 62
Equilibrium hypothesis ... 5
Erwinia carotovora var. *atroseptica* ... 52
Erysiphe graminis ... 4
Ether formation in glands, methyl ... 210
Eudesmanes ... 61
Euonymus japonicus ... 188
Euproctis pseudoconspersa ... 190
Euproctis subflava ... 191
European corn borer ... 256, 265
resistance of corn to ... 121

F

Fagara chalybea ... 175
Farnesol, biosynthesis of ... 70
Farnesyl pyrophosphate ... 68
Fatty acids ... 264
Feeding deterrents ... 247–250
Feeding stimulants ... 232–239
Flavonid(s) ... 224
oxidation states of ... 222
Food
effects of allelochemics on assimilation of ... 133
efficiency of conversion of ... 134
ingested ... 139
Fraxinus americana ... 225
Fungi ... 32
infection, potato resistance to ... 48
peptides, toxic ... 157
phytoalexins ... 3
Fungicides ... 44
Furanoterpenoids, accumulation of ... 86
Fusarium graminearum ... 39

G

Gaemannomyces graminis 262
Galactinol 43
α-Galactoside binding activity of trichloroacetate-solubilized fractions 43
Gene(s) 9, 87
  nuclear 90, 98
  -for-gene plant-parasite interactions 19
Genistein 2
Germacrene(s) 66
  conformational mobility of 75
Germacrene-A-diol 63
Germacrenediol 66
Glands, methyl ether formation in 210
Glaucolide-A 179, 181, 182
Glomerella cingulata 69, 75
  interaction with potato 72, 74
Glutinosone 64
Glyceollin 11, 72
Glycine max (soybean) phytoalexins, structures of 11
Glycosides, alkaloid 247
Gossypium 264
  barbadense 198, 204, 206
  hirsutum 158, 197, 204, 206, 239
  terpenoid content of 208
Gossypol 117, 158, 198, 259, 260
Gramicidin D. 105
Gramine 216
Green bean, (*Phaseolus vulgaris L.*) phytoalexins 3
Green bean, protection of 79
Growth
  dietary moisture effects on 147
  inhibition tests, larval 180
  regulators (juvenile and molting hormones), insect 158, 159
  toxins and 256

H

Harrisonia abyssinica 171
Headblight disease 39
Heliocide(s)
  dibromide derivative of 200
  in *heliothis* resistant cottons 199
  in plant tissue, distribution of 205
Heliopsis longipes 156
Heliothis
  resistance in cotton 197, 199
  virescens 118, 119, 133, 256, 259
  zea 121, 133, 256, 259
Helminthosporium maydis Race T (HmT) toxin 37, 90, 96
  damage to T mitochondria, mechanism of 102
HmT toxin (*continued*)
  effect on
    T cytoplasm 91–95
    mitochondria 98, 100, 101
    oxidative phosphorylation of 102
    release of malate dehydrogenase (MDH) 104
    respiratory activities of root mitochondria 99
  structures 111
  studies on the molecular structure of 108
Helminthosporium sacchari—sugarcane interaction 41
Helminthosporoside 43
High-performance liquid chromatography (HPLC) 160
Homeosoma electellum 123
Hormones, juvenile 158
Hormones, molting 159
Host
  cell resistance 56
  compounds, potato tuber 51
  interactions between *Phytophthora infestans* and potato 47
  parasite systems 9
  plant resistance
    agricultural crops 117
    insects 115, 231
    new compounds in 212
    nonagricultural plants 116
  receptor substances 18
  resistance genes 9
  specificity 129
  specific toxins 36, 108
^{1}H NMR screening techniques 203
ρ-Hydroquinone 216
13-Hydroxycapsidiol 64
2′ Hydroxy genistein 2
5-Hydroxy-1,4-naphthoquinone (juglone) 215
9-Hydroxynerolidol 63
2′ Hydroxy phaseollin isoflavin 2
Hylemya brassicae 251
Hyper postica 120
Hypersensitive area 6
Hypersensitive reaction 5
Hypocotyls
  inoculated compatible 14
  inoculated with *Phytophthora megasperma* var. *sojae*, soybean 13
  soybean 28, 31
Hypothesis, chronic effects 130
Hypothesis, equilibrium 5

I

Immunization 78
Incompatible interaction 53

Index(ices)
interrelationships of 140
parameters, nutritional 141
techniques, nutritional 134
Infection, potato resistance to 48
Infection of *Solanum tuberosum, Phytophthora infestans* 51
Inflexin 173
Infrared spectroscopy 160
Ingested food 139
Inhibitors, enzyme 132
Insect(s)
antifeedant(s)
from African plants 165
of *Spodoptera litura* in plants 185
tests 179, 181
attack, crops that exhibit resistance to 124
behavior, biologically active compounds in plants affecting 232
dietetics 130, 262
factors affecting host plant resistance to 231
growth 125, 129
host plant resistance (HPR) to 115
regulators (juvenile hormones) 158
regulators (molting horomnes) 159
Insecticide(s) 117
from cotton, natural 197
Interaction, potato-*Glomerella cingulata* 72, 74, 75
Interaction, potato-*Monilinia fructicola* 69
Ipomoea batatas 186
Isoboldine 189
Isobutylamides, unsaturated 156
Isodomedin 173
Isodon inflexus 173
Isodon shikokianus var. *intermedius* 173
Isolated roots 93
Isolation of toxic agents from plants 153
Isolubimin 62

J

Juglans nigra 218
Juglans trees, non-host *carya* 218
Juglone 215, 216
Juvenile hormones, insect growth regulators 158

K

Kaempferol 216
Kairomone 221
(−)-16-Kauren-19-oic acid 123
1-Keto-α-cyperone 65
Kievitone 2

L

Lactones, antifeedant sesquiterpene 179
Larvae, black cutworm 137, 145, 148, 149
Larval growth inhibition tests 180
Late blight, nature of 47
Leaf
disk 92, 166, 168
extracts, antifeeding activities of plant 187
infected 96
Legume insects 261
Leptinotarsa decemlineata 185, 247, 264
Lettuce leaf tissue 55
Lignan 226
Liquid chromatography 174
high-performance 160
Locusta 264
Lonchocarpus 154
Lubimin 62, 63
^{13}C NMR spectrum of 72, 73

M

Macromolecules from pathogens 39
Macromolecules from plants 41
Magnolia acuminata var. *acuminata* 227
Magnoline 216
Malate 99
dehydrogenase (MDH), effect of HmT toxin on the release of 104
oxidation 102
Malus pumila 223
Mammalian antifeedant tests 183
Manduca sexta 166, 251
Mass spectrometry 161
Medicago sativa L. 261
Melanoplus bivittalus 247
Melia
azadirachta 186
azedarach 156, 169
indica 186
Meloidogyne incognita acrita 260
Membrane(s)
depolarization 53
plasma 98
site of toxin action 97
Metabolic effects 130
Metabolites 72
plants 1
stress 78
Capsicum frutescens 64
Datura stramonium 63
Solan tuberosum 62
Solanum melongena 63
Methanol, toxic materials in 123
6-Methoxy-2-benzoxazolinone 121
2′ Methoxy phaseollin isoflavin 2
Methyl ether formation in glands 210

Methylation, effect of 122
Microsomes 95
Minerals 265
Mint 43
Miscellaneous factors 257, 258
Mitochondria
  action of chemicals on N and T 103
  chemical composition of N and T 107
  comparative biochemistry of N and T cytoplasms 103
  effect of HmT toxin on 98–101
  genetics in higher plants 110
  isolated 91, 94
  mechanism of HmT toxin-induced damage to T 102
  membranes 97
  protein 107
  respiration, T 102
  site of toxin action, evidence for 98
Moisture 143
  effects on growth and efficiency of conversion, dietary 147
Molecule(s)
  interactions 35
  parasite-produced 35
  plant-produced 38
  structure of HmT toxin, studies on 108
Molting hormones, insect growth regulators 159
Monilinia fructicola 69, 75
  interaction, potato- 69, 71
Moth
  oak 120
  resistance of sunflowers to 123
  winter 116
Mundulea 154
Mycotoxins 158
Myzus persicae 263, 265

N

1,4-Naphthoquinones 218
Nectria galligena 16
Nicotiana
  glauca 155
  sesquiterpenes 65
  stress metabolites 64
  tabacum 155
NADH (nicotinamide adenine dinucleotide, reduced) 97, 99, 102
  dehydrogenase 102
  oxidation by toxin 102
  respiration 105
Nicotine 155
Nicotinoids 155
Nigericin (NIG) 105
Nitrogen 265
Nonagricultural plants, HPR in 116
Nonpreference 116
NMR
  screening techniques, ^{1}H 203
  spectrometer 66
  spectroscopy 160
Nucleic acids 107
Nutrient(s)
  factors in plants affecting survival and development of insects 262
  indices and dietary moisture levels, interrelationships of 139
  index parameters 141
  index techniques 134
  interactions between allelochemics and 131
  physiology 129

O

Oak 99
  leaf tannins 132
  moth 120
  tree 116, 120
Oligosaccharides 30, 31
Operophtera brumata 120, 132
Oraesia emarginata 188
Oraesia excavata 188
Orixa japonica, antifeedants from 190
Ostrinia mubilalis 191
Ostrinia nubialis 247, 263
Oviposition stimulants 251, 254, 255
Oxidation states of flavonids 222
Oxidative phosphorylation 102, 105, 106
  effects of digitonin on respiration and 106
9-Oxonerolidol 63

P

Papilio polyzenes asterius 133
Parabenzoin praecox, antifeedants in 193
Parabenzoin trilobum, antifeedants of 189
Parasite
  interactions, gene-for-gene plant– 19
  -produced small molecules 35
  systems, host– 9
  virulence genes 9
Pastinaca sativa 185
Pathogen(s)
  macromolecules from 39
  protection of plants from 32
  system, plant– 10
Penicillium expansum 16
Peptides, toxic fungal 157
Periplaneta americana 215, 225
Pesticides 169
Phagostimulants 166
Phaseollidin 2
Phaseollin 2

Phaseolus
lunatus L. 4
vulgaris L. 33, 39, 263
phytoalexins from 2
protection of green bean 79
Phenolic acid 226
effects on S. *multistriatus* 225
Phloretin 216
Phoma tracheiphila 40
Phosphorylation, oxidative 102, 105, 106
Phylloxera vitifolia 115
Physiology, nutritional 129
Phytoalexin(s) 78
accumulation in plant disease resistance 27
antifungal 3, 36
bean 3
and chemicals that elicit their production in plants 1
classical 86
elicitors 1, 15
fungal 3
green bean 3
mechanism for disease resistance in plants 7
potato 3
soybean 3, 11, 27
theory 54
toxic effects of 32
Phytochemical analyses, methods of 217
Phytophthora
capsici 8
infestans 3, 48, 53, 57
infection of *Solanum tuberosum* 51
potato host, interactions between 47
megasperma 10, 27, 35, 53
elicitor, chemical nature of the 29
soybean hypocotyls inoculated with 13
Phytuberin 62, 72
biosynthesis of 73
^{13}C NMR spectrum of enriched 74
Phytuberol 75
Pieris rapae 264, 265
Pinus radiata 121
Piper futokazura, antifeedants in 193
Pisum sativum 263
Plant(s)
allelochemics 137
antifeedant(s) 165, 185, 187, 188, 191
apparency 129
attractants 240–246
biochemicals, effects of 129
chemical defense against disease in 78
disease resistance, elicitors of phytoalexin accumulation in 27
disease resistance and susceptibility 35
effects of on insect behavior 232
Plant(s) (*continued*)
effects of on insect survival and development 262
extracts, activities of 37, 186
macromolecules from 41
metabolites 1
mitochondrial genetics in higher 110
–parasite interactions, gene-for-gene 19
-pathogen system 10
phytoalexins and chemicals that elicit their production in 1
phytoalexins as a mechanism for disease resistance in 7
-produced small molecules 38
protection from their pathogens 32
repellents from African 165
resistance
to insects 115, 231
new compounds in host 212
in nonagricultural 116
response to insect growth 125
specificity, host 129
tissue, distribution of heliocides in 205
vegetative 1
Plasma membrane(s) 98
site of toxin action, evidence for the 97
Plutella zylostella 251
Pms elicitor 28, 33
Pms–soybean system 31
Pollen 93
Polygodial 172
Polyphenols, distribution of 117
Population densities 129
Populus deltoides 223
Potato 43, 124
interaction
with *Glomerella cingulata* 74
with *M. fructicola* 69, 71, 72
with *Phytophthora infestans* 47
leaves 57
phytoalexins 3
resistance to fungal infection 48
tuber tissue 50–54, 69
Pratylenchus scribneri 3
Protection, induced systemic 84
Protein, mitochondrial 107
Protoplasts 94
Pseudomonas glycinea 28
Pseudoplusia includens 140
Psila rosae 251
Psoralidin 2
Pyrausta nubilalis 185
Pyrethrum, toxicant 153
Pyruvate 99

Q

Quercetin 216

Quercus macrocarpa 223
Quiesone 64

R

Race-specific resistance 31
Raffinose 43
Reaction, hypersensitive 5
Receptor hypothesis, specific 19
Receptor substances, host 18
Repellent(s) 251–253
  African plants 165
  feeding of *Scolytus multistriatus* 215
Resistance 87, 120
  bean to *Colletotrichum lindemuthianum*, elicitation of 80, 81
  biochemical marker of toxin 42
  compatible 2
  corn 121, 122
  cotton, *Heliothis* 117, 197, 199
  disease 2, 78
    elicitors of phytoalexin accumulation 27
    phytoalexins as a mechanism for 7
    and susceptibility 35
  fungal infection, potato 48
  host cell 56, 231
  incompatible 2
  insects 4, 115
    attack, crops that exhibit 124
    behavioral factors affecting host plant 231
    developmental factors affecting host plant 231
  mechanisms of 262
  new compounds in host plant 212
  properties related to 50
  race-specific 9, 31
  sunflowers 123
Respiration
  T mitochondrial 102
  NADH 105
  oxidative phosphorylation, effect of digitonin on 106
  root mitochondria, effect of HmT toxin on 99
  succinate 106
Reversibility of toxin action 100
Rhizobium japonicum 28
Rhizobium trifolii 28
Rhopalosiphum maidis 265
Rhyncosporium secalis 42
Rice 124
Rishitin 61, 62
  ^{13}C NMR spectrum of 71
Rishitinol 62
Robinia pseudoacacia 223
Root growth 91
Root mitochondria, effect of HmT toxin on respiratory activities of 99
Rotenoids 154
Rotenone 154

S

Saccharomyces cerevisiae 28
Saponins 261
Scale of insect growth and plant's response, relative 125
Schistocerca 264
  gregaria 169, 256
  vaga 166
Scolytus multistriatus 215, 225
Screening techniques
  ^{1}H NMR 203
  plant extracts 186
  TLC 203
Seedlings 92
Sesquiterpenes
  biogenetic interrelations of 65
  biosynthesis of 69
  lactones, antifeedant 179
  nicotiana 65
  stereochemistry of bicyclic 67
  stress compounds from the solanaceae, biosynthetic relationships of 61
Shiromodiol diacetate 189
Shiromodiol monoacetate 189
Silk, corn 122
Sirex noctilio 121
Sitona cylindricollis 262, 265
Sitona hispidulus 261
Sitophilus oryzae 133
Sodium azide 105
Sojagol 11
Solanaceae, biosynthetic relationships of sesquiterpenoidal stress compounds 61
Solanum melongena stress metabolites 63
Solanum tuberosum 3
  phytophthora infestans infection of 51
  stress metabolites 62
Solavetivone 62, 65
Southern corn leaf blight disease 37
  biochemical and ultrastructural aspects of 90
  mechanism of 110
Soy 124
Soybean(s) *(Glycine max)* 10, 132, 140
  hypocotyls 13, 28, 31
  phytoalexins 3, 27
    structures of 11
  system, Pms– 31
  trypsin inhibitor 133
Species specificity, elicitor 31

Spectrometer, NMR 66
Spectrometry, mass 161
Spectroscopy, infrared 160
Spectroscopy, NMR 160
Spectrum, ^{13}C NMR 69
  2,3-dihydroxy germacrene 70
  enriched phytuberin 74
  lubimin 73
  rishitin 71
Spodoptera
  eridania 166, 180
  exempta 166
  frugiperda 180
  littoralis 166
  litura 185, 186
  ornithogalli 179
Stereochemistry of bicyclic sesquiterpenes 67
Steroids 108
Stimulants, feeding 232–239
Stress compounds from the solanaceae 61
Stress metabolites 78
Structures, HmT toxin 108, 111
Substances, bitter 156
Succinate 99
  oxidation 102
  respiration 106
Sugarcane 38, 43
  interaction with *H. sacchari* 41
Sunflower moth, resistance of sunflowers to 123
Survival of insects, nutritional factors in plants affecting 262
Susceptibility 87, 120
  plant disease resistance and 35

T

Tannic acid 119, 120
Tannin(s) 118, 120
  chestnut 132
  cotton 118, 119
  oak leaf 132
Tephrosia 154
Terpenes 256
Terpenoid(s) 260
  aldehyde components of cultivated cotton 206
  aldehydes, biosynthesis of 209
  biosynthesis of the 207
  content of *Gossypium* 208
  cotton 199
Tetranychus urticae 265
Therioaphis maculata 265
Thin-layer chromatography 160
  screening techniques 203
Tobacco 43
  budworm 211, 259
    resistance of cotton to 117
Tolerance 116
Toxin(s)
  agents from plants 153
  binding activity 37, 42
  corn silk 122
  cotton 120
  effects of HmT on
    T cytoplasm 92–95
    mitochondria 98–102
    oxidative phosphorylation 102
    release of malate dehydrogenase 104
  effects of phytoalexins 32
  evidence for the plasma membrane as a site of action 97
  fungal peptides 157
  growth and 256
  host-specific 36
  mechanism of 4
  molecular structure of HmT 108
  NADH oxidation by 102
  preparation(s) 96, 97
  production 39, 91
  purification 91
  resistance, biochemical marker of 42
  structures, HmT 111
  treatment 100, 102
Trachyloban-19-oic acid 123
Tree(s)
  citrus 40
  examined for allomones, non-host 217
  non-host *carya* and *juglans* 218
  oak 116, 120
Tribolium 261
  castaneum 133
Trichloroacetate-solubilized fractions 43
Trichoplusia ni 179, 264
Trypsin inhibitor, soybean 133

U

Ugandensidial 172
Ulmus 215
  americana twig bark 220, 224, 226
Ultrastructural studies 54, 101
Unedoside 174
Uromyces phaseoli var. *vignae* 55

V

Valinomycin 103
Veratrum 153
Verbenanceae, antifeeding activity of 191, 193
Vernonia
  flaccidifolia 179, 181
  gigantea 179
  glauca 179
  pulchella 183
Vertebrates 131

Verticillium dahliae 5, 260
Vicia fabae 8
Vitamins 264

W

Warburganal 172
Warburgia 186
  stuhlumannii 172
  ugandensis 172
Water, toxic material in 123
Weevil, boll 211
Wheat 40, 43, 124

X

Xylocarpus moluccensis 170
Xylomolin 170

Y

Yeast, brewer's 32

Z

Zanthoxylum clava-herculis 156
Zygadenus 155

A
MA